the Roll:
Table at Large
Polus ecliptic[ae] ... orienti proximus
Nord west
Nord Septentrio
Nord ost Aquila
Polus mundi Arcticus
T0373058

HOW WRITING MADE US HUMAN, 3000 BCE TO NOW

INFORMATION CULTURES

Series Editors

This book series examines how information has been produced, circulated, received, and preserved in the historical past. It concentrates principally, though not exclusively, on textual evidence and welcomes investigation of historical information both in its material forms and in its cultural contexts. Themes of special interest include the history of scholarly discourses and practices of learned communication, documentary management and information systems—including libraries, archives, and networks of exchange—as well as mechanisms of paperwork and record-making.

This series engages with vital discussions among academics, librarians, and digital humanists, presenting a vivid historical dimension by examining the roles played by cultures of information in times and places distant from our own.

Also in the series

Sailing School:
Navigating Science and Skill, 1550–1800 (2019)
Margaret Schotte

Leibniz Discovers Asia:
Social Networking in the Republic of Letters (2019)
Michael Carhart

The Maker of Pedigrees:
Jakob Wilhelm Imhoff and the Meanings of Genealogy in Early Modern Europe (2023)
Markus Friedrich

A Centaur in London:
Reading and Observation in Early Modern Science (2023)
Fabian Kraemer

How Writing Made Us Human, 3000 BCE to Now

WALTER STEPHENS

JOHNS HOPKINS UNIVERSITY PRESS
BALTIMORE

Printed in the United States of America on acid-free paper
2 4 6 8 9 7 5 3 1

Johns Hopkins University Press
2715 North Charles Street
Baltimore, Maryland 21218
www.press.jhu.edu

Library of Congress Cataloging-in-Publication Data is available.

ISBN 978-1-4214-4664-6 (hardcover)
ISBN 978-1-4214-4665-3 (ebook)

A catalog record for this book is available from the British Library.

The photographs of items from Johns Hopkins Special Collections were taken by Cameron Hanley.

Endpapers: *Mensa Isiaca*. [*Bembine Tablet of Isis*.] Engraving reproducing the 1st c. CE Egyptian bronze tablet known variously as the *Mensa Isiaca*, *Isiac Table* (or *Tablet*), or *Bembine Tablet of Isis*. Extracted and annotated by John Evelyn from Johann Georg Herwart von Hohenburg, *Thesaurus Hieroglyphicorum* (Augsburg/Munich?, 1610). Ephemeral Renaissance Collection, Sheridan Libraries Special Collections, Johns Hopkins University.

To Indian Springs School, Helena, Alabama,
in memory of Louis E. Armstrong and Howard O. Draper

. . . for thus it is written . . .

Matthew 2:5

Whence did the wondrous, mystic art arise,
Of painting speech, and speaking to the eyes,
That we, by tracing magic lines are taught
How both to embody, and to colour thought?

Needlepoint sampler of a
North American girl, 1828

Contents.

PART III

Writing in the Middle Ages

PART IV

Toward Modernity

Preface. *Homo scribens*: Humanity and Writing

The noblest acquisition of mankind is speech, and the most useful art is writing. . . . The uses of writing are too varied to be enumerated, and at the same time too obvious to need enumeration. By this wonderful invention we are enabled to record and perpetuate our thoughts, for our own benefit, or give them the most extensive communication, for the benefit of others. As without this art, the labours of our ancestors in every branch of knowledge would have been lost to us, so must ours to posterity. Tradition is so nearly allied to fable, that no authentick History can be compiled but from written materials.

—THOMAS ASTLE, 1783

Imagine our world without writing. No pencils, no pens, no paper, no grocery lists. No chalkboards, typewriters or printing presses, no letters or books. No computers or word processors, no email or internet, no "social media"; and without binary code—strings of ones and zeros that create computer programs—no viewable archives of film or television, either. Writing evolved to perform tasks that were difficult or impossible to accomplish without it; at some level, it is now essential for anything that human societies do, except in certain increasingly threatened cultures of hunter-gatherers. Without writing, modern civilization has amnesia; complex tasks need stable, reliable long-term memory.[1]

This book is about *Homo scribens*, Man the Writer, because whatever else they said about "man," most writers in the Western tradition have assumed that writing made *Homo* fully human. Am I suggesting that writing is the only skill that makes us human? Of course not.

Yet historically the idea was often implied and occasionally explicit. According to a late sixteenth-century treatise on penmanship, "Plato says that the difference which divides us humans from the animals is that we have the power of speech and they do not. I, however, say that the difference is that we know how to write but they do not."[2]

Throughout Western history, there have been other shorthand definitions of humanity in terms of some single, overarching, inherent trait. Aristotle defined "man" as the animal that laughs, *Homo ridens.* Another ancient defined humans as *Homo faber,* Man the Maker, a concept much prized in the Renaissance; a twentieth-century historian defined us as *Homo ludens,* emphasizing the cultural and intellectual significance of play. The most laudatory definition was devised by a botanist of the eighteenth-century Enlightenment, who dubbed us *Homo sapiens,* Man the Wise; later we were promoted to *Homo sapiens sapiens.* This flattering label has stuck, peremptorily declaring our superiority to all the hominins that went extinct.[3] By enshrining the epithet in anthropology and other sciences, we continue to imply that some definition of wisdom is entwined with our species' evolution. However, if *Homo scribens* is occasionally wise, we are not consistently so. Along with wisdom (however defined), writing also nurtured our habitual folly.

Neanderthals were among the elder cousins of *Homo sapiens sapiens,* coexisting with us until around thirty thousand years ago, when they disappeared as a separate species.[4] Still, a certain percentage of *Homo sapiens sapiens* mated with *Homo neanderthalensis,* so that a not insignificant number of modern humans have inherited Neanderthal DNA. Whether Neanderthals or others of our relatives laughed or played is unprovable (precisely because they did not write). But it seems likely; archaeology tells us they made things, as hominins had done since Paleolithic times. Yet writing is the one accomplishment we do not share with Neanderthals and our other ancestors.

Every age had its own ideas of how writing came about, what it was for, and what human life would be without it. For thousands of years, Sumerians, Egyptians, Greeks, Latins, Jews, Christians, and Muslims shared two projects, the creation and refinement of writing and the attempt to understand its history and meaning. Other cultures, notably in China and Central America, have long traditions of writing, to be sure, and scholars have studied them for centuries. But to do them justice here would risk tangling the emotional thread that connects the history of *Homo scribens* from Babylon to our own time. That affective evolution is coherent and compelling, from myth to method, from fireside legends of gods and heroes to scientific excavation and decryption.

WRITING AND WONDER

Throughout history, humans have regarded the art of writing with awe and even reverence. To imagine humanity without writing was not impossible, but it was in many ways difficult. The word *prehistory*, defined by the absence of written records, only entered the English language in 1836. A few years previously, in 1828, a North American schoolgirl praised writing as miraculous, "Whence did the wondrous, mystic art arise / Of painting speech, and speaking to the eyes?" This synesthetic quality, the capacity to translate information from one sense to another, had been a source of enthusiasm since the most ancient times, yet its appeal remained undiminished. Then, within twenty years, the electromagnetic telegraph expanded the definition of writing by retranslating "painted speech" into a binary system of audible pulses capable of spanning continents and oceans. In turn, the pulses were retranslatable—writable—as dots and dashes or alphabetical characters.

Two centuries after the marveling schoolgirl, we can hardly imagine her degree of enthusiasm. In our age of "digital assistants," the painting of speech—and mathematical formulas—by binary strings of ones and zeros evokes no kind of wonder. Your spoken command may look up information, transcribe it, or choose and play your favorite music; it can heat your house and pilot your automobile. Throughout five millennia, the art of writing has always been paradoxical, as mundane and practical as a pencil, yet miraculous, more stupefying in its way than end products like *Paradise Lost*, the *Divine Comedy*, the *Iliad*, or, ultimately, the Babylonian epic of *Gilgamesh*.

As if echoing the nineteenth-century schoolgirl, the science fiction writer Arthur C. Clarke has remarked that "any sufficiently advanced technology is indistinguishable from magic,"[5] and from its beginnings, writing seemed indeed magical, even god-given. Praise for letters as the foundation of civilized life developed in ancient societies as soon as records progressed beyond bare lists and inventories. On clay, papyrus and parchment, paper, stone, and metals, men—and a very few women until the Renaissance—marveled at the art of writing and celebrated its awesome magnification of memory and imagination.

To express wonder about writing was to create mythologies around it. But unless one knows where to look, praise and mythologies of writing are hard to locate among the trillions of pages produced since the fourth millennium BCE. The words for *write, writing,* and *written* are so common in all but a few languages that even electronic searches are of limited help. Formal, specialized histories of writing began developing in the late Renaissance but remained largely mythological until the nineteenth century. After 1800, praise for writing never disappeared entirely from histories, but it gradually shrank to its current almost imperceptible level. Modern histories of the art differ as drastically from mythological accounts as day from night. Look up *history of writing* today, and you will find no echo of former centuries' extravagant panegyrics or heroic mythologies. Instead, you encounter a dispassionate history of technology, from cuneiform and hieroglyphics to binary code.

And yet mythical histories are never obsolete, much less merely quaint, no matter how naive they seem in retrospect. Mythology, the common denominator of religion and literature, is *emotionally* valid even at its least factual. So the premise of this book is that the history of writing is not simply a record of what people have *done* with it, as documented by modern scholarship. It investigates how people *felt* about writing over the previous five thousand years, and for that task mythology is indispensable. To explore the emotional history of writing is to inaugurate the metahistory of the art, *the history of the history of writing,* from Babylon to the computer age.

This book is not a complete or definitive history; no history ever is, and this one does not pretend to explore every development everywhere. But it does survey millennia of continued excitement about writing, a story that concerns both general readers and those who read and write as a profession. Enthusiasts of Greek and Latin classicism; ancient Near Eastern studies; biblical studies; modern languages and literatures; ancient, medieval, Renaissance, and modern history; epic poetry and the novel; or the history of the book and printing—in short, most fields of intellectual and social history—will find in these pages stories and meditations about writing that range from the familiar to the surprising and bizarre. Unavoidably, specialists in many of these fields will wish that I had devoted more time to certain aspects or developments. But *ars longa, vita brevis*—writing has a long life span, but a single writer does not.

The history of emotions is a recent development in scholarly inquiry, but I will not attempt a more general or theoretical treatment. Instead, the following chapters trace a history of the sentiments that animated attempts to write the history of writing. Nor will I survey the *technological* development of writing from its earliest beginnings—the usual subject of histories of writing—except in the schematic "Complement" that precedes chapter one. To uncover early peoples' emotional engagement with the *technē*—or art—of writing requires us to await a moment when letters became capable of expressing nuances that imply emotion. For such reasons, the chapters that follow necessitate periodization (e.g., "the age of paradoxical optimism") that may initially seem puzzling. Moreover, since these periods are necessarily approximate, they sometimes overlap.

Chapter one, "An Age of Wonder and Discovery, 2500–600 BCE," recounts how archaic societies—Mesopotamian, Egyptian, and Homeric—envisioned the capacities of writing for tasks other than taking inventory, above all, for recording narrative. Although writing emerged as a technology for sophisticated arithmetical records by about 3000 BCE, "it was not until around 2600 BC[E] . . . that scribes began to use the cuneiform writing system to record something other than administrative transactions and lists of words. One of the scribes' new mandates was to immortalize the great deeds of the kings."[6] These centuries witnessed the development from oral to written storytelling, expanding from myth and legend toward something that begins to resemble history.

Chapter two, "An Age of Philosophy, 600 BCE–400 CE," traces the story of texts that began their careers in written form, or reached that stage quite soon, in the period of Greek and Roman classicism. The era began by accepting and expanding the possibilities of written records through poetry and historiography. Importantly, it saw the emergence of philosophy as a written search for truth rather than a description or declaration of it. The Age of Philosophy ended—or was changed fundamentally—by the weakening of imperial Roman authority in the West, its displacement by smaller "barbarian" kingdoms, and the eventual hegemony of Christian religious culture.

A conspicuous interval of this period is covered by chapter three, "Collections, Histories, and Forgeries, 300 BCE–400 CE." This period saw the rise of libraries and the consolidation of historical narratives. Homer was now envisaged as the archetypal *author* of a *text*, a single *writer* with a distinct ideological project; the triumph of written culture had suppressed all memory of how the *Iliad* and *Odyssey* gradually accumulated from several centuries of anonymous oral poets. While philosophy implied a search for truth, forgery enabled the contestation of history, and Homer's tales were an early target. Thanks to Alexander the Great's conquests, Greeks grew increasingly familiar with earlier civilizations, particularly the Egyptians, deepening their awareness of inheriting a cosmopolitan *written* culture. Among their acknowledged debts to Egypt was the invention of libraries as stable repositories of cultural memory.

Chapters four through seven cover a new epoch in the emotional history of writing, the rise of scriptures. Those singularly privileged texts, produced by the religions of Judaism, Christianity, and Islam, spoke of a sole deity. There were no goddesses: equivalents of Isis or Ishtar, Artemis, or Minerva were nowhere to be found. Though described as transcending every aspect of humanity, the divinity was treated as male and ruled societies that were thoroughly patriarchal. No longer the plural *subject* of polytheistic texts, the solitary deity was their *author*, their originator. The singularity of God, by whatever name he was invoked, underwrote the *authority* of the texts ascribed to his authorship, ensuring their superiority to other texts. In turn, the singular authority of God engendered the criterion—or rather, the need for one—of *authenticity*: How should people distinguish the single, genuine Word of God from all the competing texts that claimed the deity as their author? Chapters four through seven outline how Judaism and Christianity would be beset for centuries by this problem of *canonicity*. In contrast, the canon of Islam avoided the dilemma by containing a single holy book with no evidence that any human scribe recorded the deity's message: the only *I* in the Qur'an is Allah.

The centuries of writing activity chronicled in the first seven chapters were followed by a long period of decline and erasure. From the fifth through the eleventh centuries, by some accounts, or from the fourth through the fourteenth or fifteenth according to others, the intellectual and cultural lights that had shone so brightly before Constantine moved his imperial capital to Byzantium dimmed almost to the point of extinction. Like all historical generalizations, this one is grossly oversimplified, if not false, as historians have maintained since the nineteenth century.

In fact, from an emotional standpoint, the Dark or Middle Ages are seriously misnamed. As chapter eight establishes, the period from about 650 to 1350 CE was one of extraordinary *optimism* about writing and its capacity to make us human. In this period, most writers about writing were not aware of living in the "dark" or of being "medieval." Rather than mourn a breakdown, they habitually celebrated the triumphs of writing over oblivion, its power to prolong human memory throughout the ages. Some of the most extravagant panegyrics ever composed about the "wondrous, mystic art" date from these centuries, revealing fascination with every aspect, from its materials and tools to its benefits for science, religion, and literature.

Paradoxically, the most intense fatalism about writing was expressed in the following period. Chapter nine, "Pessimism in the Age of Rediscovery, 1350–1500," details how the Renaissance invented the barbarous Middle Ages. In the decades after Dante, Western writers began mourning a catastrophic loss of Greek and Roman culture. A biblioclasm of colossal proportions seemed to have stripped every field, from poetry and history to philosophy, of its brightest and purest gems. Not only books but all the arts had suffered irreversible setbacks, from broken statues ("idols") to vandalized temples and abandoned libraries. A great deal of this pessimism was factually based. But writers were sometimes more depressed by classical culture's cadaver than hopeful of resuscitating it. Not until the turn of the nineteenth century would there be widespread discussion about positive aspects of "Gothic" culture.

Chapter ten, "Alternating Currents, 1450–1550," explores a period that was either wracked by uncertainty or rife with possibilities, depending on who described it. With hindsight, and as some enthusiastic contemporaries foresaw, Gutenberg's press all but precluded biblioclasms comparable to the previous millennium's. Still, as the historian Eugenio Garin remarked, the period itself was "a brilliant age, . . . not a happy one." The Renaissance witnessed an increasing paranoia about religious and political subversion and an intense desire to create what historians call "a godly society."

Chapter eleven, "A Second Age of Scripture, 1500–1600," describes that godly ambition during what some have called the Age of Reformations. One of its most remarkable features was the proliferation of printed Bibles. Printing made the Bible widely available in the everyday languages that merchants and other laypersons could read for themselves, without mediation by Latin-reading churchmen. Neither literacy nor access to books was suddenly universal, but reading and interpreting the Bible became an activity that laypeople could perform. The proliferation of Bible translations fed an explosion of interpretations—and of strife; censorship institutionalized the suppression of problematic books.

Chapter twelve, "The Age of Grand Collections, 1600–1800," addresses the development of modern libraries, museums, and learned societies. Moving outside the walls of monasteries, universities, and the great houses of the noble elite, these institutions gradually spread literary, philosophical, and scientific culture among a wider public. Massive innovation affected the kinds of books being printed, as well: whole committees of scholars now collaborated on enormous projects such as the *Encyclopédie* of Diderot and d'Alembert. "Chambers of wonders," the private proto-museums of wealthy individuals, displayed remarkable natural and artistic objects, from fossils to cameos to exotic books, including illustrated catalogs of the collections themselves.

Chapter thirteen, "Skepticism and Imagination, 1600–1800," examines a particularly innovative subset of writings from the same period. Much of the lore described in previous chapters was losing its hold on scholars, for whom the autobiographies of Adam and Eve or the science of Noah and Hermes Trismegistus were no longer plausible or even interesting. As imagination drained away from such legends, it enriched literature, inspiring new genres in the vernacular languages that even servants could often read. Novels recounted the biographies and memoirs of imaginary people, while tongue-in-cheek scholarship undergirded parodies and burlesques of older genres, such as epic.

Chapter fourteen, "The Age of Decipherment, 1800–1950," narrates an unmistakable turning point in the emotional history of writing, summarizable in two words: modern archaeology. Two red-letter dates, 1822 and 1872, demarcate a half century that rang the death knell of legends first recorded in cuneiform and hieroglyphic texts. Jean-François Champollion's breakthrough in deciphering ancient Egyptian writing was matched fifty years later when George Smith extracted "the original version of Noah's flood" from *Gilgamesh*. A parade of discoveries by other excavators and scholars followed, down to the revelation in the early 1950s that the peculiar script of Linear B encoded a form of Greek that antedated Homer by six or seven centuries. Inspired by evolving archaeology, writers of fiction transformed enduring legends or created their own. An outstanding example is *The Authoress of the Odyssey*, a novelistic biography published in 1897, which identified the "real"—female—Homer.

Chapter fifteen, "The Age of Media, 1950–2020," brings the emotional history of writing to the present time. Despite the proliferating distractions of electronic media, the art of writing remains a perennial topic of novels and short stories, popularized history and serious scholarship, film and television, and even computer games. From *Fahrenheit 451* to *The Name of the Rose* and beyond, the emotional history of writing interests us more than we may realize.

Author's Note.

It is plausible that these observations may have been set forth at some time, and, perhaps, many times; a discussion of their novelty interests me less than one of their possible truth.

—JORGE LUIS BORGES, "Partial Magic in the *Quixote*"

This book is primarily intended for general readers who are interested in the ways humans have reacted emotionally to the art of writing—with admiration or, less frequently, with loathing—over the past five millennia. I have avoided encumbering it with numerous footnotes and other scholarly apparatus that would mean little to anyone other than specialists. I have provided sufficient indications to allow interested readers—whether specialists or not—to have confidence in my use of primary sources and to check them if they wish. But I have been sparing in my references to scholarly studies except where omitting them would deprive their authors of due recognition. As the quotation from Borges implies, my principal objective has been accuracy rather than originality for its own sake.

HOW WRITING MADE US HUMAN, 3000 BCE TO NOW

Introduction.

THE MYSTIQUE OF WRITING

View writing's art, that like a sovereign queen
Amongst her subject sciences [is] seen.
As she in dignity the rest transcends,
So far her pow'r of good and harm extends;
And strange effects in both from her we find,
The Pallas and Pandora of mankind.

—Quoted in William Massey,
Origin and Progress of Letters

For most of history, the epithet *scribens* would have been grossly inappropriate to describe the genus *Homo*; writing was a skill limited to a tiny elite of scribes and scholars. As the specialized technology of a guild, the art acquired a prestige, an aura, a mystique that made it seem magical, sometimes in the fullest sense of the word. Until nineteenth-century archaeology, anyone interested in the history of writing had scarcely better evidence than the Sumerians. Lacking historical perspective, but immensely proud of their craft, early scribes imagined its origin and development as superhuman, the gift of gods and heroes. Would-be historians inherited, transmitted, and embellished mythical tales about heroic or divine individuals who single-handedly invented an art imbued with a power that was sometimes tangible—that is, magical—as well as political, religious, or symbolic. Although these stories became steadily less mythical, their leitmotifs remained remarkably stable. I describe eleven of these leitmotifs below.

WRITING AS THE "WONDROUS, MYSTIC ART"

The conviction that writing was worthy of the highest admiration, a marvel so astonishing that only a god or godlike human could have invented it, permeated countless stories about it before 1800. Writing enabled memory to outlast the human voice and transcend the individual person; written thoughts could remain stable over generations or centuries. By bridging space as well as time, writing abolished isolation and created community. It could even enable interaction between the ephemeral human world and the invisible society of gods, demons, and spirits. Writing was so central to definitions of humanity that, as I note above, the concept of prehistory only emerged around 1800, while the notion that Adam, Moses, or another biblical patriarch had invented writing lingered among the religious.

INSCRIPTION AND ERASURE

Writing was a facsimile of immortality for individuals and whole societies; thus, a medieval Latin translator of Plato referred to *memoria literarum*. The phrase suggested that writing is a kind of receptacle, which *contains* memory as if it were a tangible physical object. Still, it was no secret that literary memory is not "literally" eternal because even the most durable media are overshadowed by the threat of erasure. The tension between inscription and obliteration (literally de-lettering) was and remains an omnipresent theme.

LOST BOOKS AND LIBRARIES

Lost writings are a powerful leitmotif in the emotional history of writing. The erasure of a single work seems tragic even now, but in the long manuscript age before Gutenberg, the destruction of a book could symbolize the loss of the whole world. If nothing but fragments of a text survive, the biblioclasm inevitably stimulates writers to imagine the complete whole that was destroyed. Like the armless Venus de Milo, mutilated writings have inspired nostalgic dreams of reconstitution, ranging from scholarly treatises to fantasy and kitsch. The immense Library of Alexandria was already the archetype of mass erasure during antiquity and the Middle Ages, and it still excites both scholars and nonspecialists.

REDISCOVERY

Not all lost writings are gone for good; some are merely misplaced, and startling rediscoveries have been made over the centuries. Famous recovered works that crowd scholarly daydreams include the dramatic example of an entire library belonging to Ashurbanipal, the Assyrian emperor who died in 627 BCE. Discovered in 1849–1852, it contained thousands of cuneiform tablets, many broken into tiny fragments. The trove included the epic of *Gilgamesh*, the oldest major work of world literature, containing what its first reader in two millennia christened "the original version of Noah's Flood."[1] More recently, space-age technologies have permitted unprecedented collaboration between manuscript scholars and cutting-edge scientists, who miraculously salvaged lost texts by that archetypal mathematician Archimedes.[2]

BOOKHUNTERS

Many recoveries of lost works have been owed to random good fortune, but just as frequently they were the result of deliberate searches. The figure of the Bookhunter, an Indiana Jones who traces clues and braves danger to recover priceless written treasure, was already present in ancient Egyptian myth. During the Renaissance, scholarly bookhunters transformed the ancient fables into an exciting reality; as they rediscovered landmarks of Greek and Roman culture, they laid bare centuries of dramatic stories about the history and powers of writing. Even today, the search and recovery operation is still going strong, including in cultures far older than Greece and Rome.

ANCIENT WISDOM

Biblioclasms—lost libraries and damaged manuscripts—inspire a romantic nostalgia so intense that writers have often imagined whole utopias of extinct wisdom. Sometimes hard evidence of destruction inspired these bookish fantasies, but paradoxically, daydreams of loss were often provoked by exciting rediscoveries. Until the eighteenth century, *sapientia veterum*, the wisdom of the ancients, was the scholar's imagined paradise, his (or increasingly her) Garden of Eden. Democratized literacy since 1800 has made reveries about the stupendous achievements of Egypt and Atlantis into perennial favorites of popular culture. Plato imagined Atlantis 2,400 years ago, yet modern daydreams about lost utopias, from Jules Verne and H. G. Wells to the 1985 film *Back to the Future,* differ from their ancient counterparts mainly through their anachronistic or pseudoscientific assumptions about technology and science.

FORGERIES AND FAKES

Forged texts were common in ancient Greece, and even earlier in Egypt. During the Renaissance, when genuine Greek and Roman texts and epigraphic inscriptions were being rediscovered in droves, forgery and falsification ran rife. Scholars developed techniques for detecting them, but forgers stayed a step ahead of their critics. Moreover, by the eighteenth century, novelists were employing narrative techniques—some of them dating back to ancient Greece—that blurred the boundaries between fact, forgery, and fiction in suggestive and often disturbing ways. In 1719, Daniel Defoe's *Robinson Crusoe,* subtitled "the life and strange surprizing adventures" of an Englishman from York, "written by Himself," was told so realistically that a century later many readers, including a notorious forger of Shakespeare manuscripts, still mistook it for a factual account.[3]

BOOKS OF THE DAMNED

Not all enthusiasm is positive. Whether genuine or forged, physically real or only imagined, books have at times incarnated an ideal of evil. Early Christians destroyed numbers of books they considered theologically, morally, or intellectually dangerous, including the *Book of Enoch,* which claimed to be the memoirs of Noah's great-grandfather. Other scandalous books were nonexistent or unlocatable to begin with: the mere title of a book could ignite passionate controversy, even—or especially—when no one could find copies of it. Beginning in the thirteenth century, scholars gossiped and daydreamed apprehensively about a *Book of the Three Great Impostors,* which supposedly argued that Moses, Jesus, and Muhammad were charlatans and tricksters, and their religions nothing but tissues of lies. In the skeptical eighteenth century, a book was finally forged to fit the title, to widespread disappointment.[4]

HOLY BOOKS

The opposite of damned books were sacred (*sacer* in Latin) books, which were off limits in a different way, "untouchable" because religious leaders declared them immune to all criticism. These holy books—or scriptures—are the most radically explicit example of creating *authority*—religious and political credibility—through writing. Their defenders claim that scriptures descend vertically from a god to humans, whereas modern scholars call them mere *texts* and explain their trajectories horizontally, across human history. From the *Book of Enoch* onward, legends about God and the Hebrew alphabet, including the origins of the Torah itself, were based on passages in the Hebrew Bible. The New Testament, the Qur'an, and the *Book of Mormon,* as well as numbers of would-be scriptures now forgotten, went further and described in "autobiographical" detail how they came to be written by gods or their human amanuenses.

IMAGINARY BOOKS

As the previous two categories suggest, a letter, an inscription, or even an entire book can be wholly imaginary, as thoroughly nonexistent as cloud-cuckoo-land. Paradoxically, a brief title makes the hypothetical existence of a book easier to imagine than the narrated life of a Robinson Crusoe or an Elizabeth Bennet. Conversely, it is more difficult to establish the unreality of an imaginary book than that of a unicorn or a utopia. The metahistory of writing is entwined with the history of imaginary books, and examples of full-on *mythical bibliography* are far from rare. Whether as earnest scholarly quests for literary chimeras or as satirical send-ups of learned pretense, mythical bibliography remains a major expression of the social and emotional importance of writing.

WRITING, BOOKS, AND LIBRARIES AS METAPHORS AND SYMBOLS

Various myths about the history of writing are strongly symbolic or metaphorical. At the end of Dante's *Paradiso,* his famous description of God as the ultimate book symbolized the overwhelming consequence of writing and books for Christian culture in 1320. Six centuries later, Jorge Luis Borges came to international fame through his tale "The Library of Babel" (1941). Borges describes the cosmos as an infinite library whose only inhabitants are despondent librarians searching vainly for the ultimate book that will make sense of their bibliocosm. Borges's tales and essays frequently couch the deepest philosophical truths in enigmatic narratives, glorifying language, writing, and books as convincingly as genuine primitive myths ever did, sometimes naming uncanny or savage gods as their authors.

CONCLUSIONS

The history of writing is ready for its emotional close-up: what people have done with writing is now well known, but how they *felt* about it over time remains uncharted. The celebrities of bookish myths were not only gods and humans, but also writings, and ultimately the art of writing itself. Discarded documents, when they survive, have told us much about the way people used writing, in every kind of activity from accounting to religious contemplation, poetic meditation, philosophical inquiry, and scientific research. But discarded *attitudes to writing* still await the same kind of systematic spadework that archaeologists perform on material remains of the past.

The attitudes buried in myths and legends of writing reflect times when digging in the ground was for farmers, not archaeologists. Later, scholars researched the history of writing by reading books, but they had to construct that history for themselves from scattered, sometimes enigmatic anecdotes. Like the texts of Sappho's poems or the Dead Sea Scrolls, emotional evidence about the history of writing survived in mutilated, fragmentary form. Nevertheless, that lore is as vital to the history of literature as Shakespeare's sonnets or Dickens's novels.

Generations of scholars have told us how a single author or a vaguely defined period ("the Middle Ages" or "the Enlightenment") thought about books or libraries. But aside from writing as a profession (monk, scrivener, poet, historian, journalist, novelist, etc.), little has been collected of what earlier ages thought and felt about writing *as an art*, that is, *as a whole phenomenon*, in its organic relationship to humanity and civilization. Essential evidence for the emotional history of writing is only infrequently found in revered masterworks by Homer, Dante, or Jane Austen. The best sources are often lurking in outmoded scholarship: their technical obsolescence actually makes their defunct erudition more compelling as emotional history. Hiding under the dunes of dusty bygone scholarship are stories as captivating as Percy Shelley's "Ozymandias"—a familiar poem inspired by an ancient, now-forgotten anecdote about writing.

In fact, until Shelley's time, the history of literature was considered a mere branch of the history of writing. Indeed, the two were almost indistinguishable. It was still common knowledge that the word "literature" came from the Latin term for lettering, *litteratura*. "Literature" in the 1820s referred to writings of all kinds, not just poems and fictional stories. A "man of literature" was not a poet, novelist, or university professor but, like today's "literate person," simply someone who knew how to read and write.[5]

Since 1800 the factual history of writing has been constantly rewritten as more artifacts are discovered and old theories contested or disproved. As historians of writing describe the art's technological evolution ever more precisely, they understandably reject premodern myths and legends as inaccurate—as bad science, in a word. Building on two full centuries of spectacular progress, the history of writing now rarely mentions such lore even to refute it, leaving the old myths discredited and unmourned. At most, historians may recall one or two colorful myths for an amusing contrast with the facts as now understood.

But writing has existed for over five thousand years. So what—I ask—if the history of writing has been truly historical for only 4 percent of that time? The centuries preceding 1800 contain an immense corpus of lore about writing, books, and libraries. The mythical stories are not obsolete: rather, they are the raw materials for understanding 96 percent of the *experience* of writing since Babylon. To go on recounting a triumphant story of technological progress, automatically dismissing millennia of lore as a parade of bad or obsolete science, would mean ignoring important developments in the evolution of the humanities—and of humanity. The complete history of the history of writing necessarily includes *the emotional history of writing*.

It reveals how writing made us human. Largely confined to ink and paper, vulnerable to flooding, fire, and vermin, not yet a universal skill, facing no competition from electronic screens, cinema, telephones, or even Morse's telegraph, writing was still "the wondrous, mystic art" of "painting speech, and speaking to the eyes." Even today, when we are glutted by these technical advances, a moment's reflection can remind us that, in itself, writing retains something wondrous, even mystical. It is the matrix that enshrines everything we know.

Admittedly, no single book—or lifetime—is long or rich enough to do more than outline writing's emotional history. Inevitably, given the scope of my project, I will have committed inaccuracies and omissions, but perhaps future studies will correct some of them.

• • •

Note: Thorough acquaintance with the technological history of writing is not necessary for understanding the chapters that follow. But readers desiring a brief overview will want to read "Complement," which comes next.

Complement.

WRITING AS TECHNOLOGY, FROM MYTH TO HISTORY

> If we were to write the annals of humanity and reserve the same space for each millennium, the historical period, the age of *homo scribens*, would barely fill the last page of the book. . . . [A]s for Gutenberg, he lived only a little over five centuries ago. The era on which he put his mark would thus correspond to five lines in our annals. Our contemporary mass media would appear only in the last line, with telecommunications occupying no more than a final "?".
>
> —HENRI-JEAN MARTIN, *The History and Power of Writing*

Like writers before and after Gutenberg, many modern readers remain keenly interested in questions that archaeology and other sciences are still attempting to answer about the origin of writing and its technological significance to human history. We now know that asking who invented writing is a bit like asking who invented the wheel. Writing evolved as a technology in many stages over millennia. But the perception that writing did not suddenly "happen" was not fully thinkable until the nineteenth century. Before then, if anyone asked where writing came from, when, under what circumstances, and for what purpose, the answers were mostly mythological. They closely resembled Greek and Roman stories of gods and heroes, and starred some of the same dignitaries. Writing was usually presumed to have been invented ex nihilo, from nothing, by gods or exceptional individuals such as Hermes or Moses. The story of writing had only two essential moments, before and after, and was presumed to exist here and now in about the same form its inventor gave it. It took millennia to realize that there was no "Hermes moment" when writing suddenly came into existence.

THE TECHNOLOGICAL HISTORY OF WRITING

To understand the full emotional and cultural significance of myths and legends about writing in earlier societies, we should contrast them to modern histories of writing. This is not to claim that we now know "the truth" about its origin and development; that is an impossible ideal, a moving goalpost. But since 1800, our knowledge has grown exponentially; although we will never know everything about writing, we have reliable proof of its development. That knowledge equips us to compare our own implicit cultural and emotional expectations of writing to those expressed in the mythic, prescientific stories of previous societies.

Modern scientific consensus about the history of writing has an extensive literature, kept up to date by Assyriologists, Egyptologists, classicists, and other scholars who study its relation to histories of languages, literature, books, and reading. But its main outlines can be summarized rather briefly, if we keep in mind that, like archaeology in general, it is a complex field where new discoveries are constantly being made and debated.

Archaeology indicates that people have been writing for about five thousand years. But evidence demonstrates that, like other technologies, like human beings ourselves, the art of writing evolved in a complex, time-consuming process.[1] Sumerian, almost certainly the oldest writing system, emerged by about 3200 BCE, but its characters or "letters" probably did not take on their distinctive cuneiform, or wedge-shaped, appearance for another century. This raises a fundamental question: What do we mean by writing? The evolution of cuneiform illustrates the broader fact that there are degrees of writing. According to experts, "complete" systems of writing have four characteristics: (1) their purpose is communication; (2) they consist of artificial graphic marks on a durable surface; (3) the marks are both conventional and systematic; and (4) they represent articulate speech (significant vocal sounds arranged systematically) so as to communicate thought.[2] In Sumeria, as in Egypt and elsewhere, there were phases of pre-writing, which could not convey information as fully as writing would later do.

Writing took a very long time—millennia, in fact—to evolve into its complete state. It developed in order to meet specific cultural needs that could not be satisfied in a strictly oral culture. Jerrold Cooper has made the point that the domains in which early writing was used were, in fact, invented along with writing itself—"Livestock or ration accounts, land management records, lexical lists, labels identifying funerary offerings, offering lists, divination records, and commemorative stelae have no oral counterparts. Rather, they represent the extension of language into areas where spoken language cannot do the job."[3] In Mesopotamia, where its evolution is clearest, writing was not originally *writing* but rather *counting*: it evolved gradually from systems of taking inventory, especially of livestock and grain, that probably began by 8500 BCE. These accounting systems came about because Mesopotamian societies were growing ever larger. Scattered hamlets were developing into the first cities, and growth required increasingly complex systems of organization. Bureaucratic structures were needed to administer the division of labor, the distribution of goods, and taxation.

Taking inventory by writing words and numerals evolved from the use of tokens or tally marks to count (the latter survive in the convention of four vertical lines crossed by a diagonal to signify the number 5.) The evolution was extremely gradual: five thousand years may have been required to move from the simplest means of tallying to taking inventory with the most primitive, pre-cuneiform writing system. It currently seems that writing first developed in the city-state of Uruk, just as Chaldean legend affirms.[4] Modern archaeologists consider Uruk one of the first cities, emerging about 4000 BCE from the amalgamation of two more primitive Bronze Age settlements. The site had been settled for centuries by farmers and herders before it underwent urbanization. This rough chronology suggests that the urban development of Uruk and the evolution from counting to written literature may run parallel to each other.

Roughly five thousand years separate us from the emergence of complete writing. A previous progression in Mesopotamia, from tally marks to complete writing, probably took as long or longer. Little clay tokens were used by 8500 or 8000 BCE to account for commodities; the variety of tokens suggests that particular goods were represented by characteristic shapes and markings, and this differentiation probably also developed gradually. Eventually, ways of classifying and keeping track of tokens became necessary; according to one theory, the solution was the *bulla*, a hollow ball-shaped clay envelope. Along with the bulla, complex token shapes and markings emerged by about 3500 BCE, four centuries before the arrival of cuneiform "complete writing." A bulla would presumably contain tokens for a single commodity or for several different commodities involved in a single transaction: for example, the transfer of livestock or woven cloth to a temple treasury, or the exchange of such items between individuals. There was one problem, apparently: like a piggy-bank, the bulla originally had to be broken to see the tokens inside. So the shapes of the tokens were sometimes impressed into the wet clay exterior of the bulla, evidently so that breaking would be unnecessary except in certain circumstances.

At some point, it apparently occurred to someone that there was needless duplication in having tokens that both rattled around inside a bulla and were represented on its surface. Thus, within a century or thereabouts, the bulla may have evolved into a tablet showing numerical notation as well as symbols of goods. It has been noted that the early clay tablet had approximately the same shape as the bulla, and thus the same ergonomic relation to the human hand. As yet, however, there is no agreement whether cuneiform letters evolved in any direct way from tokens; as with natural selection, evolution from counting to writing may have proceeded by leaps. The exact relationship is still unclear and likely to remain so.[5] But on both bullae and tablets, numbers were represented by pushing a rounded stylus-tip into clay at various angles.

For the items being counted, the earliest known tablets do not show cuneiform writing but rather *pictograms*, rudimentary sketch-like pictures (a human foot, say, or a fish) incised with a pointed stylus.

The invention of the cuneiform (wedge-shaped) stylus tip facilitated uniformity of pictograms: knowledgeable scribes could build images by combining multiple strokes of varying depth and angle. Several developments led from pictographic, or picture-writing, to complete cuneiform writing, that is, from the representation of ideas (sheep, barley, numbers) to the transcription of spoken language in all its phonetic and grammatical complexity.[6] The earliest pictograms were *logograms*, representations of entire words. Obviously, a collection of logograms is not a text, since it cannot transcribe the complexities of speech. To evolve into full representations of speech, logograms underwent two transformations. On the one hand, as pictogram or visual representation, the logogram gradually lost its *iconicity*, that is, its recognizability as the picture of a human foot or a fish. The change from pointed to wedge-shaped stylus-tips gave pictograms the possibility of abstraction and standardization. A cuneiform pictogram looked less like a fish than most freehand drawings, but cuneiform pictograms of the same concept looked more alike than did freehand drawings. Stylization also increased the possibilities for symbolism. There are parallels in our modern globalized culture; the pictogram we recognize as a human heart is an abstract, stylized symbol that has almost nothing to do with cardiology, but it regularly symbolizes the concept "love."

Even more important, the possibility of representing sound emerged from the creative misuse of pictograms. You cannot draw a picture of the sound *ah*. But you can create a *rebus*, a picture of a thing which represents the sound of its name.[7] In Sumerian, two parallel wavy lines represented the thing *water* but also the sound *ah*, the spoken word for water. A comparable English example is *eye*; as a pictogram it can represent the concept "I" because the two words are *homophonic*, pronounced exactly the same. The evolution from a simple picture of the eye to its creative misuse as "I" is known as the transition from a *logographic* system, in which the picture has no necessary phonetic content (it could be vocalized as *eye, oeil, occhio, oculus,* etc., according to the language of the user), to a *logosyllabic* system, in which the picture represents a syllable of a particular language, such as *eye/I*. Logosyllabic systems greatly reduced the number of signs.

Determinative signs were also developed. These specialized signs meant nothing in isolation but gave contextual guidance when combined with other signs. Determinatives were particularly important for disambiguating: in our modern analogy, determinatives might signify whether the heart pictogram should be read literally, as a noun ("heart"), or figuratively, as a verb ("to love").[8] In the logo-syllabic phase, determinatives would signify the difference between homophones, say, anatomical "eye" versus first-person pronoun "I." In Sumerian there were further possibilities—in fact, necessities—for signifying various classes of objects, personages, actions, etc. There were determinatives for cities, gods, and other semantic classes, including wooden objects.

The final phase of development, phonetic writing, depended on having a set of signs that *only* represent sounds, the smallest units in a language. In this way, any word—including one from another language—could be represented by its mere sound or sounds. It was a great improvement over *syllabic* writing, which represented syllables of a particular language. The earliest Greek alphabet, which evolved just in time to record the Homeric epics, had sixteen signs. By contrast, the Greek Linear B syllabary system, which had gone extinct several centuries before, had sixty basic signs.[9] In alphabetic writing, speech finally becomes the immediate content of writing: alphabetic writing represents spoken sounds, which represent words, which represent things or concepts. (But the actual *sounds* of English words are more varied than the twenty-six vowels and consonants of the English alphabet. The sound of the letter *a* in *father* is the phoneme /ɑ:/, whereas the phoneme /æ/ represents the *a* in *cat*. So, since the late nineteenth century, the International Phonetic Alphabet has developed over 160 signs to account for such differences in phonemes.) A final development is silent reading, which no longer depends on speaking the signs aloud: a cluster of signs is immediately recognizable to the eye (or the fingertips, in Braille) as a word, whether it be *cat* or *catastrophe*.

Alphabets remain true to the rule observed in other characteristics of writing: they were not created ex nihilo by a Hermes-equivalent. They were instead based on preexisting elements. *Alpha* and *beta*, *aleph* and *beth*, are words representing the first two letters of the Greek and Hebrew alphabets, respectively. *Alpha/aleph* and *beta/beth* evolved from pictograms for "ox" and "house," which evolved into letters in the Phoenician alphabet that became the parent of both Greek and Hebrew writing. Phoenician writing, which stabilized around 1050 BCE, was itself an offshoot of the Proto-Sinaitic or Proto-Canaanite alphabet attested by 1700 BCE; this system derived in turn from an even earlier (ca. 2000 BCE) Egyptian system that indicated how to pronounce hieroglyphic signs.[10]

But Greek and Hebrew writing did not evolve in the same way. Although the Hebrew letter *beth* is still a word that signifies "house," to Greek speakers the letter *beta* never represented houses but simply the sound *b*. Moreover, there were degrees of phoneticism: in the development from Egyptian to Phoenician to Hebrew, no system had signs for vowels; letters represented consonants only, while the vowels needed for pronunciation were supplied by speakers or readers according to customary rules.[11] Since Greek had a smaller system of consonantal sounds than Phoenician, the Greek alphabet adapted leftover consonants from the Phoenician alphabet to create letters for vowels—like other Greek vowels, *alpha* was a repurposed Phoenician consonant.[12]

So writing systems were not created ex nihilo: their various features preexisted them for varying lengths of time. Yet the crucial elements tended to coalesce "punctually," at a certain point in time. Once the core of a system was in place, refinements evolved gradually. In a foreshadowing of modern historical understanding, the ancient Greeks credited the legendary hero Cadmus with creating their basic alphabet from Phoenician letters; later Greeks told of gradually adding (and sometimes abandoning) individual letters over time. Scholars have suggested that a single, bilingual "Cadmus" might have existed: a Phoenician scribe ("Cadmus") could have met an illiterate Greek bard (a "Homer") in a cosmopolitan mercantile town. Over a certain period, the Phoenician adapted his alphabet to transcribe Greek sounds, while the poet performed (repeatedly, no doubt) the oral tales that became the *Iliad* and *Odyssey*. Perhaps, but perhaps not. Writing never developed fast enough to record its own history: by the time serviceable alphabets evolved, oral tales had mythologized and simplified the origin of writing. And history itself would not be adequate to the task, until it moved beyond repeating stories to the detective work of nineteenth-century archaeology and cryptography.

One of the side effects of complete writing (especially alphabetic) was the ability to represent context. Before writing, all communication was through speech and gesture. If one was counting sheep, one knew the context of the inventory because one was standing among the sheep. When ambiguity arose in strictly oral communication, it had to be resolved by reference to context ("we are/were counting sheep"). By itself, partial writing actually made contextualization more complex: context had to be supplied by oral commentary when it was not already evident ("we used these particular marks because we were counting sheep"). Since their context cannot be reconstructed, the earliest surviving pictographic "texts" are almost undecipherable: the only people who could supply the context died thousands of years ago. Thus, the evolution of writing might be expressed as a progression from 100 percent context (i.e., no writing or drawing, just spoken words and pointing) to smaller and smaller proportions of nonwritten contextualization.

Although its origins were in bureaucratic operations, writing eventually outgrew the pragmatic limitations of its ancestry, gradually demonstrating its usefulness for organizing other sorts of information besides inventory, from land surveying to codification of laws. It is estimated that by about 2600 BCE, imagination—the use of writing for nonbureaucratic purposes—enabled ancient people to record poems and stories that moderns can recognize as literature. Now, thanks to five thousand years of gradual evolution in writing systems, a competent novelist or historian using a couple dozen alphabetic signs can transport me mentally, in total silence, to almost any setting and situation by merely accumulating descriptive details, no matter how different her society is from mine. I need only be a competent reader of her language or have access to a good translation.

Six thousand years, 8500–2600 BCE, may seem a long time to progress from taking inventory to writing a poem or story, but this impression is deceptive. First, not all "literature" is written. Preliterate societies preserve poems and stories for centuries; the example of Homer shows that orally transmitted lore can be both complex and extensive. If written literature took six millennia to develop, oral recitation must have seemed adequate to transmit story and ritual during that period. Oral "literature" and writing coexisted in parallel before they converged to create written literature. Unlike complex record-keeping, complex literature could develop extensively before being written down; it could rely on narrative sequence, along with rhythm, assonance, and other linguistic features to assist human memory in preserving the story. Homer's *Iliad* and *Odyssey* were not written down until the eighth or seventh century BCE, but they recorded events from four or five centuries earlier.

Only writing can preserve literature in the absence of living performers. Although we will never know exactly how *Gilgamesh* sounded, either before or after being written down, writing kept it alive, though dormant, for over three millennia. Even if original voice recordings had somehow once existed, their fragile media would not have survived.

The features that allowed oral literature to be transmitted without writing seem to have influenced the chronological development of written literature. We moderns assume that prose is more natural, and therefore older, than verse, which we consider "artful" and "artificial," but this is a distortion created by our long-term immersion in written culture. Historically—and in all known cultures—prose was an innovation, a *later* development than verse. In the twentieth century, field work conducted among illiterate bards demonstrated that narratives in verse could be transmitted for centuries independently of writing, as had long been suspected for the poetry of Homer.[13] Prose, however, is not merely at home in a written environment: it cannot exist in complete isolation from writing. "[Prose] has not always been present in all cultures, even in all literate ones. Historical evidence shows that it is verse that precedes prose. . . . There is an epoch in [all known literary] traditions in which there is no prose, and apparently never had been. There is then, subsequently, a time in which prose appears. And the appearance of prose does not at all coincide with the appearance of writing. It is subsequent to the appearance of writing. There are many cultures that have verse and have writing, but that have no prose."[14]

As is clear from this summary, the actual or "true" history of writing is a long and still incomplete story of technology—a history of signs, the instruments used to inscribe them, and the practical possibilities for interpreting them. Most of the developments explored here were not even suspected before 1800. In the absence of hard data, imagination, emotion, and tradition were forced to write the history of writing.

PART I

Pagan Antiquity

CHAPTER ONE.

An Age of Wonder and Discovery, 2500–600 BCE

It can no longer be doubted that the Sumerians, after giving humankind the most precious technological and cognitive tool ever invented, namely, writing, by that token gave us the first highly articulated system of literature.

—HERMAN VANSTIPHOUT, *Epics of Sumerian Kings* (2004)

Archaic antiquity, the times that were already ancient to Plato, was an age of discovery, fully comparable to the age of Columbus. But unlike the "New World," writing was never waiting to be encountered out beyond some formidable geographic boundary. It had to be invented. Yet in Latin, *inventio* means "discovery," not "invention" in our sense. Unlike us, ancient humans often imagined that writing had in some sense always been "out there," preexisting somewhere like a kind of America, waiting to be encountered by humanity, ready to reveal things of ultimate importance. But at least one account of a human's *inventing* writing ex nihilo is extremely ancient.

ORIGINS

Archaeology suggests that writing was invented first in Sumeria—southern Mesopotamia—spreading somewhat later to Akkadia in the north. Appropriately, Sumer and Akkadia transmitted the earliest complete myths about the discovery of writing.

About four thousand years ago, a Sumerian scribe recorded a story known as "Enmerkar and the Lord of Aratta." Doubtless proud of the craft he practiced, the scribe repeated a myth that was probably several hundred years old. The story concerned Enmerkar, the legendary first king of Unug [Uruk], a city-state in ancient Iraq, who would have reigned around 2800–2700 BCE. Uruk may be the city that Genesis 10:10 calls Erech, saying it was founded by the mythical Nimrod, the "mighty hunter before the Lord" who also built the Tower of Babel. Enmerkar supposedly invented writing while competing with the Lord of Aratta, king of a prestigious, wealthy, faraway, and possibly imaginary city.[1] Each king wanted to prove his own and his culture's superiority to the other. So they sent a courier back and forth between the two cities, having him deliver orally increasingly complex and difficult messages, including threats, riddles and other challenges.

Since writing had not yet been invented, the courier had to recite everything from memory. Finally Enmerkar entrusted a message to the courier that was too complex for him to memorize. The king's message "was very grand," we are told, "its meaning very deep." But "the messenger's mouth was too heavy; he could not repeat" the message. So the resourceful Enmerkar "patted some clay and put the words on it," creating the first cuneiform tablet. "Before that day, there had been no putting words on clay. But now, when the sun rose on that day—so it was."[2]

Enmerkar's invention solved the herald's memory problem. On hearing his rival's message, the Lord of Aratta was impressed by its complexity and sophistication. When he commanded the messenger to explain how he remembered it, the man simply handed him the lump of clay. "The Lord of Aratta took from the messenger / the tablet (and held it) next to a brazier. / The Lord of Aratta inspected the tablet. / The spoken words were mere wedges—his brow darkened."[3] Enmerkar's invention was utterly unheard of, a revolution in communication. Eventually, writing would make living human messengers optional or obsolete.

Enmerkar's quarrel with the Lord of Aratta and his invention of writing illustrate the Janus-faced paradox of mundane practicality and wonder-working brilliance that characterizes ancient meditation on the art of writing. In fact, writing arose gradually in Sumeria, not as a tool of political rivalry but to meet the bureaucratic needs of inventory. That bureaucracy was in service to a theocratic state, administering religious ceremonies and the taxation that funded them. Despite what looks like a celebration of "genius," Enmerkar's legend echoes the theocratic context of his invention: he needed gold and precious stone to build a temple for the goddess Inana. The Lord of Aratta had plenty of these opulent materials, but a prolonged drought was starving his people; they needed grain, which Uruk had in great abundance. Threats and militaristic boasting had gained nothing for either of the rulers. So, since Enmerkar was the special favorite of Inana, the goddess inspired him to propose a mutually beneficial exchange with Aratta, thereby inventing trade, as he had invented writing. The story may seem utopian now, but international cooperation was not its point. Instead, it demonstrated "the technological superiority of Sumer over the mere owners of raw though precious materials," such as Aratta. Knowing and using writing was Enmerkar's unanswerable boast, the proof of his state's supremacy: "large-scale trade is seen to depend on writing, which simply implies Sumerian." The scribes responsible for this text "coined a term concordant with 'cuneiform,'" and "hinted at the indubitable fact that writing was invented for economic, not intellectual reasons."[4]

Although Enmerkar was celebrated as writing's heroic inventor, in another, apparently previous, Mesopotamian tradition, there was no human originator of writing or anything else. This legend portrayed primeval humans as hopeless incompetents, who had to be taught even the simplest things by a mysterious semidivine being. One late version, recorded by a third-century BCE author named Berossos, asserts that

> *in the very first year* there appeared from the Red Sea [the Persian Gulf] in an area bordering on Babylonia a frightening monster, named Oannes. . . . It had the whole body of a fish, but underneath and attached to the head of the fish there was another head, human, and, joined to the tail of the fish, feet like those of a man, and it had a human voice. Its form has been preserved in sculpture to this day. [emphasis added; figure 1.1]

FIGURE 1.1. Mesopotamian priest. His garb imitates the physique of an Apkallu, as in Berossos's description of Oannes. *A Second Series of the Monuments of Nineveh: Including Bas-reliefs from the Palace of Sennacherib . . . from Drawings Made on the Spot, during a Second Expedition to Assyria.* Ed. Austen Henry Layard. London: J. Murray, 1853. Plate 6, "Fish God. Nimroud." Sheridan Libraries Special Collections, Johns Hopkins University, 913.352 L426M 1852 FOLIO V. 2 C. 1.

Berossos relates that this being

> spent its days with men, never eating anything, but *teaching men the skills necessary for writing, and for doing mathematics and for all sorts of knowledge:* how to build cities, found temples, and make laws. It taught men how to determine borders and divide land, also how to plant seeds and then to harvest their fruits and vegetables. In short, *it taught men all those things conducive to a settled and civilized life. Since that time nothing further has been discovered.* At the end of the day, this monster Oannes went back to the sea and spent the night. It was amphibious, able to live both on land and in the sea.
>
> Later, other monsters similar to Oannes appeared, about whom Berossos gave more information. . . . *Berossos says about Oannes that it had written as follows about the creation and government of the world and had given these explanations to men.* [emphasis added][5]

According to this text, humanity has never discovered anything for itself in all the millennia of its existence; worse, we are incapable of discovery. All knowledge was imparted to our earliest ancestors by this strange being. Oannes's gifts transformed us from beasts to humans.

But there is an enigma: Why should this author, who evidently reveres writing, say that it was transmitted by a monster? Berossos obviously considered Oannes to be the creator of human society. The mystery is solved if we notice that the story was repeated over many centuries. The crucial change occurred when Christian writers took over the tale. Berossos had written in Greek, and so Greek Christians of the fourth century CE, wanting to proclaim the superiority of their religion over paganism, transmitted the tale of Oannes only to ridicule it as a bizarre distortion of the truth about humanity's place in the cosmos, which the one true God had revealed to Moses in the Book of Genesis.

Berossos was literate because he was a priest, and the reverence he showed for writing reveals a priestly snobbery. He presumes an order of societal worth according to class or caste, so the first art he mentions is writing, and the last is agriculture. His presentation of

Oannes's teachings conflicts with modern conceptions of human societal development, which consider writing a late and very gradual acquisition, made necessary in part by the prior invention of agriculture. But Berossos assumes that writing preceded humanity, so that it, not agriculture, is the first and most necessary skill for humans; for Berossos, people needed writing just to *survive*. Without writing, he implies, the first people could not even remember Oannes's teachings, and would have either perished or remained beasts. The inadequacy of preliterate memory and the exaltation of writing as the art of not forgetting was also the point of "Enmerkar and the Lord of Aratta," but that story made writing a human invention. Was it a newer, "secular humanist" revision of writing's traditional history?

In addition to bestowing all knowledge "conducive to a settled and civilized life," Oannes also revealed cosmology and other aspects of prehuman reality. Prehuman realities are also narrated in the Book of Genesis, but the two accounts differ significantly. Genesis did not name a source but simply recounted that "in the beginning God created heaven and earth." Berossos identifies Oannes as his source for the earliest history and cosmology: "His *Babyloniaca*, then, becomes a direct descendant of the actual records deposited . . . before the Flood at Sippar, for the implication is that everything that Berossus has reported up to the Flood derives from these very tablets."[6] Thus, Berossos makes Oannes the ultimate informant for *written* history, whereas the Bible does not mention writing until Moses leads the Israelites out of Egypt. The late appearance of writing left Jews and Christians to imagine a far different scenario from Berossos's: God was the source of knowledge about cosmology and early human history, but he communicated orally until giving Moses the tablets of the Decalogue. The long interval between creation and the Ten Commandments made Moses the only identifiable source for information about archaic times, as well as the logical transmitter of writing "vertically," from the deity to humans.

Although Berossos does not describe him as a god, Oannes's amphibious lifestyle suggests the superhuman ability to live anywhere, under any conditions. Since water is the primal element in Babylonian creation myths, Oannes's adaptability may also reflect the antiquity of his story. Scholars speculate that Oannes was an *Apkallu*, one of the beings that Sophie Cluzan describes as "mythical sages from antediluvian times, who emerged from the sea to reveal science, arts, and technical knowledge to humanity. . . . Owing to their technical knowledge, they were . . . the privileged intermediaries of the gods. In the first millennium BCE, they were considered the mythical ancestors of literate people, keepers . . . of secrets."[7] Berossos mentions the *Apkallu* and gives them names that echo those on a cuneiform tablet from twenty centuries before his time.[8] These features indicate that his account probably condenses a long tradition of writing about civilization and the importance of writing for not forgetting it.

DEATH AND RENAISSANCE OF CULTURE

Writing has equal importance in Berossos's account of the period around the Great Flood. In Mesopotamian mythology, the Deluge demarcated the primeval world from the "modern," postdiluvian world. Berossos tells of a King Xisouthros, who, like Noah, is warned by a god that a flood will destroy the world, and is commanded to save human and animal life in a boat. However, Xisouthros receives a radically different message from Noah's: the god instructs him to safeguard writing, in addition to preserving life.[9] Unlike Genesis, Berossos does not simply presume that people who survive the Deluge will automatically reconstruct their culture, nor does he suggest that Babylonian civilization is prone to impiety, an implication that Jews and Christians saw in the story of Babel (Genesis 10:8–10). Berossos declares outright that civilization is good, and that its rebirth after the Flood requires writing. But because he explicitly mentions the necessity of preserving books, his vision of humanity is even more pessimistic than the morality tale of Genesis. Writing is defined as the art of not forgetting, and to lose it is to forfeit humanity.

In fact, Berossos distinguishes clearly between strategies for saving life and those for safeguarding civilization. In Berossos's Flood-story, written for readers in the hegemonic Greek-speaking culture during the age of Alexander the Great, he Hellenizes the Sumerian god Enki as "Kronos." The god "appeared to Xisouthros in a dream and revealed that on the fifteenth of the month Daisios [April–May] mankind would be destroyed by a great flood. He then ordered him to bury together all the tablets, the first, the middle, and the last, and hide them in Sippar, the city of the sun [i.e., of the sun-god Shamash]. Then he was to build a boat and board it with his family and best friends . . . provision it with food and drink and . . . take on board wild animals and birds and all four-footed animals."[10] "All the tablets" constitute a whole library, containing the encyclopedic memory of antediluvian Sumeria. Significantly, Kronos/Enki commands Xisouthros to safeguard the books even before building the boat. It may seem illogical that the god does not simply order the king to take the books aboard the boat, but this would have been physically impossible. The books were essentially brickbats, and a load of them would be far more ballast than the already groaning vessel was likely to need. In fact, the curious phrase "the first, the middle, and the last" implies that these records were at least in part chronological. Since Berossos declares that 430,000 years elapsed between the beginning of history and the Great Flood, we might imagine that many more than half a million tablets were necessary to contain even the most important information. The fact that Xisouthros's "Ark" does not rescue knowledge as well as life remained essential during the long Jewish and Christian afterlife of Flood-myths. Despite their familiarity with lighter materials—papyrus, animal skins, and eventually paper—when Jews and Christians retold the Flood-myth of Genesis, they did not imagine that Noah rescued writing by taking books into the Ark. Instead, like Berossos, they imagined someone who inscribed antediluvian knowledge on a heavy medium like stone or metal or brick, and left it outside the Ark to fend for itself. (See chapters seven and following.)

After the Flood subsided—a fact Xisouthros ascertained in the same way Noah did, by releasing birds from his boat—he "broke open a seam on a side of the ship" and the company disembarked. Then Xisouthros "prostrated himself in worship to the earth and set up an altar and sacrificed to the gods"—again like Noah. However, unlike Noah, Xisouthros thereupon became a god. He immediately vanished, and despite their searches, his friends could not locate him, until "the sound of a voice that came from the air gave the instruction that it was their duty to honor the gods, and that Xisouthros, because of the honor he had shown the gods, had gone to the dwelling place of the gods, and that his wife and daughter and the steersman had enjoyed the same honor." In the final development of his tale, Berossos again stresses the importance of books and writing: "The voice then instructed them to return to Babylonia, to go to the city of Sippar, as it was fated for them to do, to dig up the tablets that were there, and to turn them over to mankind. . . . After they understood all this, they sacrificed to the gods there and went on foot to Babylonia."

The return to Sippar initiates the first Renaissance in human history, a total rebirth of civilization from oblivion: "And those who had arrived in Babylonia dug up the tablets in the city of Sippar and brought them out. They built many cities and erected temples to the gods and again renewed Babylonia."[11] This passage implies that Xisouthros deserved deification as much for saving "all the tablets" as for rescuing life. However, Xisouthros himself does not decide that saving the books would be a good idea: the god's commands suggest that, like knowledge and writing, foresight and providence are divine attributes, not rational human precautions.

We cannot know when writers first began to reflect on the nature and history of writing; but it seems possible that Berossos's story of Xisouthros came down to him from the world's oldest bibliographic myth. John Dillery notes, "While [Berossos] was writing in Greek—something no Near Eastern intellectual had yet done—he conformed to conventions that were in some cases more than two thousand years old" by his time. Berossos's reference to safeguarding "all the tablets, the first, the middle, and the last" (or "the beginnings, middles, and ends of all knowledge" in another translation) has suggestive parallels in earlier texts.[12] Cuneiform traditions intimate that this feature must have been found explicitly in Sumerian tradition. For one thing, Berossos calls his Noah-figure Xisouthros,[13] a name that transliterates the Sumerian name Ziusudra, an archaic Noah-name meaning "Life of extended days," which appeared in a tale about the Great Flood around 2600 BCE.[14]

Did Berossos transmit a tale from nearly five thousand years ago, in which Ziusudra saved writing as well as life? It seems likely from this evidence. However, Berossos was representing a conquered, "barbarian" culture to a cosmopolitan Greek-speaking audience. As the representative of an ancient and bookish culture, he was arguing for its superiority, and rebutting the triumphalist Greek superiority complex. So, as a Babylonian patriot, he *might* have created Xisouthros's primordial renaissance from whole cloth "only" 2,300 years ago rather than transmitting lore that was twice as ancient. In either case, Berossos's dramatic stories vividly solicit our wonder at the power of writing to withstand erasure and safeguard fragile human culture.

Xisouthros's myth differs from Noah's because Berossos assumes that the Great Flood was above all a threat to the benefits of human culture; it was a menace to the gift of a deity not, as in the Book of Genesis, the punishment sent by an angry god to blot out human sinfulness. And unlike Yahweh, the god who alerted Xisouthros had no control over the Flood, only superior knowledge of it. By burying the books, Xisouthros saved human civilization from a purely natural threat, not from theological or moral disgrace. (A closer parallel to Genesis is in *Atrahasis*, another Babylonian Flood-story, when one of the gods drowns the world because noisy humanity, created by the gods to spare themselves menial drudgery, keeps him awake at night!)[15]

Xisouthros's library conforms to a pattern that persists throughout history. A book, an encyclopedia, or, as with Xisouthros's cache of tablets, a library containing all knowledge, is so important to human civilization that foresight (divine here, but human in later mythologies) must preserve it from destruction in the Flood. The antediluvian book is an archetype, a story of absolutes: it embodies supreme, total, or crucial wisdom, threatened with apocalyptic destruction but essential to human survival and saved by triumphant prudence and artifice. As Berossos demonstrates most vividly, the codex or library of all knowledge is a powerful symbol of the eternal struggle between memory and oblivion, inscription and obliteration, human civilization and bestial ignorance. Ashurbanipal himself, the king whose rediscovered library brought *Gilgamesh* to light, boasted that "I have learnt the craft of the sage Adapa, the hidden secret of all scribal learning. . . . I have examined cuneiform signs on stones [dating] from before the Flood, whose [meaning] is sealed, inaccessible, and confusing."[16]

XISOUTHROS'S AFTERLIFE

Ironically, Berossos's myth of the rescued primeval library was itself almost erased. His *Babyloniaca* was not much read in antiquity; five centuries later, the Christian Eusebius of Caesarea (d. 339/340) knew Berossos only indirectly, through quotations by intervening writers. Until a late sixteenth-century Protestant scholar discovered excerpts from Eusebius in a ninth-century Byzantine chronicle, traces of Berossos's work had all but vanished. Like Xisouthros's books, Berossos's seemed fated to be a lost library. But serendipity blessed them both, as it blessed another ancient Mesopotamian tale, the epic of *Gilgamesh*, which emerged in the mid-nineteenth century from its long hibernation in the sands of Iraq.

Gilgamesh celebrated one of the earliest kings of Uruk, a contemporary of Enmerkar, who would have lived circa 2800–2700 BCE. The epic was probably assembled a millennium later, between 1900 and 1600 BCE.[17] *Gilgamesh* survives in the Akkadian language, but numbers of its episodes were prefigured several hundred years earlier, in short Sumerian poems about "Bilgamesh."[18] From Sumerian to Akkadian, the hero's story maintained a common core of heroic adventures.

Some texts are *metaliterary*: they are writings about writing as well as events. Tablet 1 of the standard Middle Babylonian redaction of *Gilgamesh*, dated between 1300 and 1100 BCE, provides a striking example of metaliterary writing.

> He [Gilgamesh] saw what was secret, discovered what was hidden,
> he brought back a tale from before the deluge.
> He came a far road, was weary, found peace,
> and set all his labours on a tablet of stone.
> . . . .
> [See] the tablet-box of cedar,
> [release] its clasp of bronze!

[Lift] the lid of its secret,
[pick] up the tablet of lapis lazuli and read out
the travails of Gilgamesh, all that he went through.[19]

Although most of us read silently now, we can still "hear" a voice coming from the text, as if it were being read aloud. The voice emanating from *Gilgamesh* describes a tablet and exhorts us to examine the secrets it reveals. But which tablet? Not the splendid one engraved on semiprecious stone: *Gilgamesh* was preserved on much humbler clay tablets and excavated from the Library of Ashurbanipal in the mid-nineteenth century (see plate 1).

Whether the tablet of lapis lazuli ever existed, mentioning it and its decorative box implied that Gilgamesh deserved such a durable, monumental presentation because his adventures were all "firsts"—in fact, some were also "lasts." He was the only living man to visit the afterlife and recover secrets from before the Great Flood. Gilgamesh learned the ultimate secret, that death really is unavoidable, from the Babylonian "Noah," whom the epic calls Uta-napishtim rather than Xisouthros. Ironically—and maddeningly for Gilgamesh—Uta-napishtim/Xisouthros was the exception who proved the rule of human mortality because the gods deified him as a reward for saving all life. Although *Gilgamesh* does not specify who invented writing, it does imply that the art matters so much that the hero's archetypal adventures, including his brutal defeat by ultimate realities, are worthy to be preserved on rare and precious lapis lazuli.

Naram-Sin was another renowned king, celebrated on a famous stele from about 2230 BCE. An Old Babylonian poem poses as his autobiography.

[Open the foundation-box] and read well the stele
[That I, Naram-Sin], son of Sargon,
[Have written for] all time. . . .
Whoever you may be, governor, prince, or anyone else,
Whom the gods shall name to exercise kingship,
I have made a foundation box for you,
I have written you a stele. . . .

> Behold this stele,
> Listen to the wording of this stele. . . .
> Let expert scholars tell you my stele.
> You who have read the stele.[20]

In contrast to *Gilgamesh*, this poem gives us the actual, autobiographical voice of Naram-Sin, who foretells the entrance of another agent in the evolving drama: the scholar or historian who perpetuates the memory and fame of exalted rulers like Naram-Sin. This imagined scholar is no counter of sheep or baskets of grain: his inventories are lists of kings, the institutional memory of the society.

Naram-Sin's autobiography reveals that some tablets or smaller stelae were treated as cornerstones, time capsules meant to be buried and consulted at some future date for the wisdom they contained. The *Stele of the Vultures*, a larger, self-standing Sumerian monument, dated to about 2460 BCE, seems more familiar: it commemorated a military victory of the city-state Lagash. The stele informs us first of all that it has a proper name of its own, which echoes the name of a god.

> The stele, / its name / is not a man's name; it is: "Ningirsu, / Lord, Crown of Lumma, / is the life of the Pirig-eden Canal." / The stele of the Gu'eden— / beloved field / of Ningirsu / [which] Ennatum / for Ningirsu / returned to his [the god's] hand— / he [Ennatum] erected it.[21]

Notice that "its name / is not a man's name," as if the scribe wanted to imply that the stele was somehow divine, despite being a human artifact. The *Stele of the Vultures* is a *self-referential* artifact: it claims to be the original of which it speaks, whereas both Naram-Sin and *Gilgamesh* refer to an original tablet that the reader has to imagine.

In Berossos's story of Xisouthros, the providential god is not said to have invented the knowledge he orders the king to save—Oannes revealed it long ages before. But neither is it clear that Oannes invented writing. Nor does early Sumerian myth always consider writing a human or divine *invention*. In the poem "Inana and Enki," writing is a sort of transcendent possession, almost a concrete thing, which can belong to only one proprietor and cannot be shared. It is "one of the hundred or so basic elements, or 'essences'—the Sumerian term is *me*—of civilization. These *me*'s reside with Enki, the god of wisdom and intelligence and Inana's father, in the ancient cult center of Eridug." Writing is a fact, like rocks or trees, rather than a created artifact. So the goddess Inana, "coveting them, endeavors to acquire the *me*'s for her city, Uruk, by getting her father intoxicated—a common ploy in Sumerian literature—and duping him into giving them to her. Succeeding in her plot, Inana loads the *me*'s, including that of the scribe's craft (Sumerian *nam-dub-sar*) into the Boat of Heaven bound for Uruk."[22] Much later, Berossos thinks of Xisouthros's buried and disinterred books as repositories of information, but they also function allegorically, somewhat like Inana's *me*'s.

INVENTORS, MALE AND FEMALE

In a human alternative to the transcendent *me*'s, Enmerkar, the first king of Uruk, was supposed to have invented cuneiform. *Gilgamesh*, Berossos, and the *Stele of the Vultures* also imply that writing exists primarily for human purposes, whatever its origin. But like "Inana and Enki," other myths defined writing as an attribute of the gods. A Sumerian poem called "The Marriage of Sud" does not include any mention of refined, cultural one-upmanship between narcissistic rival kings as in "Enmerkar and the Lord of Aratta," or their heroic exploits, as in *Gilgamesh* and the *Stele of the Vultures*. Instead the poem communicates that writing is what keeps the entire society alive: it is sacred because it is about communal bonds.

"The Marriage of Sud" recounts how Enlil, the Sumerian god of wind and storms, sought and won the hand in marriage of Sud, a "foreign" goddess. After their suitably passionate wedding night, Enlil "naturalized" his bride into his own regime and changed her name from Sud to Ninlil, echoing his own name. He also "decreed her fate," putting her in charge of several related areas of human life. "*The scribal art, the tablets decorated with writing, the stylus, the tablet-board, / To compute the accounts,* adding and subtracting, the blue measuring-rope, the . . . measuring-rod, the marking of the boundaries, the preparation of canals and levees, / *Are fittingly in your hands.*

The farmer repays you the favor in the fields."[23] As grain-goddess, Ninlil personifies all the various activities that kept the human society fed and functioning, from the farmer who tilled the field, to the scribe who recorded the crop. Yet there seems to be no mention of human or divine inventors.

But we do hear of one historical inventor connected with writing. The earliest identifiable author of literary writings was a Mesopotamian princess, Enheduanna, daughter of the powerful King Sargon of Akkad, and aunt of Naram-Sin. Enheduanna lived from 2285 to 2250 BCE, and was high priestess of the Moon-god Nanna. Like Gilgamesh, Enheduanna holds several "firsts," but hers are writerly and literary: She is the earliest poet to be known by name; her forty-two temple hymns were the first known collection of such poems, and constitute one of the earliest attempts at systematic theology. She made the first extant claim to poetic originality, asserting in one poem, "The compiler of the tablets was En-hedu-ana. My king, something has been created that no one has created before."[24] But did she physically inscribe words on clay with a stylus? Ancient authors were not necessarily their own scribes. So Enheduanna may have composed orally and memorized her verses until she could dictate them to a scribe, as Roman poets, and even the blind John Milton, did many centuries after her. And her scribes were not necessarily male: "Finds from the site of the city of Sippar have yielded tablets with the names of four female scribes, trained, it would seem, by their own scribal families."[25]

Yet, given the rarity of women recognized as writers throughout history, it is unsurprising that by the Old Babylonian period (ca. 1500 BCE) the tutelage of writing had parted ways from the "feminine" patronage of grain and fertility. It passed to a male god, Nabu, scribe of the great god Marduk and patron of both writing and wisdom. By about 1000 BCE, Nabu was one of the most powerful gods of the Babylonian pantheon, co-regent with Marduk in some myths.[26]

ARCHAIC EGYPT

Whether Egyptian or Mesopotamian writing is more ancient, evidence suggests a comparably long history of writings about writing.[27] A major difference between them was the scarcity of stone in Mesopotamia and its abundance in Egypt. Many surviving ancient Egyptian texts are inscriptions on monumental tombs. Some contain the earliest extant autobiographies, so that, as in Mesopotamia, writing is the art of not forgetting remarkable individuals. A Fifth Dynasty tomb (ca. 2450–2300 BCE) reads: "The elder Judge of the Hall, Hetep-her-akhet, says: I made this tomb on the west side of a pure place, in which there was no tomb of anyone, in order to protect the possession of one who has gone to his *ka*. As for anyone who would enter this tomb and do something evil to it, there will be judgment against them by the great god. I made this tomb because I was honored by the king, who brought me a sarcophagus."[28] This inscription is not, strictly speaking, self-referential: unlike the Mesopotamian examples, it speaks of the tomb rather than about itself or its stone "page."

However, another Old Kingdom text, the "Instruction Addressed to Kagemni," clearly describes its own ratification or canonization ceremony. "The vizier had his children summoned, after he had understood the ways of men, their character having become clear to him. Then he said to them: 'All that is written in this book, heed it as I said it. Do not go beyond what has been set down.' Then they placed themselves on their bellies. They recited it as it was written. It seemed good to them beyond anything in the whole land."[29] As in other early Egyptian cases, the import of the text is social, a matter of ethical, moral, or religious etiquette. For our purposes, the more interesting point is the metaliterariness of the passage, the social authority it accords to written words.

Miriam Lichtheim dated the language and style of the "Instruction" to the Sixth Dynasty (ca. 2300–2150 BCE). It was not inscribed on stone but found instead on papyrus of a relatively late date. Was her dating correct? Only Egyptologists would know. But when self-referential texts in later times refer to their own dating, they often provoke the suspicion of fraud. A slab of granite from the New Kingdom (Twenty-fifth Dynasty, ca. 710 BCE) bears an inscription claiming to be the copy of a very old document preserved, like the "Instruction," on papyrus. "This writing was copied out anew by his majesty in the house of his father Ptah-South-of-his Wall, for his majesty found it to be a work of the ancestors which was worm-eaten, so that it could not be understood from beginning to end. His majesty copied it anew so that it became better than it had been before, in order that his name might endure and his monument last in the house of his father Ptah-South-of-his Wall throughout eternity."[30] Judging from this colophon, the original text would have been composed between 3000 and 2040 BCE, inscribed on organic material, and damaged by insects and great age. But was there an original text at all? From antiquity to modern times, the claim that a text was copied from a badly damaged original is often the telltale sign of forgery. In this case, if a fragile original ever existed, it was "erased" less by vermin and age than by the durable granite that replaced it, containing a majestic and possibly fraudulent record of "his majesty" and his ancestors. Yet a humiliating comeuppance awaited this inscription: like its papyrus "original," the text shows many gaps, and its midsection "has been almost completely obliterated." Why? because in much later times the proud monument was demoted to a millstone, eroded by the pressure and friction of its menial task.[31]

Since before Plato, Westerners have recognized that in ancient Egyptian mythology, the god Thoth invented writing and was the patron of scribes. In short, Thoth embodied the value Egyptian culture attached to writing. In "The Eloquent Peasant," an Egyptian text from circa 2040–1650, a man repeatedly petitions a magistrate to redress wrongs done him by a noble. In his eighth petition, the peasant exclaims, "Do Justice for the Lord of Justice / The justice of whose justice is real! Pen, papyrus, palette of Thoth, / Keep away from wrongdoing!" Thoth is the magistrate's patron in both his roles, as judge and as writer. At last the peasant receives justice, after the king's high steward orders all his petitions "read from a new papyrus roll, each petition in its turn."[32] Here writing functions as the art of not forgetting the law and morals.

We could expect a profession protected by such a divine patron to take pride in its trademark "pen, papyrus, and palette." But scribes occasionally took this conviction to extremes. A papyrus from circa 1186–1077 BCE asserts that scribes and the books they write are the only guarantors of immortality—not so much for anyone they celebrate as for themselves. A modern scholar remarks that "the claim that only writers are immortal is astonishing" because it clashes with the received idea of Egyptian culture, first because "the vast majority of Egyptian literary works were produced anonymously." Even more surprising is the writer's disregard for the Egyptian belief in the afterlife, "a *transformed* existence after death for which the buried corpse was merely the point of departure."[33] Here the names of "learned scribes . . . have become everlasting" since they did not entrust their names to the conventional means of remembrance. "They knew not how to leave heirs, / Children to pronounce their names; / They made heirs for themselves of books."[34] Their reed pens and writing boards take the place of children, the surface of stones the place of wives. Even if they were eunuchs, scribes left an inheritance.

But not all the scribes' "relatives" are inanimate: "People great and small / Are given them as children," because "their name is pronounced over their books, / Which they had made while they had being." The text then makes the strongest claim for writers—and thus for writing—that we have yet seen: "good is the memory of [the books'] makers, / It is for ever and all time!" Not only corpses, but houses and tombs and stelae crumble, families and clans die out, we are told, "but a book makes him remembered / Through the mouth of its reciter." Thus, "better is a book than a well-built house, / than tomb-chapels . . . / Than a stela in the temple!" Books "act as chapels and tombs / in the heart of him who speaks [the scribe's] name." Therefore, the poem recommends, "be a scribe, take it to heart, / That your name become as theirs," because "better is a book than a graven stela." Though scribes may have no children of their own, "the children of others are given to them / To be heirs as their own children," by reading their books: "Death made their names forgotten / But books made them remembered!"[35] Here, finally, books enshrine the art of not forgetting writers.

This claim, that the writer immortalizes himself as well as the person he praises, was to become the gravitational center of writings about writing in later times.

THE *BOOK OF THOTH*

There is compelling recent research concerning the lore of writing in ancient Egypt. Papyrus fragments inscribed at various times during the Alexandrian and Roman periods (ca. 330 BCE–400 CE) come from a hypothetical, reconstructed *Book of Thoth*. Fragments in the cursive hieratic and demotic scripts (each faster and easier to write than the hieroglyphic system) concerning scribal craft and lore appear to reflect a coherent body of thought about the role of writing and seem compatible with ideas in texts centuries older. The surviving fragments may even be the remnants of a single, much older text. According to its translators, "the Book of Thoth provides a very rich insight into Ancient Egyptian thinking about the writing system and the act of writing itself," and "manifestly aimed to give to the engaged reader a deep understanding about the symbolic and religious aspects of writing."[36] The book is framed as a dialogue between a master and a disciple and has the characteristics of an initiatory ritual.

Richard Jasnow notes that in the *Book of Thoth*, "the ever-popular [Egyptian] themes" of capturing birds and fish in nets appear to symbolize the activity of scribes. The scribal master is "he who has netted the bas [birds]," while the disciple declares, "I desire to be a bird-catcher of the (hieroglyphic) signs of Isten (= Thoth)."[37] Jasnow adds that "the chief sacred shrine of Thoth in Hermopolis was called the 'House-of-the-Bird/Fish-net,'" and that "images of netting and birds connected with writing may well have an association with this ancient designation of Thoth's temple."[38] Jasnow contends that "the designation of the Master as . . . 'He-has-netted-the-*bas*' perhaps not only refers to the *bas* as the sacred [book]rolls, but also figuratively describes the act of writing, where the *ba* birds represent the individual written signs or sounds."[39] In later Mediterranean cultures, trapping or fettering was an evocative description of active memory, while nets and other textiles symbolized writing as texts.

ARCHAIC GREECE

The oldest reference to writing materials in a European language occurs in Homer (7th–8th c. BCE). Egyptian lore emphasized the personal and professional virtue of scribes, but the *Iliad*'s only mention of writing imputes mysterious, perhaps magical powers to the art, and enmeshes it in a story of treachery. While the Mesopotamian and Egyptian writings discussed above publicize virtue, Homer shows writing concealing vice. Like the biblical Joseph, the hero Bellerophon repulsed a queen's attempt to seduce him and was slandered by her for attempted rape. Rather than kill Bellerophon as his wife demanded, the devious king of Argos sent him to the queen's father, another king, hoping to "pass the buck" of assassinating him. The king of Argos commanded Bellerophon to deliver "murderous symbols, / which he inscribed in a folding tablet, enough to destroy life, / . . . to his wife's father, that he might perish." Instead of killing Bellerophon, the queen's father devised even more indirect traps for the hero than her husband had, sending him on successive "missions impossible" against the dreadful monster Chimæra, a powerful tribe of men, and the warlike Amazons, before finally "[spinning] another entangling treachery" by having his bravest vassals ambush Bellerophon. After failing at every attempt, the king divided his kingdom with Bellerophon and betrothed another of his daughters—a sister of the deceitful queen—to him.[40]

The tale of Bellerophon implies a negative or uncomprehending view of writing and the secrets it can conceal. The story appears to be about the infancy of Greek writing, but the actual history of Greek writing is far more complicated. The poems of Homer—whoever he, she, or they were—are as stratified as the site of Troy itself. The Mycenaean civilization Homer describes ended suddenly in a Mediterranean-wide "Bronze Age collapse" during the twelfth century BCE. The significance of certain technologies became confused or lost in the collapse. Notoriously, rather than fighting from chariots, Homer's heroes use them mainly as taxis to and from the front lines. Another cultural loss was the Mycenaeans' complicated syllabic writing system, Linear B.

The story of Bellerophon is Homer's sole reference to writing; in the *Iliad* and *Odyssey*, no one reads mythological and legendary stories. Instead, illiterate "singers of tales" perform them orally.[41] Indeed, Homer's own Trojan stories were transmitted orally by bards during a "dark age" of four or five centuries when Greek culture apparently forgot how to write. Bernard Knox observes that "the word Homer uses—*grapsas*" literally means "scratching" and only later became "the normal word for writing."[42] The Greeks only rediscovered writing through contact with the Phoenician alphabet in the late eighth or early seventh century BCE, just in time to record the Homeric poems. In the interim, the connection between the instructions on Bellerophon's tablets and his "mission impossible" apparently mystified the oral poets whose tales Homer inherited. For these bards, mention of the tablets was a fossil, like fighting from chariots: they seem to have thought the "symbols" were magical and somehow contained a power "murderous . . . enough to destroy life" (*Iliad* 6.178).

Perhaps, in some version of the story that was lost centuries before the *Iliad*, the tablets contained a simple utilitarian message: "Kill the bearer of this letter." Yet the only remaining Linear B texts are inventories, essentially lists of nouns and numbers, inadequate even for composing a simple demand for murder. But could "Homer" the writer have been thinking of the new Phoenician-derived writing? Was he simply using magic as a metaphor, nothing more than poetic license? Ultimately, the historical circumstances are irretrievable; what remains in the text is the uncanny suggestion of magical secrecy, treachery, and attempted murder.

Homer's reference to Bellerophon's tablets cannot be considered self-referential. The *Iliad* is not accounting for its own origin but referring to writing outside of and preexisting its own story. This broader kind of reference, metaliterariness, later became widespread in fictional literature, especially in historical novels couched as autobiographies or epistolary exchanges, such as *Robinson Crusoe*, *Tristram Shandy*, *Les Liaisons dangereuses*, and thousands of later ones (see chapters twelve and thirteen).

The archaic Mesopotamian, Egyptian, and Greek examples show a striking awareness of writing as an art—a human activity that has rules and can be taught—and its implications for human culture during formative stages of its development. The fascinated consideration of writing as both a technology of social organization and something more philosophical or symbolic seems to have been strong in Mesopotamia and Egypt. But unless far more texts of Linear B or a breakthrough in understanding the even older Linear A writing system should emerge, one would have to conclude that such awareness was all but missing from archaic Greece. The sole Homeric example ultimately connects the history of reflection on the art of writing to the history of magic, where the use of symbols (both graphic and object-based) to perform wonders remained constant even in literate contexts.

CHAPTER TWO.

An Age of Philosophy, 600 BCE–400 CE

Philosophy is the attempt to give an account of what is true and what is important based on a rational assessment of evidence and arguments rather than myth, tradition, bald assertion, oracular utterances, local custom, or mere prejudice. As with many of the arts and sciences that make up Western civilization and culture, philosophy was first defined as such by the Greeks around the fifth century BCE. However, evidence suggests that many of the problems, concepts, and approaches that became known as philosophy in Greece originated in other places and times.

—RICHARD POPKIN, *The Columbia History of Western Philosophy* (1999)

Classical antiquity was the age when Greece and Rome gave writing the variety and versatility that caused later ages to marvel. As we saw, the Homeric texts do not refer to writing except in the story of Bellerophon. Instead they claim inspiration by the Muses, the "daughters of memory," those godlike guardians and transmitters of what we could call the collective memory of the Hellenes. A great number of texts emerged during the centuries that followed the *Iliad*'s debut in writing (ca. 750–650 BCE). Hesiod, who lived around the time when the Homeric poems were written down, is considered the first Western poet to include details of his own life in his writing. He may have personally written down his poems or dictated them, though whether or to what extent he used writing cannot be established.[1] He says that the Muses turned him from a shepherd into a poet and "breathed into me / divine song, that I might spread the fame of past and future."

However, they warned him—and us—that although they knew "how to tell the truth" itself, they did so only "when we wish." Moreover, they warned that they knew "how to tell many lies that pass for truth."[2] As for the difference between truth and lies or fiction, there was no guarantee—and, in a strict sense, no notion—of accuracy or consistency. When later Greeks looked for a human origin of their poetry before Homer and Hesiod, they attributed it to the mythical figures Orpheus and Musaeus. Eventually (5th–4th c. BCE), written poems were attributed to Orpheus, and some of the shorter ones (3rd–2nd c. BCE) survive.

Philosophy, the "love of wisdom," was the most characteristic invention of post-Homeric society. The goal of philosophy, whether it investigated humanity or the natural world, was always in some sense truth. Philosophers of the sixth and fifth centuries BCE are conventionally called pre-Socratic, on the understanding that Socrates (d. 399 BCE) was a revolutionary thinker. Socrates and his disciple Plato railed against their contemporaries the Sophists, or "wise men," whom they accused of selling techniques of argument rather than truth, but all of them paid more attention to rhetoric, ethics, and metaphysics than had the pre-Socratics. These early thinkers had focused on the physical world or cosmos, speculating about its origins and makeup in a critical or rationalistic manner, without depending on mythological explanations.

Writings of the pre-Socratics survive only in fragments preserved by later philosophers, and there is some question whether Thales, the earliest pre-Socratic, wrote at all; no work of his was cataloged in the famous Library of Alexandria, in Egypt. Thales famously predicted an eclipse that occurred in 585 BCE, perhaps basing his knowledge on Babylonian astronomy (though later Greeks thought he learned it in Egypt, an apprenticeship they commonly attributed to their philosophers). According to Aristotle, Pherecydes of Syros, around 540 BCE, wrote on the gods and the world's origin but did not "say everything in mythical form." Pherecydes also seems to have been a transitional figure for having written in prose rather than verse, as earlier philosophers had done.[3] At the other end of the Age of Philosophy, one of the last great figures was Hypatia, an expert in mathematics and astronomy. She taught Platonic and Aristotelian philosophy in Alexandria, until she was murdered by a Christian mob in 415 CE.[4]

The Age of Philosophy produced far more writings, on many more subjects, than the Archaic age. Like epic poetry, philosophy had begun in oral culture. Theater—comedy and tragedy—also depended fundamentally on oral presentation to the public, but by the later sixth century BCE, plays were being recorded in writing. The iconic tragedies of Aeschylus, Sophocles, and Euripides are all from the fifth century, when, as one historian observes, prose and poetry "snowballed"; copyists, booksellers, book collectors, and private libraries appeared by about 400 BCE.[5]

The growth of writing and reading inspired the discipline of *philology*, a "love of words or discourse" as a set of *written* processes that could be understood historically as well as philosophically. Philology began in earnest in the third and second centuries BCE when scholars in the great libraries of Alexandria and Pergamum undertook to create standard texts of Homer and the fifth-century Greek tragedians. The *Iliad* and *Odyssey* posed a particularly complicated task, since their transmission had originally been entirely oral.

Because the Homeric stories were not written down until the infancy of the Greek alphabet, after 800 BCE, no single authoritative "original" text of either poem existed. The two epics had to be assembled, choosing from a welter of episodes, in stages that can only be conjectured. Altogether, the development from oral tales to written epics took about seven centuries to complete. Dividing the *Iliad* and *Odyssey* into "books" of comparable length—twenty-four for each poem, "numbered" by the letters of the Greek alphabet—was simply the most visible phase of the enormous project that created the two poems.

Assembling the *Iliad* and *Odyssey* did not put an end to the process. Because the Homeric poems were copied and recopied by hand over many generations, scribal errors frequently crept in, as anyone understands who ever copied more than a few lines by hand.[6] Scribes in all ages sometimes complained openly, on the very copies they penned, that transcription was mechanical, soul-killingly dull, and painful to the eyes, the hands, and the body as a whole. Desks designed for comfortable writing were a late medieval invention.[7] (Reading as well would long present what now seem obstacles to comfort: even spaces separating words were not universal before the twelfth century. Moreover, some ancient texts had been written *boustrophedon*, "as the ox plows," reversing direction from one line to the next, alternating between left to right and right to left.)[8] Mistakes frequently garbled the sense of a passage, and they accumulated over the centuries. Errors multiplied in a perverse kind of chain reaction: a scribe who was unskilled or momentarily distracted could miscopy a word, or worse yet, accidentally skip a word or a whole line, leaving a gap, or lacuna, in the text. The reader, and the next scribe, must hazard a guess—educated or not—at the originally intended sense. The safest procedure for restoring an author's meaning was to compare, or "collate," two or more copies, hoping they did not all contain the same errors. Even so, manuscripts containing the correct wording might not have survived, or could be unavailable to the copyist-editor. Trying to guess the original meaning might risk actually falsifying the passage.

THE MUNDANE MYSTIQUE OF CLASSICAL WRITING-LORE

In archaic Greece, as in Mesopotamia, writing had been valued for the supremely mundane purpose of accounting and inventory. However, unlike in Mesopotamia, the reflection on writing would remain largely unromantic in the Greek and Roman world, even after large bodies of mythology about the gods were written down. Except for legends concerning Egyptian divinities, notably Isis and Thoth (Hermes), Greeks and Romans considered the origin and nature of writing to be human, not divine. For the ancient Greeks, by contrast to the modern world, oral culture was a supreme, public, human value; they prized bardic storytelling (even after Homer was written down), theatrical representation, political oratory, and philosophical discussion. Writing was a mere instrument, a technology subservient to this oral culture, valued mainly for preserving the outlines and possibilities of spoken discourse. This "mundane mystique" of writing may have also reflected Greeks' awareness—and Romans' after them—that their alphabets were imported "horizontally" from other cultures rather than revealed "vertically" by superhuman figures like Oannes and Thoth. Writing was of this world, not transmitted from an otherworld, as Mesopotamian and Egyptian scribal elites had defined it. Indeed, writing systems that preceded the alphabet required a long and difficult scribal apprenticeship that nourished the sense of a superhuman origin. Not until writers of Greek and Latin developed familiarity with Judaism and Christianity, did writing as an attribute of the gods or their gift to humans become an emotionally significant concept for them.

Post-Homeric mythology about the human history of writing, particularly the dependence of the Greek alphabet on the Phoenician model, has been borne out by modern scholarship. Herodotus, who died in 425 BCE, is celebrated as the "father of history." In his usage, *historiē* meant "inquiries" or "researches," a quest for knowledge by humans. His overall view of writing had mythological features, but the agents who interested him were human; he wrote that Phoenicians led by Cadmus settled in Boeotia, "and they transmitted much lore to the Hellenes, and in particular, taught them the alphabet which, I believe, the Hellenes did not have previously, but which was originally used by all Phoenicians." He noted that the Ionians called letters *ta Phoinika*, "the Phoenician things."[9] He was aware that "with the passage of time, both the sound and the shape of the letters changed," and remarked that "after making a few changes to the form of the letters," the Hellenes "put them to good use." Herodotus claimed to have seen inscriptions dating from the transitional phase, when the Greek alphabet still showed its Phoenician roots: "I myself have seen these Kadmeian letters at the sanctuary of Ismenian Apollo in Boeotian Thebes. These letters, which are engraved on three tripods, look for the most part like Ionian letters."

Like later Greeks and Romans, Herodotus was attentive to the history of writing materials. He noted that "the Ionians have called papyrus scrolls 'skins' (*diphthera*) since long ago" because "when papyrus was scarce, they used the skins of goats and sheep instead." In fact, he concluded, "even in my time many barbarians still write on such skins."[10] Herodotus equated the move from animal skins to papyrus with cultural sophistication, but for him as for later Greeks, the term barbarians—*barbaroi*—did not primarily indicate sanguinary ferocity (after all, papyrus was an Egyptian commodity!). Rather, *barbarian* expressed a negative linguistic and cultural judgment: the word mimicked the sound of "babbling" in languages other than Greek, or of Greek spoken by nonnatives.

Nonetheless, Herodotus's assessments of barbarian cultures were remarkably positive. His anecdotes about non-Greek writings are suffused with a sense of wonder and suggest one reason why later authors accused him of telling tall tales rather than histories. Because he claimed to depend on native informants, Herodotus's assessments of barbarian writings are less factual-sounding than what he wrote about the Hellenes. One of his explanations remains a commonplace in Western culture. Unaware that ancient Egypt actually had three systems of writing (hieroglyphic, hieratic, and demotic), Herodotus related that the Egyptians "write with two different scripts, one called 'sacred,' the other called 'public.'" His notion survives in the terms *hieroglyphic* and *demotic,* Greek for "sacred carvings" and "popular" writing.

Archaeology has validated the graphic distinction, but for Herodotus it confirmed his admiring claim that "of all peoples, they [Egyptians] are the most religious."[11] Much of his enthusiasm for Egypt derived from his belief that its religion was extremely ancient. He even wrote that "the names of the gods came to Hellas from the barbarians . . . and derive specifically from Egypt."[12] One of his more striking examples, which continued to influence the study of mythology as late as the seventeenth century, was Herakles (Hercules). Contrary to his countrymen's belief, he proclaimed, "the Egyptians did not take the 'name of Herakles' from the Hellenes, but . . . the Hellenes . . . took it from the Egyptians," who "recognize a certain Herakles, a god of great antiquity." By contrast, "the Hellenes tell many different naïve stories, and their myth of Herakles is especially foolish."[13] According to his Egyptian informants, he said, the Egyptian Herakles lived "17,000 years prior to the reign of Amasis"—an Egyptian ruler who died in 526 BCE, a century before Herodotus's birth.[14] Herodotus's fantastic timeline was unremarkable at a time when biblical chronology, with its brief timeline, had not penetrated Greece. Christianity was five centuries in the future, and even the Hebrew Bible was unknown to Greeks. Their references to the birth and even death of gods often reflected the belief that gods were originally very ancient humans, "divinized" by descendants grateful for their benefactions (as Romans later divinized their emperors). Given this mundane mystique, Herodotus had no need to think of writing as a gift of the gods.

Extensive libraries and systematic bibliographers like those of Ptolemaic Alexandria were still two or three centuries in the future, so for information about non-Greeks, Herodotus says he relied on the priests who served Egyptian temples and trusted their objectivity. He relates that at Memphis, priests of Hephaestus (the Egyptian god Ptah) admitted unchauvinistically that Phrygian civilization was older than Egyptian. He also respected priests at Egyptian Thebes and Heliopolis, "for of all the Egyptians, the Heliopolitans are said to be the most learned in tradition."[15] Sometimes, as here, he seems to cite oral traditions rather than written sources. But his claim to be writing history rather than myth reflected the value he attached to temple documents. In another discussion of the gods, he says the Egyptians calculate that the god Dionysus was born 15,000 years before King Amasis, adding that "the Egyptians claim they are absolutely certain of the truth of these calculations, since they have always counted and always recorded the years."[16] Elsewhere, he says that priests of Hephaestus at Memphis informed him about Min, the legendary founder of their city and first king of Egypt, and that "after this king, the priests related from a papyrus roll the names of 330 kings of Egypt." As far as Herodotus was concerned, it was true, if "it is written."

He was especially respectful of monumental inscriptions as historical sources. He found inscriptions by a later king, Sesostris, interesting both as history and for their peculiar, gendered aspect. "[He] assembled a large army with which he marched through Asia, conquering every nation that he encountered. Now, for those . . . who struggled bravely and fiercely for their freedom, he set up pillars in their lands with inscriptions declaring his name, his native land, and how he had subdued them with his might. And for those whose cities he took easily and without a fight, he inscribed the same words on the pillars as he did for the courageous peoples, but he added an image of female genitals, wishing to publicize their impotence." Herodotus says the majority of these scabrous monuments "are evidently no longer in existence," but he himself "saw some of them in Palestinian Syria that had the previously mentioned inscription and depiction of the genitals of a woman."[17]

Whether he always "told the truth" or not, and even if his written sources are sometimes exaggerated or even imaginary, Herodotus's notion that writings are the most reliable source of history became commonplace among later historians. Columns, pillars, and stelae fascinated him as enduring sources of historical evidence. Narrating the exploits of Darius, he says that after the Persian king had gazed on the Bosphorus, he "set up on its shore two pillars of white stone, one engraved with Assyrian writing and the other with Greek, listing all the peoples who had contributed troops to the army that he was leading."[18]

Herodotus's claim to rely on trustworthy, historical, written Egyptian sources motivates his detailed contradiction of Homer's *Iliad* and *Odyssey*.[19] According to Herodotus, after fleeing with Paris from her husband, Menelaus, Helen never arrived at Troy. Instead, Proteus detained the lovers in Egypt. In the *Odyssey* (4.477–79, 582–87), Proteus was a shape-shifting Egyptian god, but Herodotus calls him a human king based in Memphis. King Proteus discovered the extent of Paris's crimes from the Trojan prince's servants; he was tempted to kill him but sent him on his way, keeping Helen in Egypt throughout the war. So the catastrophic war was fought in vain, since the Trojans failed to convince the Greeks that Helen was not in Troy. Herodotus relates that after the war, Menelaus sailed to Egypt, reclaimed Helen from Proteus and started for home, though not until he had sacrificed two Egyptian children to obtain favorable winds. This assertion debunks the Greek Troy-saga, but also ironically mirrors Agamemnon's sacrifice of his daughter Iphigenia to gain favorable winds for beginning the Trojan expedition. "Afterward, when the Egyptians found out what [Menelaus] had done, he fled straight to Libya in his ships, hated and pursued. But where he went from there the Egyptians did not know. They said that they knew some of these details from their own inquiries, and others from what had happened to them directly, which of course they knew with certainty."[20]

Herodotus's first-person accounts of his research supplant Homeric myth with strictly human testimony, allegedly far older than Homer's, and passed down in writings by priestly guardians of religion. "When I asked the priests whether the Hellenes' account of what happened at Troy was fictitious, they claimed that they knew the truth of this affair, because Menelaos himself was the source of their information."[21] So barbarians, not Homer, transmitted the real truth about the war, which they learned firsthand from a chief conqueror of Troy. Herodotus went further, indicting Homer's tales as inconsistent with human psychology, logical behavior, and the bare facts of Trojan history:

> That is what the Egyptian priests said, and I agree. . . . [I]f Helen had been in Troy, the Trojans would certainly have returned her to the Hellenes, whether Alexandros [Paris] concurred or not. For neither Priam nor his kin could have been so demented that they would have willingly endangered their own persons, their children, and their city, just so that Alexandros could have Helen. . . . In fact, since the kingship was not even going to devolve upon Alexandros, he could not have hoped to control matters in Priam's old age. It was Hektor, both older and more of a man than Alexandros, who was to inherit the crown . . . and he would never have entrusted affairs of state to a brother who committed injustices.[22]

Herodotus did not condemn Homer for a liar, as later Greeks and Romans did. Instead, he conjectured that the bard's poems showed he "had also heard this version of the story. But since it was not as appropriate for epic composition as the version he adopted, he rejected it." Later ages accused Homer of writing dishonest history, but Herodotus was more sophisticated, treating him as an editor respectful of the difference between history and poetic myth, a distinction that Aristotle later made. Herodotus's assumption that Homer was a writer rather than an illiterate bard endured largely unchallenged until the eighteenth century.

To modern readers familiar with Greek myth, Herodotus's tale could seem like a joke or a forgery. But he apparently told it so as to provide a plausibly motivated *human* story, a history stripped of gods and marvels, and based on authoritative archival sources. His respect for Egyptian religion as a guardian of historical truth showed through his assertion that "a divine force arranged matters" to "clearly demonstrate . . . that when great injustices are committed, retribution from the gods is also great."[23]

PLATO'S HISTORY OF WRITING

Like Herodotus, other classical writers habitually showed a pragmatic, utilitarian orientation toward writing, particularly in their attention to the history of writing materials. Plato, who lived about three quarters of a century after the "father of history," included factual-sounding references to the use of tablets of various sorts. Some, on durable materials, were used for contracts, public records, and even personal documents, such as wooden tablets proving honorable discharge from military service.[24] In the *Statesman*, Plato mentions "laws . . . inscribe[d] on tablets of wood and of stone," while in the *Laws* (5.541c), he refers to "written records inscribed on [tablets] of cypresswood being laid up in the temples as a memorial to times to come." Further on (*Laws* 6.753c), he speaks of nominating candidates by means of tablets on which each voter inscribed "the name of his nominee, his father, his tribe, and the ward to which he belongs, [along with] his own name with the same particulars." A letter attributed to Plato updates the dangers Homer had referred to: the writer declares he will continue his discussion "in riddles, so that in case something happens to the tablet 'by land or sea . . .' he who reads may not understand." The writer cautions that it is "a very great safeguard to learn by heart instead of writing" because "it is impossible for what is written not to be disclosed."[25] This ironic twist on the art of not forgetting turns it into an untrustworthy practice that can overrule and negate the writer's intention, revealing secrets to the wrong reader.

Some tablets were like modern notepads. Usually made of wood, they formed little trays of wax that could be inscribed with a pointed stylus; its blunt other end erased the writing by smoothing over the wax. In addition to notebooks and rough drafts, waxed tablets were used for brief letters. Bernard Knox observes that Homer's word for

Bellerophon's tablets, *pinax*, was used by later Greeks for waxed tablets.[26] Although the *Iliad* does not mention wax, evidence suggests that waxed tablets were used soon after Homer. On one luxurious tablet from the seventh century BCE, the early Etruscan alphabet, which was adapted from the Greek, is inscribed around the ivory frame, perhaps for the benefit of an affluent beginning writer (see plate 2). It is unlikely that Plato referred to wax-covered tablets in the *Laws*; perhaps he was thinking of them in the *Theaetetus* (section 191c–196), though his actual words refer to "blocks" of wax bearing impressed seals rather than writing on tablets. When describing schoolmasters teaching children to write, he referred to slates (*Protagoras* 326d); being both durable and erasable, they continued in use until modern times. Such media as slates and wooden tablets seem to be implied by the Greek word for writing, *graphein*, apparently related to a word for scratching, as we saw in the myth of Bellerophon.

For longer texts, where ease of consultation and relative durability were both required, Greeks of the literate age used animal skins or papyrus rather than tablets. Being made from a plant that was not indigenous to Greece, papyrus had to be imported, so it is plausible that writing on skins preceded the use of papyrus, as Herodotus assumed. Unlike Herodotus, however, Plato did not compare the relative merits of skins and papyrus as writing media.[27] But he told a richly evocative story that depends implicitly on the use of papyrus. Plato recounted that the Athenian lawgiver Solon (d. 558 BCE) once visited a temple library in the Egyptian city of Sais on the Nile, the river renowned as source of the papyrus reed. Like Herodotus, Solon learned in Egypt that Greeks had no idea of the true age of "ancient history." Plato tells that during Solon's conversations with the Egyptian priests, he made the gaffe of parading his own historical knowledge, and was swiftly ridiculed by a priest.

> [Solon] started talking about Phoroneus—the first human being, it is said—and about Niobe, and then he told the story of how Deucalion and Pyrrha survived the flood. He went on to trace the lines of descent of their posterity, and tried to compute their dates by calculating the number of years that had elapsed since

> the events of which he spoke. And then one of the [Egyptian] priests, a very old man, said, "Ah, Solon, Solon, you Greeks are ever children. There isn't an old man among you." On hearing this, Solon said, "What? What do you mean?" "You are young," the old priest replied, "young in soul, every one of you. Your souls are devoid of beliefs about antiquity handed down by ancient tradition. Your souls lack any learning made hoary by time."[28]

In other words, Greek "history" is myth, based on an *absence* of written sources (a primary meaning of *muthos* is "speech"). As Berossos would do several decades later, Plato here established a direct relationship between the survival of writing and the continuity of civilization. The old Egyptian priest explained that the fabulous tales of Deucalion and Phaeton were merely distorted reflections of the distant past. He observed that it was improbable that someone named Deucalion had actually built a boat to ride out a flood, or that a child of the sun god ever lost control of his father's chariot and scorched the earth.

The old priest continued, presumably in a more conciliatory tone, that the Greeks were rustic and childish because, although they were an ancient people, their cultural memory was never old. Like Berossos, the old priest blamed nature, not the wrath of a god, for such calamities: floods and wildfires were caused by influences of the heavenly bodies.[29] Again like Berossos, the priest disqualified oral transmission as a means of preserving history because even written records were vulnerable to natural cataclysms. Underneath the Greeks' childish myths lay the reality that throughout the past, their record-keeping had been frequently interrupted by floods and fires. Greece's characteristic climate and terrain were eternally subject to extremes of heat and wet, so Greek civilization was doomed by topography and weather to be short-lived, always dying in its infancy, perpetually condemned to start over afresh. When floods occurred, only people living on high ground avoided being washed away and drowned; conversely, great conflagrations spared only the inhabitants of wet, low places. Whatever cultural memory Greek writings had built up after any single cataclysm would eventually be erased by another disaster. In the

Laws, Plato elaborates further on the destruction of civilizations by floods: for "an immense period of time," "untold tens of thousands of years," "the many generations of men who led such a life were bound, by comparison with the age before the Deluge or with our own, to be rude and ignorant in the various arts."[30]

Plato's *Timaeus* drew a scientific contrast between the Greeks' tragic loss of cultural memory and the optimal preservation of Egyptian temple archives. If the superiority of Egyptian to Greek culture was due to its greater antiquity, this resulted from Sais's fortunate terrain and climate. Thanks to the drainage of the Nile Delta, Sais was protected when the river flooded, but the delta's abundance of water also protected the city from conflagrations. Sais was therefore a uniquely bibliophilic environment: inside the almost miraculous haven of its archives, historical records were safe from the obliteration that menaced them everywhere else. Thus the priests had preserved accurate records of Egypt's interminable history. But—an unhoped-for bonus to Solon and Athens—Sais also warehoused records of foreign peoples, revealing that their histories were astonishingly longer than the foreigners themselves could suspect. As the old priest explained, "Of all the events reported to us, no matter where they've occurred—in your parts or in ours—if there are any that are noble or great or distinguished in some other way, they've all been inscribed here in our temples and preserved from antiquity on."[31]

Plato stages a contrast between civilization and savagery by dramatizing the eternal conflict between writing and erasure, inscription and obliteration, memory and oblivion. His story alludes to the fragility of writing, not only in general but presumably with special reference to papyrus, the preferred writing support for both Egyptian and Greek record-keeping. Like Berossos's myths of Oannes and Xisouthros, the story of Solon's Egyptian trip emphasizes that a civilization's written memory prevents not only its loss of identity but its complete annihilation. Despite the vulnerability of writing, Plato's myth is a paean to that art as the only guarantor of human continuity, a repository of memory that may be glorious but is constantly threatened. This paradox was the feature of Plato's story that would have the broadest implications for the history of culture.

Plato's Egyptian tale is more self-consciously writerly than his usual accounts of Socrates's discussions with his pupils. Somewhat like early modern novelists, he gives the story a strong imaginary premise, yet claims it is based on true historical records: it is, in essence, a historical novella about the meaning of history itself. It is fully as positive about the benefits and powers of writing as Berossos's stories of Oannes and Xisouthros. As we saw, the Babylonian writer declared spectacularly that writing had existed for hundreds of thousands of years, a claim that itself may have been as ancient as *Gilgamesh*. Plato refers to records from somewhat more recent times, supposedly written "only" a handful of millennia before his time. However, unlike the myths of Berossos, Plato's has the feel of an original composition, invented, as his myths often were, to prove a point. In many ways it resembles Herodotus's revisionist story of Helen and the Trojan War.

Herodotus's and Solon's trips to Egypt, whether historical or invented, reflect a twinned interest in exotic places and primordial times that had taken firm root in Greek intellectual life before Plato. Greek legends often related that major figures of its culture had journeyed to Egypt and learned history and other wisdom from the already ancient civilization on the Nile. Even Homer was fabled to have made the trip; in fact, according to one late account, he was actually an Egyptian, and another made him a Babylonian.[32] Although Egyptians and Babylonians were "barbarians," Greek legends credited them with historical knowledge about the ancient Hellenes that Greeks themselves had forgotten because it was obliterated by catastrophes.

Like Herodotus, Plato differs from Berossos in one crucial respect: he was interested mainly in human history and purely human achievements. While the Babylonian assumed that all knowledge had been transmitted vertically from the superhuman Oannes, Plato's story of Solon launched a horizontal myth. Rather than present a feckless humanity receiving arts, crafts, or sciences through otherworldly revelations, Plato imagined the human transmission of historical knowledge from one age or society to another. We could plausibly defend this vision as humanistic; indeed it captured the imagination of many European humanists in the fifteenth and sixteenth centuries, and was influential even before then.

Like Herodotus, Plato tells a revisionist history, although he does so tongue in cheek. As if to counter the old Egyptian priest's humiliating accusation of Greek historical illiteracy, Plato slyly, playfully enlists writing in the service of Greek pride. He insinuates that, although the Egyptian priests dismiss Greek civilization as perpetually infantile, they acknowledge that Athens's history is longer and prouder than Egypt's. With a wink to the reader, Plato recounts that, in the optimally preserved records at Sais, Solon discovered that Athens had once been more powerful and important than Egypt. Documents on file in the priestly archive chronicled how the original, primeval Athens had valiantly saved Mediterranean civilization nine thousand years before Solon's birth, when it repelled an invasion mounted by Atlantis, an advanced and aggressive civilization located on an immense continent off the western shores of Europe. Modern Greeks, wrote Plato, knew nothing of this: not only had their own records been repeatedly destroyed by floods and fires, but Atlantis disappeared into the ocean shortly after the Athenian victory, leaving no trace.

Socrates's disciple Critias claims to possess the notes Solon took when he visited the priestly archives of Sais. According to Critias, Solon had been so impressed by what he learned from the Egyptians that he wrote it down and entrusted the manuscript to an ancestor of Critias. Critias thus claims a written chain of custody for the story he tells about the Greeks' lack of ancient writings. Had Solon not journeyed to Egypt, the Athenians would never have known the grandeur of their own ancient history, since their entire civilization had been destroyed and rebuilt many times over. Ironically, Atlantis itself would have vanished from Greek cultural memory if not for the notes that Solon brought back from the archives of Sais, which, Plato imagines, enabled Socrates's circle to learn the "true history" of civilization and the causes for Greeks' ignorance of it. It sounds very much as if Plato were one-upping Herodotus's claim to have learned historical truth from barbarian records.

Like Berossos, Plato had no notion of a universal flood similar to the biblical Noah's. Even Deucalion's flood must have been localized, since it apparently spared Sais, and anyway the *Timaeus* treated that flood as a mythologized distortion of history. Although they agreed that writing was fragile, Plato and Berossos differed markedly on the nature of its enemies. Berossos and the Mesopotamian myths he depended on imagined only one means of catastrophic obliteration: a flood. As they knew, clay tablets actually benefit from fire, which turns them into durable terracotta, whereas unfired tablets of dried clay would dissolve. (Many centuries later, discoveries in the Iraqi desert were to establish that difference quite dramatically. The fire that destroyed Nineveh in 612 BCE ravaged the palace containing the Library of Ashurbanipal, but it was a boon to the library's holdings, which the fire preserved.)[33] Conversely, the usual Greek and Egyptian supports for writing—papyrus, parchment, wooden or waxen tablets, even stone and metal—were supremely vulnerable to fire; thus Plato included fire as the other archenemy of writing. For him, the preservation of culture and civilization took place on the razor's edge, as it were, in a "Goldilocks zone" of climatic equilibrium midway between too wet and too hot. After Plato, *obliteratio* or de-lettering became an archetypal, obligatory motif in Western bibliographic myths.

AGAINST WRITING: PLATO'S SUPPOSED PREJUDICE

Even in antiquity, not all authors celebrated writing's ability to preserve memory. The range of emotions about writing in past ages is impressive, for not everyone had positive feelings about it. Although classical texts characteristically display a "prosaic" or humanistic attitude toward writing, Plato's dialogues went beyond complacency, embracing ambivalence. His *Phaedrus* tells a counter-myth, a second, theological romance about writing which, on its surface, rebuts the humanistic tale of Solon's Egyptian researches. In *Phaedrus*, Socrates relates a story from "once upon a time" that explains how writing was invented in Egypt—a civilization quite as old as Mesopotamia, and more familiar to Greeks. The god Theuth (i.e., Thoth) once went on a regular spree of invention, says Socrates, creating numbers and arithmetic, geometry, board games, dice, and, "above all else, writing." The god visited the King of Egypt to show off his brainchildren. Among his discoveries, he was proudest of writing and enthusiastically touted it as "a potion for memory and for wisdom," which "will make the

Egyptians wiser and will improve their memory." Supremely unimpressed, the king emphatically disagreed. He declared that

> since you are the father of writing, your affection for it has made you describe its effects as the opposite of what they really are. In fact, it will introduce forgetfulness into the soul of those who learn it: they will not practice using their memory because they will put their trust in writing, which is external and depends on signs that belong to others, instead of trying to remember from the inside, completely on their own. You have not discovered a potion for remembering, but for reminding; you provide your students with the appearance of wisdom, not with its reality.[34]

The king accused the god of inventing an unnecessary crutch, a prosthesis that would create the very handicap it was intended to relieve. Writing is not only unnecessary but debilitating, causing memory to atrophy. Knowledge not already in the knower's memory is of no real benefit, thought Socrates.

In *Timaeus* and *Phaedrus,* Plato produced polar opposite assessments of the value of writing: Was it a boon or a bane to memory? An instrument of human providence, or the gift of a deluded god? Ironically, the *Phaedrus* starts on a positive note: much like Berossos, it describes writing as the gift of a superhuman figure. But writing comes in for severe criticism; it is so different from memory that Socrates compares it to a handicapped interlocutor who lacks any mental flexibility and can only repeat the same words to everyone, whether they understand or not.

In recent times, historians of writing and other scholars have showered so much attention on Plato's counter-myth in *Phaedrus* that it has all but eclipsed the *Timaeus*'s positive tale. But *Phaedrus* tells less than half the story of Plato's attitudes to writing. After all, he was a writer, and Socrates was not (so far as anyone knows). But the Socrates who tells the story of "Theuth" might not be an accurate representation of the Athenian philosopher who died in 399 BCE.[35] In any case, the *Timaeus* implicitly refutes the *Phaedrus*'s deprecation of writing. Socrates's disciples, who tell the story of Solon, emphasize the social dimension, moving from their master's fixation on personal wisdom and contemporary society to transgenerational concerns about memory and cultural continuity. Plato's apparently contradictory fables about writing actually assess both sides of humanity's relation to it. To emphasize the *Phaedrus* at the expense of the *Timaeus*, as some scholars have done, is to confuse the social value of writing (historical continuity) with its significance for individuals.

The philosopher Jacques Derrida scorned this confusion: "Only a blind or grossly insensitive reading" of the *Phaedrus* "could . . . have spread the rumor that Plato was simply condemning the writer's activity." The god Theuth's description of writing as a *pharmakon* (potion, remedy, medicine, or elixir) for forgetfulness stresses only the positive connotations of the Greek word for *drug*, but *pharmakon* can mean "poison" as well as "remedy." A drug can kill as well as cure. "The stated intention of Theuth being precisely to stress the worth of his product," he presents the *pharmakon* "from a single one, the most reassuring, of its *poles*."[36] When the king objects that "the *pharmakon* of writing is good for *hypomnēsis* (rememoration, recollection, consignation) and not for *mnēmē* (living, knowing memory)," he dramatizes both the power and the limitations of writing.

The ambivalence of Plato's *pharmakon* may reflect the primitive distrust of writing that Homer portrayed through the "murderous symbols" on Bellerophon's tablet. Derrida remarks on Plato's "mistrust of the mantic and magic, of sorcerers and casters of spells," claiming that "Plato is bent on presenting writing as an occult, and therefore suspect, power." Derrida further relates Plato's suspicion of the occult to his animus against painting, optical illusions, and *mimesis* (the artistic imitation of reality) in general. Socrates made this distrust explicit and notorious when he excluded poets from his *Republic*: "One and the same suspicion envelops . . . the book and the drug, writing and whatever works in an occult and ambiguous manner open to empiricism and chance, governed by the ways of magic and not the laws of necessity."[37]

Socrates's aversion to writing and the ambivalence toward it between *Timaeus* and *Phaedrus* can be seen as symptoms of the transition from an oral Greek culture to a literate one, a process that reached a watershed moment during Plato's lifetime. The classicist Eric Havelock postulated that a fundamental change in Greek philosophy was traceable to the development of writing, causing "a larger intellectual revolution, which affected the whole range of the Greek cultural experience."[38] Havelock wrote, "All human civilisations rely on a sort of cultural 'book,' that is, on the capacity to put information into storage in order to reuse it. Before Homer's day, the Greek cultural 'book' had been stored in the oral memory. . . . Between Homer and Plato, the method of storage began to alter, as the information became alphabetised [written down], and correspondingly the eye supplanted the ear as the chief organ employed for this purpose."[39] The infamous decision to exclude poets from Plato's *Republic* was the key to Havelock's thesis. Havelock read the *Republic* less as the political program for a future utopia than as "an attack on the existing educational apparatus of Greece," in which "the poets are central."[40]

"Poetry" for Plato had nothing to do with our postliterate idea of it, wrote Havelock. Poetry was not yet the aesthetic production of an individual *writer* but rather the oral medium that transmitted a traditional, completely unwritten culture. The poets Plato referred to were not creative geniuses *writing* for a sophisticated, literate audience but illiterate bards and, by Plato's time, rhapsodists. Some rhapsodists were literate, but their *oral* recitations of epic poetry constituted the encyclopedia—that is, expressed the total outlook—of their largely illiterate culture.

Plato attacked this poetry as a haphazard, unsystematic, unsophisticated instrument for transmitting culture. Worse yet, poetry seduced its listeners, stimulating their passions through song and proposing terrible models of conduct. Living only three or four centuries after the *Iliad* and *Odyssey* were written down, Plato was commenting on an ancient, traditional culture that was still largely unreflective. He was judging that culture from the perspective of a subculture—call it philosophy—that developed over the previous two centuries, culminating in his own decision to *write* it. The goal of philosophy was not to conserve and transmit a single society, as the bards had done, but to understand humanity and the world at large. The task called for analytical thinking not respect for ancestral precedent and tradition. Plato's own program "is conveyed in the Greek term *episteme* for which our word science is one possible equivalent." His curriculum should equip his student "to define the aims of human life in scientific terms and to carry them out in a society which has been reorganised upon scientific lines."[41]

According to Havelock, however, Plato misconstrued the purpose of Greek poetry: he expressed "the astonishing presumption" that it had been "*conceived and intended* to be a kind of social encyclopedia" that should educate its listeners in approximately the same way as philosophy.[42] This was unfair as well as unhistorical. As Havelock declared, "The poet as a possible claimant to fill this role becomes an easy target; we feel too easy. He should never have been placed in such an inappropriate position."[43] Judging oral poetry by the standards of written philosophy, Plato concluded that, by his time, poetry was "doing a very poor job" of educating citizens. Socrates eshewed literacy, as had his philosophical predecessors, who transmitted their thought orally and even in verse. Plato, despite his ambivalence toward writing, made philosophy the writable product of an increasingly literate society, beginning a trend that would canonize Homer as a writer, an author, until the eighteenth century.

Since the 1960s Havelock's work has inspired broader meditations on orality and literacy by philosophers, including early proponents Walter Ong and, notoriously, Marshall McLuhan. The latter proposed that literate, "linear" culture culminated with Gutenberg's invention of movable type but was transformed again by electronic communication. McLuhan described a process by which, in the transition from print to television to computers (and, we may now add, the internet and "smart" telephones), the "global village" increasingly relies on the richer sensory resources of oral transmission within a generally literate culture.[44]

Havelock's revisionist perspective on the *Republic* and Derrida's on the *Phaedrus* suggest that the *Timaeus*, which is thematically related to both dialogues, may have been similarly distorted over time. In *Timaeus* and its companion-piece *Critias*, Plato's myth of Atlantis became one of the most influential stories ever told. As Western societies became increasingly literate, the fiction of a primeval written source for the tale clearly added to Atlantis's appeal. Westerners no longer believe that the myths of Zeus, Athena, and the other Greek gods are literally true or possible, but Atlantis continues to fascinate readers and filmgoers, whether naive or sophisticated, who daydream about archaic, utopian societies, rivaling or surpassing our own. Solon supposedly learned that, notwithstanding the Atlanteans' aggression toward Europe, they had enjoyed an ideal, virtuous society, regulated by laws decreed by Poseidon and preserved on a column of precious metal in the god's sanctuary (*Critias* 119c).

Plato's description of Atlantis has inspired centuries of utopian fiction and pseudo-historical speculation. Imagining entire lost societies seems particularly to have "grabbed" writers of the Renaissance who were sensitized by their recovery of writings from pre-Christian Greek and Roman culture. It can be shown that Renaissance writers were moved by the *Timaeus*'s depiction of inscription and erasure, but their successors have increasingly neglected the topic. Ironically, as literacy spread in modern times, Plato's emphasis on writing and the Egyptian archives was progressively displaced by fascination with more spectacular ancient technologies, such as those that produced the pyramids.

Writing philological fictions about writing, as in *Timaeus*, necessarily begins from books, libraries, and erudition—and often ends there—whereas writing a utopian fiction like *Critias* allows an author to embrace every conceivable aspect of society in a lively, novelistic way, suggesting a complete alternate reality. From Thomas More's *Utopia* (1516) onwards, "Solon's" fragmentary description of Atlantean society inspired works that were both novelistic and philosophical, imagining alternatives to the social realities of writers' own times. Francis Bacon's *New Atlantis* (1627), describing an ideal society on an island in the Pacific Ocean, was directly inspired by its Platonic namesake. Others carried the utopian sociological and technological themes into the realm of science fiction, creating such early classics as Tommaso Campanella's *City of the Sun* (1602) and Swift's *Gulliver's Travels* (1726/1735). Utopias beyond the Earth even appeared, in Cyrano de Bergerac's *States and Empires of the Moon* (1657), his *States and Empires of the Sun* (1662), and in Voltaire's *Micromegas* (1752).

DIODORUS OF SICILY, WRITING, HISTORY, AND LIBRARIES

But along with Herodotus and Plato, several writers of classical Greece and Rome remained important in the emotional history of writing, from their own times until after 1800. During these centuries, they were quoted and invoked as authorities by almost everyone who speculated about the history of writing. The Sicilian Greek Diodorus Siculus (ca. 60–30 BCE) is in some ways more crucial than Herodotus or Plato for Greco-Roman ideas about the history of writing. Diodorus wrote a history of the world so extensive that he called it *Bibliotheke*, or *The Library*, as if he were trying to realize the "universal" history of Plato's Egyptian priests. Yet his deepest meditation on writing comes almost as an aside, not at "the beginning of time," but when discussing the history of Greeks in their early Sicilian colonies. He tells that around 600 BCE, the city of Thurium (modern Catania) chose as its lawgiver one Charondas, who went about his task with enthusiasm and thoroughness comparable to Solon's projects for Athens. "He, after examining the legislations of all peoples, singled out the best principles and incorporated them in his laws; and he also worked out many principles which were his own discovery."[45]

Diodorus was particularly impressed by Charondas's egalitarianism, a rather striking exception to the societies we have so far seen: "Charondas also wrote another law which is far superior . . . and had also been overlooked by lawgivers before his time. He framed the law that all the sons of citizens should learn to read and write, the city providing the salaries of the teachers; for he assumed that men of no means and unable to provide the fees from their own resources would be cut off from the noblest pursuits."[46] Diodorus then launched into a panegyric about the benefits of writing that remained unequaled until the time of Pliny the Elder a century later: "In fact the lawgiver rated reading and writing above every other kind of learning, and with right reason; for it is by means of them that most of the affairs of life and such as are most useful are concluded, like votes, letters, covenants, laws, and all other things which make the greatest contribution to orderly life. What man indeed could compose a worthy laudation of the knowledge of letters?"[47] Diodorus accepts his own challenge with alacrity:

> For it is by such knowledge alone that the dead are carried in the memory of the living and that men widely separated in space hold converse through written communication with those who are at the furthest distance from them, as if they were at their side; and in the case of covenants in time of war between states or kings the firmest guarantee that such agreements will abide is provided by the unmistakable character of writing. Indeed, speaking generally, it is writing alone which preserves the cleverest sayings of men of wisdom and the oracles of the gods, as well as philosophy and all knowledge, and is constantly handing them down to succeeding generations for the ages to come.

While writers before and after him intoned similar praises of writing, Diodorus, as if responding to Plato's *Phaedrus*, proclaimed that "while it is true that nature is the cause of life, *the cause of the good life is the education which is based upon reading and writing*." He credits Charondas with a kind of class consciousness, the realization that although writing benefits humanity as a whole, it can exclude and debase individuals if they are illiterate. "And so Charondas, believing as he did that the illiterate were being deprived of certain great advantages, by his legislation corrected this wrong and judged them to be deserving of concern and expense on the part of the state."[48]

This awareness, and the effort to put it into practice, says Diodorus, made Charondas not merely a wise legislator, but the greatest of all. "He so far excelled former lawgivers who had required that private citizens when ill should enjoy the service of physicians at state expense, that, whereas those legislators judged men's bodies to be worthy of healing, he gave healing to the souls which were in distress through want of education, and whereas it is our prayer that we may never have need of those physicians, it is our heart's desire that all our time may be spent in the company of teachers of knowledge."[49] To phrase this somewhat anachronistically, Diodorus implies that reading and writing are essential to the "care of the self" as well as the welfare of the society.[50] As we shall see, the notion that reading is "healing for the soul" is Diodorus's best-known contribution to the emotional history of writing.

The early books of his *Bibliotheke* are an extended encomium of history as the noblest use of writing. Like Plato's Egyptian priests, Diodorus set out to write the history not merely of his homeland but of all civilizations. For him, as for Herodotus, the universal archives of the Egyptians were not imaginary but a historic example to be praised and imitated. Accordingly, Diodorus saw himself as continuing a great and idealistic tradition: "It has been the aspiration of these writers [of universal histories] to marshal all men, who, although united one to another by kinship, are yet separated by space and time, into one and the same orderly body." Because their art overcomes spatial and temporal isolation, he says, historians are "ministers of Divine Providence," a Stoic concept that was later modified by Christianity. "For just as Providence, having brought the orderly arrangement of the visible stars and the natures of men together into one common relationship, continually directs their courses through all eternity, apportioning to each that which falls to it by the direction of fate, likewise the historians, in recording the common affairs of the inhabited world as though they were those of a single state, have made of their treatises a single reckoning of past events and a common clearing-house of knowledge concerning them."[51] Here is an early example of ambitious comparisons between the order of the cosmos and the organization of a book, such as we will see in later chapters: for Diodorus, a universal history describes a human cosmos.

Despite this idealistic, ecumenical depiction, Diodorus recognized that universal history was not the original form of historical writing. By nature, he said, historians are inclined to patriotism, and every nation claims to be the most ancient and accomplished: "Not only do Greeks put forth their claims but many of the barbarians as well, all holding that it is they who were autochthonous and the first of all men to discover the things which are of use in life, and that it was the events in their own history which were the earliest to have been held worthy of record."[52] Here we can almost hear an echo of the Egyptian priest's sardonic chuckling at Solon's naivety about Greek antiquity.

The question of precedence cannot be resolved, says Diodorus, precisely because writing is a relatively new art. "Now as to who were the first kings we are in no position to speak . . . for it is impossible that the discovery of writing was of so early a date as to have been contemporary with the first kings. But if a man should concede even this last point, it still seems evident that writers of history are as a class a quite recent appearance in the life of mankind."[53] As a new art, writing is necessarily still evolving; historiography is merely its latest phase. Implicit here is the idea that writing is a human invention, not a gift of the gods.

Though he does not explicitly say so until this point in his book, Diodorus assumes throughout, like Herodotus, that the study of history can only be undertaken thanks to the advantages offered by the written word. In opening his work, Diodorus emphasized that the study of history benefits all peoples. Its purpose should not be to establish cultural precedence, but rather to impart vicarious experience, a saving of both time and pain: "The understanding of the failures and successes of other men, which is acquired by the study of history, affords a schooling that is free from actual experience of ills." Diodorus offered "long-suffering" Odysseus as proof that reading offers a less strenuous education than the school of hard knocks: "The most widely experienced of our heroes suffered great misfortunes" owing to his lack of historical knowledge, but we can learn from his example because reading is "a schooling, which entails no danger, in what is advantageous."[54] The idea that history repeats itself, inherited from Thucydides, became a cliché, but Diodorus offered it as a still-fresh insight.[55]

Unlike Herodotus, Diodorus considers Homer a historian rather than a spinner of myths; and history, he says, is more effective than myth in "reinforcing morals toward goodness and virtue." Thanks to writing, history offers a kind of immortality as an incentive to virtue: "For all men, by reason of the frailty of our nature, live but an infinitesimal portion of eternity and are dead throughout all subsequent time; and while in the case of those who in their lifetime have done nothing worthy of note, everything which has pertained to them in life also perishes when their bodies die, yet in the case of those who by their virtue have achieved fame, their deeds are remembered forevermore, since they are heralded abroad by history's voice most divine."[56]

Diodorus is saying that the most comprehensive history will offer the greatest range of examples, and thus the most abundant incentives to virtue.

> If a man should begin with the most ancient times and record to the best of his ability the affairs of the entire world down to his own day, so far as they have been handed down to memory, as though they were the affairs of some single city, he would obviously have to undertake an immense labour, yet he would have composed a treatise of the utmost value to those who are studiously inclined. For from such a treatise every man will be able readily to take what is of use for his special purpose, *drawing as it were from a great fountain.*[57]

Universal history offers yet another benefit, by minimizing information overload. "In the first place, it is not easy for those who propose to go through the writings of so many historians to procure the books which come to be needed, and in the second place . . . because the works vary so widely and are so numerous, the recovery of past events becomes extremely difficult of comprehension and of attainment."[58] Diodorus reveals that "information overload" was not discovered by the computer age, or even by the spread of printing in the 1500s. In addition, he was troubled by a paradox that later writers often overlooked: information overload is only problematic once information *under*load—the inaccessibility or destruction of sources—has been overcome.

Because he deprecated patriotic one-upmanship, Diodorus intended to offer the best history of historiography that he could assemble. "Since Egypt is the country where mythology places the origin of the gods, where the earliest observations of the stars are said to have been made; and where, furthermore, many noteworthy deeds of great men are recorded, we shall begin our history with the events connected with Egypt."[59] Without mentioning the *Timaeus*, he was as impressed as Plato by the record-keeping of Egyptian priests. The Egyptians, he says, "have preserved to this day the records concerning each of these stars over an incredible number of years, this subject of study having been zealously preserved among them from ancient times." They claim the Chaldeans of Babylonia learned astrology from them, he relates, but he does not attempt to settle the question.[60]

From at least the time of Herodotus, ancient Greek authors acknowledged their culture's debts to Egyptian writing, and Diodorus gives an especially full catalog.

> Many customs that obtained in ancient days among the Egyptians have not only been accepted by the native inhabitants, but have aroused no little admiration among the Greeks; and for that reason, those men who have won the highest repute in intellectual things have been eager to visit Egypt, in order to acquaint themselves with its laws and institutions, which they considered to be worthy of note. . . . Now it is maintained by the Egyptians that it was they who first discovered writing and the observation of the stars, who also discovered the basic principles of geometry and most of the arts, and established the best laws.[61]

After repeating the Egyptians' attribution of these arts to Thoth, Diodorus, like Herodotus, says that many of the Greeks' gods, and their primeval heroes, such as Herakles, were originally Egyptian. He implies that Herakles was a writer, and says he "set up his pillar in Libya."[62] Such inscriptions prevent our being misled when we encounter Egyptian heroes bearing the same names as famous Greeks: "The son of Alcmenê, who was born more than ten thousand years later . . . in later life became known instead as Heracles . . . because, having avowed the same principles as the ancient Heracles, he inherited that one's fame and name as well."[63]

To this unbelievably ancient age belong two of the primeval gods, Isis and Osiris. Diodorus subscribed to the euhemeristic idea that the gods were simply humans who had been "divinized" by their descendants because of their exceptional benefactions. He devotes more space than Herodotus to Isis and Osiris, describing them as great inventors and avid travelers, who each, like Herakles, erected a pillar to commemorate their own beneficence to humanity.

> Some historians give the following account of Isis and Osiris: The tombs of these gods lie in Nysa in Arabia. . . . And in that place there stands also a stele of each of the gods bearing an inscription in hieroglyphs. On the stele of Isis it runs: "I am Isis, the queen of every land, she who was instructed of Hermes, and whatsoever laws I have established, these can no man make void. I am the eldest daughter of the youngest god Cronus; I am the wife and sister of the king Osiris; I am she who first discovered fruits for mankind; I am the mother of Horus the king; I am she who riseth in the star that is in the Constellation of the Dog; by me was the city of Bubastis built. Farewell, farewell, O Egypt that nurtured me."
>
> And on the stele of Osiris the inscription is said to run, ". . . I am Osiris the king, who campaigned over every country as far as the uninhabited regions of India and the lands to the north, even to the sources of the river Ister, and again to the remaining parts of the world as far as Oceanus. I am the eldest son of Cronus, and being sprung from a fair and noble egg I was begotten a seed of kindred birth to Day. There is no region of the inhabited world to which I have not come, dispensing to all men the things of which I was the discoverer."

Isis and Osiris lived so long ago that their monuments have been largely erased: "So much of the inscriptions on the stelae can be read, they say, but the rest of the writing, which was of greater extent, has been destroyed by time."[64] Despite claiming elsewhere that writing is comparatively new, Diodorus implies here that written commemoration must have existed in remote antiquity.

In more recent ages, says Diodorus, the famous sages of Greece were all educated to some extent in Egypt:

> As evidence for the visits of all these men, [the Egyptian priests] point in some cases to their statues, and in others to places or buildings which bear their names; and they offer proofs from the branch of learning which each one of these men pursued, arguing that all the things for which they were admired among the Greeks were transferred from Egypt.

> Orpheus, for instance, brought from Egypt most of his mystic ceremonies, the orgiastic rites that accompanied his wanderings, and his fabulous account of his experiences in Hades.

From Orpheus, Homer got the idea that Hermes was a psychopomp, or guide for souls to the underworld, but Orpheus himself originally got it from Egypt.[65]

Despite all his praise of Egypt and his sense of the cultural historian's obligation to avoid partisanship, Diodorus could not altogether refrain from discussing Greek claims. But his history of writing among the Greeks is somewhat more complex than that of Herodotus. He credits mythical poets, including Linus and Orpheus, with writing down primeval stories of the god Dionysus that they heard in Libya and Nysa, the Arabian town where he located the stelae of Isis and Osiris. One of these legendary figures, Pronapides, was "the teacher of Homer and a gifted writer of songs"; another "lived at the same time as Orpheus" and was a brother of the Trojan king Priam.[66]

These stories do not contest the Phoenicians' invention of writing, but Diodorus records others that do. Some authors say that "the Syrians are the discoverers of the letters, the Phoenicians having learned them from the Syrians and then passed them on to the Greeks, . . . these Phoenicians . . . sailed to Europe together with Cadmus and this is the reason why the Greeks call the letters 'Phoenician.'" In fact, some say "that the Phoenicians were not the first to make this discovery, but that they did no more than to change the forms of the letters, whereupon the majority of mankind made use of the way of writing them as the Phoenicians devised it, and so the letters received the designation we have mentioned above."[67] What to make of this information may be nothing more than that Diodorus, attempting fairness, was diligently recording the various versions he had encountered and that these legends in the fifth book of his *Library* were not intended to contradict the praise of Egypt in the first book.

A CULTURAL CHAUVINIST?

However, Diodorus also included legends that flatly contradicted his, Plato's, and Herodotus's homage to Egypt as the birthplace of civilization. He tells us that some say that "to the Muses . . . it was given by their father Zeus to discover the letters and to combine words in . . . poetry." Discussing the Heliadae, or offspring, of Helios, Diodorus says that these "children of the Sun" lived on the island of Rhodes, and thus were quintessentially "Greek." So, "besides having shown themselves superior to all other men, [they] likewise surpassed them in learning and especially in astrology."[68] At this point, Plato's story of Solon among the Egyptian priests comes in for a significant revision. One of the Heliadae,

> Actis, sailing off to Egypt, founded there the city men call Heliopolis, naming it after his father; and it was from him that the Egyptians learned the laws of astrology. But when at a later time there came a flood among the Greeks and the majority of mankind perished by reason of the abundance of rain, it came to pass that all written monuments were also destroyed in the same manner as mankind; and this is the reason why the Egyptians, seizing the favourable occasion, appropriated to themselves the knowledge of astrology, and why, since the Greeks, because of their ignorance, no longer laid any claim to writing, the belief prevailed that the Egyptians were the first men to effect the discovery of the stars.[69]

Following in Plato's footsteps (though without his irony), Diodorus reverses the traditional roles of Greece and and Egypt in the genealogy of knowledge, but declares that the Phoenicians also benefited from Greeks' ignorance of their own achievements. "Likewise the Athenians, although they were the founders of the city in Egypt men call Saïs, suffered from the same ignorance because of the flood. And it was because of reasons such as these that many generations later men supposed that Cadmus, the son of Agenor, had been the first to bring the letters from Phoenicia to Greece . . . since a sort of general ignorance of the facts possessed the Greeks."[70] Diodorus expands Plato's ironic hint that since Athens predates Sais, any glorification of Egypt's culture is indirect praise of Athens.[71] But Diodorus strongly implies that "reasons such as these" indicate that the Greeks really were cheated of their claims to antiquity and cultural preeminence.

Diodorus may have been unable to shake off Greek patriotism, and his history of writing may ultimately be discontinuous and self-contradictory, but the deepest significance of his inconsistencies is to demonstrate the enormous importance of writing for the knowledge of history and thus for the continuity of civilization. Perhaps he was sincere earlier when he declared his neutrality toward quarrels over cultural antiquity and precedence. But he provides the strongest evidence so far that the invention of writing was considered the invention of civilization, not simply the means of not forgetting it.

CHAPTER THREE.

Collections, Histories, and Forgeries, 300 BCE–400 CE

> Now, notwithstanding these great and manifold benefits, which men have all along received from this curious and wonderful invention, "It is very remarkable (says one of our celebrated penmen) that writing, which gives a sort of immortality to all other things, should be, by the disposal of divine providence, so ordered, as to be careless in preserving the memory of its first founders."
>
> —WILLIAM MASSEY, 1763

THE EGYPTIANS AS INVENTORS OF LIBRARIES

Although he claimed, like Herodotus, to base his descriptions of Egypt partly on his personal experiences there, Diodorus also quoted an earlier Greek traveler to Egypt, Hecateus of Abdera (4th c. BCE). Unfortunately, aside from what the *Bibliotheke* preserved, Hecateus's *Aigyptiaka* is completely lost, adding another layer of irony to Diodorus's depiction of cultural continuity. Even more ironically, Hecataeus bequeathed him one of the most enduring legends about the history of writing.

> Not only do the priests of Egypt give these facts from their records, but many also of the Greeks who visited Thebes [three centuries ago] and composed histories of Egypt, one of whom was Hecateus, agree with what we have said.
>
> Ten stades from the first tombs, he says . . . stands a monument of the king known as Osymandias. . . . And it is not merely for its

> size that this work merits approbation, but it is also marvellous by reason of its artistic quality and excellent because of the nature of the stone, since in a block of so great a size, there is not a single crack or blemish to be seen. The inscription upon it runs: "King of kings am I, Osymandias. If anyone would know how great I am and where I lie [i.e., in my tomb], let him surpass one of my works."[1]

Diodorus's story of Osymandias was to have profound influence on the historiography of writing. In 1455, the Italian humanist Poggio Bracciolini translated much of the *Bibliotheke* into Latin. From then until 1818, when Percy Shelley published his famous sonnet "Ozymandias," European readers thought of the ancient pharaoh not as Shelley's hubristic despot but as the founder of the earliest library (figure 3.1).

Diodorus quotes Hecateus on the stupendous complex of buildings now known as the Ramesseum—Osymandias being a throne-name of the powerful pharaoh Ramesses II (d. 1213 BCE). Diodorus says that Hecateus paid particular attention to "the sacred library, which bears the inscription 'Healing-place of the Soul.'[2] Other translators have rendered Diodorus's phrase *psuches iatreion* as "medicine for the soul" or "the soul's dispensary."[3] Diodorus's quotation of this peculiar phrase recalls his praise of Charondas's Catanian law mandating universal male education as "healing for souls."[4] Thanks to Poggio's translation, Diodorus's name became permanently attached to the "motto" of Osymandias in the scholarly imagination and was perhaps the most widespread encomium of writing known to early modern scholars. Whatever Diodorus thought about the true inventors of writing, centuries of Europeans saw his homage to Osymandias as the highest expression of traditional Greek admiration for writing, books, and libraries.

Osymandias appears nowhere else in surviving ancient literature: if not for the *Library*, nothing of his legendary achievement would have reached us. Diodorus's quotation from Hecateus is ironic in yet another way: the older writer is emblematic of a vast and shadowy library of ancient authors whose works are almost completely lost and

FIGURE 3.1. "The first library, erected by Osimandias." The scribe's text depicted reads, "Medical chamber for the soul," translating Diodorus of Sicily's mention of the inscription *Psuches Iatreion.* Print. Rijksmuseum. Legaat van de heer S. Emmering, Amsterdam. Public domain, Wikimedia Commons.

have reached us only as scattered quotations, little patches quilted into the fabric of later writers' books. Diodorus named his *Library* well, for it reminds us that tantalizing survivals of ancient texts were constantly mentioned in writings about writing, from Sumeria onwards. There is nothing so fascinating as a fragment, and this is particularly true of textual survivals. "Fragments of Ancient Greek Historians" is the title of two large, modern collections that extricate even the tiniest passages of lost ancient texts from works written later and examine them for clues of all kinds.[5] Not only historians but a hefty percentage of all ancient authors who wrote in Greek survive only or primarily in such collections of fragments: epic poems, pre-Socratic philosophy, and other genres are represented.[6] Diodorus's title indicates he would have recognized these collections as libraries, but for us they are libraries of lost as well as rescued manuscripts. Erudite scholars in all sorts of disciplines still pore over the fragments in these libraries, just as passionately as Shelley's admirers daydream over his "Ozymandias." Like Plato, Diodorus foreshadows those novelists—as well as many outright forgers—who have kept company with learned daydreamers since antiquity.

STARRING THE LIBRARY OF ALEXANDRIA

What if we could travel back in time and see all the lost books of antiquity gathered in one place and time, as they were before they perished? In effect we would be visiting the fabled Library of Alexandria. For centuries, this collection, gathered by a successor to Alexander the Great in the Greco-Egyptian city that bore his name, was considered a Platonic ideal, the library that came as close as possible to owning every book.[7] In Alexandria, the library as building and institution was only part of the Mouseion, the "Museum" or House of the Muses, which apparently foreshadowed university campuses housing think tanks. It was a complete environment for scholars, financed by the dynasty of the Ptolemies, and so thoroughly under their patronage that it was sometimes compared to a bird cage. The modern commonplace of the university as "ivory tower" is not an inappropriate comparison.

Such was the appeal of the Ptolemies' project to collect "all books" that, by the second century CE, the Roman writer Aulus Gellius speculated that it had contained 700,000 scrolls. Scholars have questioned this figure, but even now, some nonspecialist reference books cite it as factual.[8] The legends surrounding the Library's destruction are comparably exaggerated: it is unlikely that Julius Caesar destroyed significant parts of the library when he burned his ships to escape the troops of Cleopatra's brother. Modern scholars also speculate that the books destroyed were probably not the originals housed in the Mouseion, but copies of them stored in warehouses at the port and being readied for export. In fact, there were apparently several "destructions of the Library of Alexandria," all but the last of which were partial. A famous legend relates that when Muhammadan armies conquered Alexandria in 641, the Caliph Omar ordered that the library scrolls remaining from previous destructions should be burned as fuel to heat the city's public baths, a supply that lasted for months. Omar's reasoning was supposedly that all the books either agreed with or contradicted the Qur'an, and were thus either superfluous or heretical. (Strikingly similar nonchalance—though more radical—was supposedly shown by a crusading Catholic abbot in southern France, who in 1209 ordered his troops to exterminate an entire town that sheltered Albigensian heretics, proclaiming that "God will recognize his own.") Whether Omar's legend is true or not, it was probably not originally a Christian slander, since we hear it from a Muslim source.[9] At any rate, Luciano Canfora relates that well before Omar's time, the Library of Alexandria earned its place among "the melancholy experiences of the war waged by Christianity against the old culture and its sanctuaries: which meant, against the libraries."[10] Christian zealots targeted books as well as "idolatrous" works of art: iconoclasm was reinforced by biblioclasm.

The mythology of the Alexandrian library is as ancient as Osymandias's "healing-place of the soul," but surviving historical evidence for it is abundant. In both cases, the concept of "library" is anachronistic, driven by the later history of words and institutions. Despite the evidence of fragment collections like those previously mentioned, the two-thousand-year process of mythmaking has doubtless exaggerated the size of the Alexandrian collection. However large the Library's holdings may have been, scholars caution that the term *bibliotheke* did not ordinarily indicate a vast collection like those of the British Library, the Bibliothèque Nationale, the Vatican Library, or the Library of Congress. The Greek word for "library" was originally the term for one or more bookshelves (*bibliothekai*). The fabled primeval library of Osymandias was probably not even a room, but rather a single bookshelf.[11] "Library," like any other concept, changes over time and place becoming prone to exaggeration and idealization.

The destruction of the Alexandrian library is the best known biblioclasm in antiquity but hardly the only one. The Library of Ashurbanipal, as we saw, was destroyed in the seventh century BCE, a good four centuries before the Mouseion was founded. Like Ptolemy Soter, the legendary founder of the Alexandrian library, Ashurbanipal apparently intended his collection to contain a copy of every book in existence, though neither man could know quite what that ambition entailed. Both libraries shared two essential, modern-seeming features: the deliberate campaign of collection-building and an organized catalog of their holdings. But despite being newer than the library at Nineveh, the Alexandrian collection well deserves its fame as the poster child of biblioclasm. The fire that gutted Ashurbanipal's library preserved a large percentage of his clay tablets; the same cannot be said for the papyrus rolls of Alexandria.

A number of recent books, including Canfora's *The Vanished Library*, illustrate just how thickly the history of writing, books, and libraries is encrusted with myth and legend.[12] The rise of archaeology since Shelley's "Ozymandias" allows modern scholars to question the old literary myths, even while partially depending on the same literary sources. That reliance allows moderns to indulge, albeit unconsciously or ironically, the mystique of writing. Our ultimate aim is the same one that drove pre-nineteenth-century scholars: to understand better the history of writing, books, and libraries. But the centuries since Champollion have made possible a more down-to-earth assessment of that history. While the element of nostalgia in modern studies is often strong enough to resemble fiction, the account they render is hypothetical, not mythical, a "possible world" not a "lost" world.

The point, for us, is not demythification or whether stories about lost libraries contain elements of historical truth. Rather, what counts is our awareness that myths and legends about writing are most numerous and romantic when they concern masses of books that have perished, especially in the distant past. Like paleontology, they evoke the pathos of extinction and the thrill of recovery and reconstruction. The recent "refoundation" of the Library in modern Alexandria will go a long way toward either de- or remythologizing the ancient museum of the Ptolemies, depending on its orientation to the tourist trade.[13] Likewise, the two myths of Ozymandias—either Shelley's hubristic despot or Diodorus's inventor of libraries—which are otherwise so divergent, intersect at the point where wonder locates both the magnificence of human achievement and its inevitable destruction by time, physics, and folly. All three myths remind us that our concerns with human survival are entwined with our dread that, even beyond our individual fragility, the "codex" of human civilization could be easily erased by warlords, religious fanatics, or the elements. Beneath fears of obliteration and oblivion lies our conviction that amnesia is the archetype of all human losses.

EGYPT, LAND OF BOOKS

As we saw, the god Thoth was revered by Egyptians as the inventor and patron of writing four thousand years ago. Although Socrates belittled Thoth's invention in Plato's *Phaedrus*, later philosophers who wrote in Greek did not share this attitude. Aware of the Mesopotamian and Egyptian reverence for writing, they adopted Thoth as the personification of writing's antiquity and divinity. In his treatise *On the Mysteries of the Egyptians*, the Neoplatonic philosopher Iamblichus (d. 325 CE) adopted the persona of "Abamon," a modern Egyptian priest, and declared that "Hermes [Thoth], the god who presides over rational discourse, has long been considered, quite rightly, to be the common patron of all priests; he who presides over true knowledge about the gods is one and the same always and everywhere. It is to him that our ancestors in particular dedicated the fruits of their wisdom, attributing all their own writings to Hermes."[14] This phrasing shows an early awareness that misattribution of authorship does not always betray a desire to deceive; acknowledging a belief about the origin of an idea can be a simple homage.

Iamblichus/Abamon repeats traditional Greek references to Egypt as the primeval source of knowledge and civilization: "Pythagoras and Plato and Democritus and Eudoxus and many other of the Hellenes of old" were "granted suitable instruction by the [Egyptian] scribes of their time."[15] Somewhat like Herodotus, "Abamon" claims to transmit wisdom originally inscribed on Egyptian monuments: "if you put forward a philosophical question, we will settle this . . . by recourse to the ancient stelae of Hermes, to which Plato before us, and Pythagoras too, gave careful study in the establishment of their philosophy."[16] In another passage, he declares that a "whole gamut" of questions "has been covered by Hermes in the twenty thousand books" he wrote, "according to the account of Seleucus, or in the thirty-six thousand, five hundred and twenty-five, as Manetho reports."[17] The 20,000 books symbolize the vast wisdom of Hermes, while 36,525 is a hundred times 365.25—the number of days in the solar or Julian year—symbolizing an ideal correspondence between the cosmos and the corpus of Hermes's works. ("Abamon" again emphasizes multiples of a hundred when he declares that Hermes "has handed down . . . a hundred treatises giving an account of the empyrean gods and a number equal to this about the aetherial ones, and a thousand about the celestial ones.")[18]

The emphasis on mysteries in Iamblichus's title, on exaggerated numbers—whether of books or centuries separating him from Hermes—and on esoteric languages and writing-systems communicates wonder by alternating between precision and imprecision. Manetho, one ostensible source of Iamblichus's numerology, was a Greek-speaking Egyptian who flourished in the mid-third century BCE, half a millennium before Iamblichus, but the Julian year was introduced two centuries after Manetho's death. So the text "Abamon" quotes was the work of a much later Jewish or Christian author, who attributed his own ideas to the old Egyptian. This Pseudo-Manetho claimed fantastically to have excerpted extremely ancient chronicles from "the monuments . . . in the sacred language and inscribed in hieroglyphic characters by Thoth, the first Hermes." "Manetho" had been unable to read these archaic documents and so was forced to depend on translations made "after the deluge from the sacred [Egyptian] language in hieroglyphics into the Greek language and disposed into books by Agathodaimon, son of the second Hermes . . . in the inner sancta of the temples of Egypt."[19] The mention of "the deluge" betrays the Jewish or Christian outlook of the forger, since the point of Plato's *Timaeus* had been that the drainage of the Nile Delta made it impossible for Egyptian writings to suffer erasure from a deluge.

Like Iamblichus and Pseudo-Manetho, Alexandrian and Roman writers attributed numbers of improbable or contradictory feats to the "Egyptian Hermes," Thoth. Pseudo-Manetho's attempt to combine a mass of heterogeneous legends led him to hypothesize two Hermeses, and other writers imagined a whole series of them.[20] What remained constant was the notion that some divine or at least godlike personage had revealed momentous truths long ages previously. Jewish, Christian, and Muslim writers would compare—and sometimes connect—Pseudo-Hermes's revelations to those of Moses.

A ROMANCE OF OBLITERATION

Between antiquity and early modern times religious and philosophical texts and technical treatises on magic, astrology, and alchemy attributed a vast body of revelations to Hermes. The most famous of these appear in the *Corpus Hermeticum*, a late-antique collection. These quasi-religious, semi-philosophical texts "can be understood as responses to . . . the very complex Greco-Egyptian culture of Ptolemaic, Roman, and early Christian times."[21] But despite strong resemblances to Christianity, these works lost currency among Western Christians. Only the *Asclepius* was translated into Latin; the others remained in Greek until discoveries in the fifteenth century brought back many ancient writings that had gone missing from Western Christendom.

Asclepius endured throughout the Middle Ages because it depicted an incredibly ancient civilization, which practiced a religion stranger than anything described about Egypt in Plato or in the Hebrew Bible. "Hermes himself" describes primeval Egypt as "an image of heaven" and "the temple of the whole world." Until now, Egypt was a literal heaven on Earth. In fact, "everything governed and moved in heaven came down to Egypt and was transferred there." However, laments "Hermes," Egypt's archaic religion is dying; Egyptians have become impious and will ultimately abandon it.

Scholars of forgery call this "Hermetic apocalypse" a *prophetia ex eventu*, or pseudo-prophecy after the fact: the forger imagines his own belated age being prophesied by the primeval Hermes. The god laments that in the distant future, "it will appear that the Egyptians paid respect to divinity with faithful mind and painstaking reverence—to no end. All their holy worship will be disappointed and perish without effect, for divinity will return from earth to heaven." Foreigners will occupy Egypt and will prohibit "reverence, fidelity, and divine worship." The *Asclepius* imagines ("predicts") that Egypt, its people, and its religion will pass away, as in fact they have: "This most holy land, seat of shrines and temples, will be filled completely with tombs and corpses. . . . Egypt will be widowed, deserted by God and human . . . and the number of the entombed will be much larger than the living. Whoever survives will be recognized as Egyptian only by his language: in his actions he will seem a foreigner."[22]

By "predicting" that Egypt will preserve only enigmatic traces of primeval monuments, *Asclepius* assembled a romance of annihilation and fragmentation: "O Egypt, Egypt, of your reverent deeds only stories will survive, and they will be incredible to your children! Only words cut in stone will survive to tell your faithful works, and the Scythian or Indian or some such neighbor barbarian will dwell in Egypt."[23] The pathos of this prophecy from an imaginary deep past captured the Renaissance imagination and prefigured the desolate "nothing beside remains" of Shelley's "Ozymandias." The role of such writing is to evoke, without actually describing, a distant, utopian past, far more magnificent than the dreary present, a past in which writing actually was powerful, even magical. *Asclepius* lacked the sardonic political edge of "Ozymandias": the Hermetic apocalypse offered no "upside," no well-deserved fall of a tyrant, no positive morality. It was an elegiac daydream of destruction, an exercise in imagination that began from a squalid present and time-traveled backward toward primal, Edenic perfection, only to be yanked back with a thump to where it began.

The appeal of prophecy is eternal in both history and literary imagination. The idea that "It is written," that everything has been foretold, appears in Ovid's *Metamorphoses* when Jupiter tells Venus that, in the dwelling of the Three Fates, she will find "the records of all that happens on tablets of brass and solid iron, a massive structure, tablets which fear neither warfare in the heavens, nor the lightning's fearful power, nor any destructive shocks which may befall, being eternal and secure. There shalt thou find engraved on everlasting adamant thy descendant's fates."[24] We tend to see these "books of eternity" or fate as purely symbolic, and Ovid probably agreed with us. But the passage reflects a fundamental truth about the residual power and mystique of writing: if, "It is written," then "it," whatever it is, must be true. This principle of authority is the ultimate form of the mystique of writing. And, by the converse of this logic, if something is true or real—that is, has happened—then in some sense it was written, "from all eternity." Prophecy is the history of the future, words waiting to be fulfilled or incarnated by events and human actions. And in magic, particularly in natural, nondemonic, ritual magic, when writing is present, its role is not merely to foresee but to implement: writing does not simply describe the world but changes or causes it.[25]

Ovid's "archives of the world" describe history as un- or ahistorical. It is not history but a form of eternity, things that simply *could not avoid happening*. It is an age-old dream, but there is another fantasy, powerful since the Sumerian times recalled by Berossos, of a written memory that would eternalize contingency, those things that might have not happened but did. And this memory, too, would be impervious to change, unerasable and "fearing no cataclysm." As if inscribed on diamond tablets rather than clay or papyrus, it would last "for all eternity." It would far outdo even the archives of Sais or Sippar, not only in durability but also for its relevance to our fundamental concerns. Its subject would be *us, our* ancestry, our own history, and our posterity. Even if no one remained to read it, we could take comfort in its existence, for, "It is written." This would be the ultimate form of history. But we shall have to wait several centuries to encounter it again.

The Roman poet Horace (d. 8 BCE) was acutely aware of literary tradition. His *Art of Poetry* (*Ars poetica*), a longish poem on the craft, was, with Aristotle's *Poetics*, the most influential ancient set of instructions but was livelier and pithier than the *Poetics*. Many of the hoariest catchphrases about poetry, and literature in general, derive from the *Ars poetica: ab ovo, in medias res, ut pictura poesis, deus ex machina,* and *quandoque bonus dormitat Homerus* ("sometimes good old Homer dozes off" and contradicts himself). Horace's interest in literary theory reflected his belief that literature—good *writing*—was the key to a kind of personal immortality. Excellent poets make a culture's heroes immortal. Writing to a friend, he declared: "Many a brave man lived before Agamemnon, but all lie buried unwept and unknown in the long night, because they lacked a sacred bard. In the grave there is little to distinguish unrecorded valour from forgotten cowardice. I shall not pass you over in silence, unhonoured by my pages; nor shall I allow jealous oblivion to erode your countless exploits, Lollius, without fighting back."[26] Somewhat like the Egyptian scribe in chapter one, Horace taught that the great poet will himself become immortal, for written words never die—so long as they are excellent, scribes being common in Rome. Congratulating himself on his poetic achievement, Horace boasted that "I have finished a monument more lasting than bronze, more lofty than the regal structure of the pyramids, one which neither corroding rain nor the ungovernable North Wind can ever destroy, nor the countless series of the years, nor the flight of time. I shall not wholly die, and a large part of me will elude the Goddess of Death. I shall continue to grow, fresh with the praise of posterity, as long as the priest climbs the Capitol with the silent virgin."[27] The first words of this passage (*Exegi monumentum aere perennius*) are among the most famous Horace wrote; Horace's works have indeed been carefully preserved and have inspired countless poets. By referring to the priest and the vestal climbing the Capitoline Hill in Rome, he implied that the Roman Empire would outlast the pyramids—but slyly ("prophetically") suggested that his words might outlast both.

Less quoted are words on the other side of the Horatian coin. In one poetic epistle, the poet addresses his book and chides it for wanting to escape prematurely to the booksellers' quarter, like a handsome young slave itching to roam the world. He predicts it will eventually be hurt, like a slave losing his erotic charm: "You will find yourself packed into a corner . . . you will be loved in Rome till your youth leave you." Like a grubby old man, the premature book will be miserably exiled to North African or Iberian provinces, "when you've been well thumbed by vulgar hands and begin to grow soiled . . . stammering age will come upon you as you teach boys their ABC."[28] Schoolbooks have immortality of a sort, perhaps, but not a dignified one.

Nor was writing always the ultimate value. Roman writers, like their Greek predecessors, considered public speaking a paramount skill; some considered memory more important than writing. They devised ways of composing and remembering speeches by using mnemonic devices. The "memory palace" ordered the segments of an oration by associating them with objects contained in various rooms of a house being visualized by the mind's eye. Delivering the speech involved mentally walking through the house, "observing" the objects and recalling the segments associated with them.[29] The rhetorician Quintilian's *Institutio oratoria*, or *Education of an Orator* (ca. 95 CE) had little room for the romantic kinds of writing lore related by historians and philosophers. But he discussed the importance of orthography for proper pronunciation, which he expressed in a highly suggestive metaphor: "Within the limits prescribed by usage, words should be spelt as they are pronounced. For the use of letters is to preserve [*ut custodiant*] the sound of words and to deliver [*reddant*] them to readers as a sacred trust [*velut depositum*]: consequently they ought to represent the pronunciation which we are to use."[30] Quintilian's words imagine writing as a strongbox or treasury: the sounds of words, like coins or other valuables, are safeguarded by being locked away in writing, so that their meanings can be "paid out" at their full value to lawful recipients. Still, "the limits prescribed by usage" hint that time may weaken the strongbox.

WRITING THE HISTORY OF WRITING

Traditions concerning the history of writing were available to Latin readers in the grandfather of encyclopedias, Pliny the Elder's *Natural History*. Pliny had an insatiable curiosity for every sort of knowledge.

Fittingly, he died like a modern tornado chaser while too closely observing the eruption of Vesuvius in 79 CE, a fate his nephew recorded in a famous letter.[31] But Pliny was as much a reader as a writer: he carefully listed all the many authors whose books he pillaged for his encyclopedia, on a huge range of topics from botany and zoology to medicine, magic, and human history. He gleaned important lore about writing not only from the early history of Rome but from Greece, Egypt, and Mesopotamia, as well.

Unlike Plato or the Jews, Pliny saw no need for a creator-god, and assumed that the world and humanity had existed forever. Likewise, he asserted, the use of writing was *aeternum*—very ancient or even eternal—at least among the Assyrians. But he omitted—if he knew about them—such mythological trappings as Oannes and the *Apkallus*: only the human history of writing interested him. But his chronology was as ample as Berossos's: he quoted "an authority of the first rank" who claimed the Babylonians recorded astronomical observations for 730,000 years "on baked bricks," while conceding that Berossos counted only 490,000 years. Pliny admitted that some Latin authors thought writing was invented in Egypt by Mercury (or Hermes, i.e., Plato's "Theuth"). On a more historical scale, he affirmed that King Cadmus adapted the Phoenician alphabet to create sixteen Greek letters, but he then quoted Greek authors who asserted that the Egyptians had letters 15,000 years before Cadmus. Pliny also said the Pelasgians, an Aegean people who antedated the Greeks (and whom some identified with the Etruscans), had brought the alphabet to Latium.[32]

HISTORY OF WRITING MATERIALS: WRITING AND THE HUMANITY OF LIFE

Pliny devoted more attention than previous writers to the history of materials, and demonstrated, at times explicitly, that writing is what makes us human: we are *Homo scribens*. He quoted the famous polymath Marcus Varro (d. 27 BCE), a contemporary of Julius and Augustus Caesar, on the chronology of materials: "First of all people used to write on palm-leaves and then on the bark of certain trees." Books took their Latin name, *libri*, from this tender inner bark (*arborum libris*). Afterwards, "folding sheets of lead began to be employed for official muniments, and then also sheets of linen or tablets of wax for private documents." But Pliny reserved the bulk of his attention for the history and manufacture of papyrus, asserting that it was the material "on which the immortality of human beings depends [*qua constat immortalitas hominum*]" because "our civilization or at all events our records [*humanitas vitae . . . certe memoria*] depend very largely on the employment of [papyrus]."[33] *Humanitas vitae* means literally "the humanity of life," but it also implies *humanitatis vita*, "the life of humanity."

Papyrus was so important that Pliny devoted seven more chapters to it. Like Herodotus, he considered papyrus more important than parchment. (Papyrus was made from the pith of the stalks of the papyrus plant. Parchment was made from the skins of sheep and cattle in a process that differed from the creation of leather; a prized grade of parchment was made from the skins of immature animals.) Shortages of papyrus could be disastrous. "As early as the principate of Tiberius" (14–37 CE) a reduced supply "led to the appointment from the Senate of umpires to supervise its distribution," for "otherwise life was completely upset [*in tumultu*]."[34] Herodotus had assumed parchment was used before papyrus, but Pliny says Varro argued that papyrus came first: "When owing to the rivalry between King Ptolemy and King Eumenes about their libraries, Ptolemy suppressed the export of [papyrus], parchment was invented at Pergamum," the city that lent its name to that material (*pergamena* in both Greek and Latin).[35] Some older writers, said Pliny, argued that papyrus was used in Latium four or five hundred years before Alexander, and even during the Trojan war. This he found truly puzzling: "why, if [papyrus] was already in use . . . has Homer stated that even in Lycia itself wooden tablets . . . were given to Bellerophon" rather than a letter on papyrus?[36] Pliny does not specifically mention the Romans' use of unwaxed wooden tablets, although large numbers of them have been excavated in Europe since the early 1970s, including over two thousand at Vindolanda, a Roman fort near Hadrian's Wall in Northern England. Containing personal and administrative documents, the tablets—made from thin slices of birchwood and alderwood—had been incompletely burned when the fort was abandoned in 105 CE. So Pliny likely knew of such tablets.

Ironically, perhaps, his chapters are markedly less mythological than what most medieval and Renaissance authors relate about the history of writing. Like the rest of his encyclopedia, writing-lore was based on his voracious reading and seems to represent an ancient consensus, although he quotes many authors who survive only as fragments. Even in the eighteenth century, the best historical discussions of writing began by repeating his chronology of materials. His treatment of the papyrus reed was so influential that William Massey chose a full-length engraving of the plant as the frontispiece to his *Origin and Progress of Letters* (1763), taking it as a symbol of the entire phenomenon of writing (figure 3.2; see also chapter twelve).

THE LOST AND FOUND BOOKS OF NUMA POMPILIUS

From the Renaissance through the eighteenth century, Pliny's discussion of papyrus would be found useful for dating the history of writing materials and even detecting literary forgeries. In fact, Pliny himself may have suspected forgery in one case involving papyrus. He recounted a notorious incident from two centuries before his time that dramatically reveals both the importance of writing in Latin culture and a sophisticated philological awareness among Romans.

> There are important instances that make against the opinion of Marcus Varro in regard to the history of [papyrus]. Cassius Hemina, a historian of great antiquity [ca. 146 BCE], has stated in his *Annals*, Book IV, that the secretary Gnaeus Terentius, when digging over his land on the Janiculan [Hill of Rome], turned up a coffer that had contained the body of Numa, who was king at Rome, and that in the same coffer were found some books of his.... [This happened] 535 years after the accession of Numa [i.e., in ca. 181 BCE]; and the historian says that the books were made of [papyrus], which makes the matter still more remarkable [*maiore etiamnum miraculo*], because of their having lasted in a hole in the ground.[37]

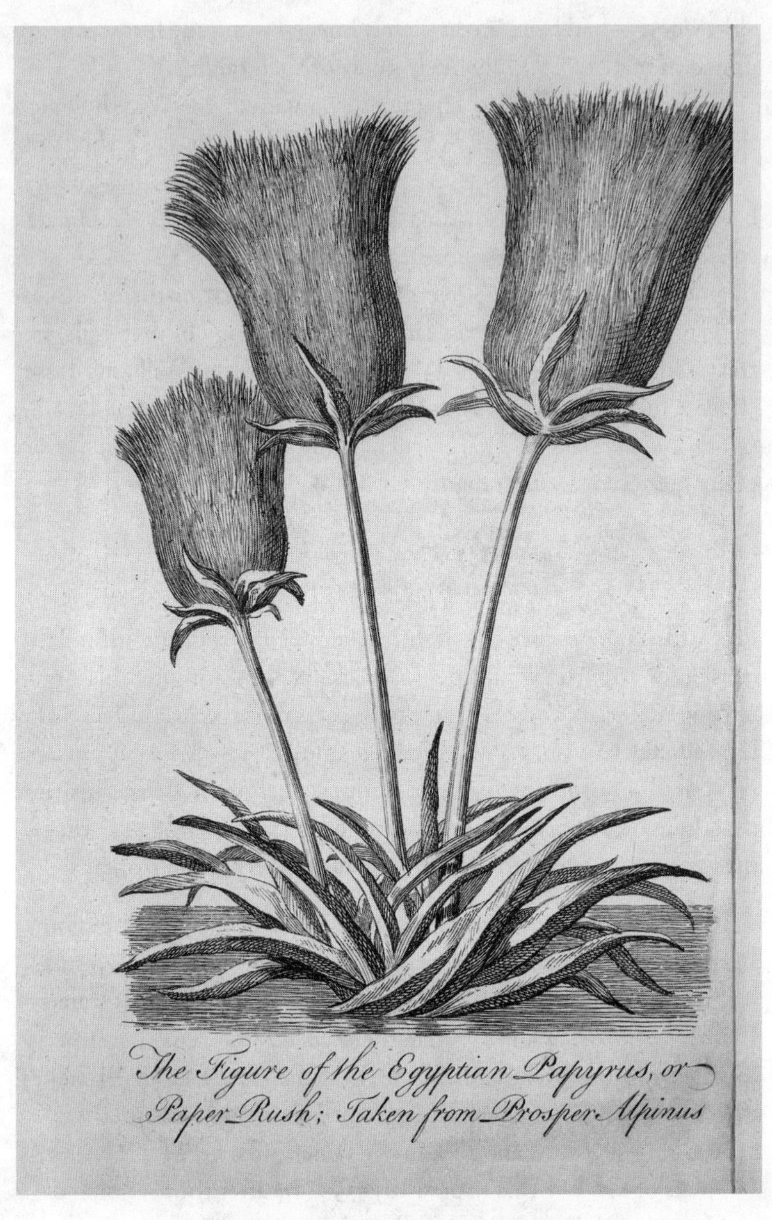

FIGURE 3.2. The papyrus reed. Frontispiece of William Massey's *Origin and Progress of Letters*. London: Printed for J. Johnson, 1763. Sheridan Libraries Special Collections, Johns Hopkins University, Z105.M41 1763 c. 1.

Unlike modern archaeologists, Pliny probably had no idea that papyri could survive for millennia in the arid sands of Egypt, but his sources were skeptical that documents buried underground could last half a millennium in the wetter climate of Rome.

> Consequently on a point of such importance I will quote the words of Hemina himself: "Other people wondered how those books could have lasted so long, but Terentius's explanation was that about in the middle of the coffer there had been a square stone tied all round with waxed cords, and that the three books had been placed on the top of this stone; and he thought this position was the reason why they had not decayed, and that the books had been soaked in citrus oil [*citratos*],[38] and he thought that this was why they were not moth-eaten."[39]

Most scholars assume that Pliny must have accepted this explanation. Indeed, he does not seem perturbed by the fact that Terentius was a professional scribe or that the discovery was made on Terentius's own land. Yet Pliny's exclamation *maiore miraculo*, while not quite meaning "a greater miracle," does seem skeptical.

That Terentius was a forger, and that Pliny suspected forgery are both likely, but ultimately unprovable. The incident was so remarkable that other versions of it circulated long before Pliny, as he indicates. The accounts sometimes disagree on details, making it difficult to indict Terentius or anyone else as *the* forger. However, all the accounts indicate that authors were alert to evidence that someone wanted to head off suspicion of the "discovery." In his *History of Rome*, Livy (d. 17 CE) mentioned not one but two coffers, "each about eight feet long and four broad," which advertised their significance even before they were opened: "Each chest had an inscription, in Latin and Greek letters: one to the effect that Numa Pompilius . . . king of the Romans, was buried there, the other, that the books of Numa Pompilius were inside."[40] Like Pliny's version, Livy's implies that someone was anxious about the credibility of the find. Both coffers' lids were "fastened with lead," leaving witnesses apparently puzzled that the chest bearing "the inscription about the buried king" was empty, "with no trace of a human body or anything else." Someone—whether at the scene or afterwards—declared that the corpse and "everything else" had been destroyed by "the wasting action of so many years." Yet the books were "not merely whole, but looking absolutely fresh." Being contained in a separate coffer, they had no need to compete for space with an imaginary corpse or the rectangular stone mentioned by Pliny: the coffer simply contained "two bundles tied with waxed rope, containing seven books each." Although Livy did not record any witnesses' puzzlement over the storage precautions, his version makes the condition of the books seem even more implausible than Pliny's does.

Other details in Livy's version betray the stagecraft of a forger anxious to deny authoring the books. The landowner was apparently absent at the dramatic moment of discovery; instead, "*husbandmen* were digging up the ground." Although the coffers were buried in an agricultural field, they were not discovered until workers dug to "*a greater depth than usual.*"[41] Moreover, in Livy's version, the landowner did not open the chests immediately after discovery but only after seeking "the advice of his friends." If the landowner was indeed a scribe, this real or feigned reluctance would shield him from suspicion. Livy also seems more attentive than Pliny to the subversive ideological implications that contemporaries had seen in the "discovery" and appears skeptical of the books' authenticity. He quotes an annalist who said the Greek philosophical books "were Pythagorean, confirmation of the common belief . . . that Numa was a pupil of Pythagoras" and remarks that this ascription was "arranged by plausible invention [or rather, lie: *mendacio probabili*]." Yet Livy also says that the Greek books expounded "a system of philosophy which might have been current at [Numa's] time," an opinion that seems to come from yet another source. A modern commentator conjectures that unlike Pythagorean philosophy, "the system expounded was simple and practical rather than theoretical and academic."[42] But we should not exaggerate Livy's philological skepticism. His main reason for doubting Numa's contact with Pythagoras was nativist or patriotic: "It was Numa's native disposition, then, as I incline to believe, that tempered his soul with noble qualities, and his training was not in foreign studies, but in the stern and austere discipline of the ancient Sabines, a race incorruptible as any race of the olden time."[43]

Pliny found a number of conflicting accounts regarding the subjects treated in the "books of Numa." One claimed there were "seven volumes of pontifical law and the same number of Pythagorean philosophy"; another that "there were twelve volumes of the *Decrees of Numa*." Yet another specified "twelve volumes *On Matters Pontifical* written in Latin and the same number in Greek containing *Doctrines of Philosophy*," while the prolific scholar Varro "says that there were seven volumes of *Antiquities of Man*," a subject that Varro himself famously wrote about.[44] Whatever their contents, Roman authorities considered "Numa's" books socially disruptive and destroyed them almost immediately.

Livy implies they were objectionable on religious grounds. Since Numa was considered the founder of both Roman law and Roman religion, this was a matter of the utmost seriousness.[45] But at bottom, Roman historians were fascinated by the scientific incongruity of papyrus remaining intact and legible after being buried for over half a millennium. And well might they be suspicious: as a *scriba publicus*, the landowner would have access to the technical knowledge and writing materials necessary to forge the documents and perhaps, if Pliny's version is accurate, to invent the "square stone" and other precautions, either as a story or in reality.

THE BOOKS OF NUMA AS INSPIRATION FOR FORGED HISTORIES

For six hundred years after the writers that Livy and Pliny mentioned as their sources, the "discovery of Numa's books" was discussed by classical authors, including Valerius Maximus (ca. 14–37 CE) and Plutarch (d. 120 CE), and by the Christian apologists Lactantius (d. ca. 320 CE) and Saint Augustine (d. 426 CE).[46] For centuries afterward, "Numa's books" directly or indirectly inspired both literary forgeries and the scholars who eventually debunked them.

Forgery is the ultimate expression of a culture's conviction that writing is powerful or, as Pliny remarked, essential to our humanity. Ironically, forgeries involve a double debunking, since their avowed or tacit aim is overturning a consensus about history. Two of the most influential forgeries, the alleged "histories" of Pseudo-Dares and Pseudo-

Dictys, claim, as Herodotus had done, to debunk Homer's account of the Trojan War on the basis of reliable eyewitness testimony. In the Latin Middle Ages, "Dares and Dictys" enjoyed a considerable fortune since, outside Byzantium, Homer would not be available in the original Greek until the 1400s. For readers who would never know the *Iliad*, "Dares and Dictys" functioned as a perverse substitute for Homer. They so poisoned the wells that, as late as the seventeenth century, even some readers of Homer defended them as genuine proof of his inaccuracy or deceit. The two forgeries made ostentatious references to the history of writing and thus played on literate prejudices in favor of written history and against myth, so authors frequently accepted their "evidence" that Homer had systematically misrepresented the Trojan War. Their pugnacious attitude toward the oldest genuine Greek text is the broadest of several ironies surrounding these two false "histories."

The apparently older text, from the fourth century CE, presented itself as the *Diary of the Trojan War*. The complete text of the *Diary* survives only in Latin, but a fragment in Greek was found among the famed papyri excavated from the garbage dump at the extinct Egyptian city of Oxyrhynchus in 1899–1900. "Dictys of Crete" was supposedly a warrior of Agamemnon, and his text mimics the kind of historical evidence Pliny had discussed. "Dictys, a native of Crete from the city of Cnossos and a contemporary of [Agamemnon and Menelaus], knew the Phoenician language and alphabet, which Cadmus brought to Achaea. He accompanied the leaders [who] chose him to write down a history of this campaign. Accordingly, writing on linden [-wood] tablets and using the Phoenician alphabet, he composed nine volumes about the whole war."[47] The wooden tablets ironically remind modern readers that Homer's *Iliad* mentioned tablets given to Bellerophon several generations before the Trojan War. If Dictys's *Diary* had been genuine, it would obviously be a firsthand, day-by-day account of the entire war, whereas Homer only described the few days when Achilles angrily boycotted the fighting. If Homer's poetic account, supposedly written several hundred years after the war, is not only truncated but unreliable, then where differences exist, Dictys's eyewitness account must prevail.

But how did a text so ancient survive for so long? Why should its early readers assume the text was not written yesterday—was not, in fact, a forgery? Answer: because, like the books of Numa, it was "rediscovered" under unusual circumstances. Supposedly, the original tablets had been placed in Dictys's own tomb, enclosed and protected by a special box. "Time passed. In the thirteenth year of Nero's reign an earthquake struck at Cnossos and, in the course of its devastation, laid open the tomb of Dictys in such a way that people, as they passed, could see the little box. And so shepherds who had seen it as they passed stole it from the tomb, thinking it was treasure. But when they opened it and found the linden tablets inscribed with characters unknown to them, they took this find to their master."[48]

The strongest feature of the preface is the abundance of detail the writer adopts to win our confidence in the authenticity of the story. Details in fact reveal a distinct literary and political program behind the creation of the *Diary*. As the italicized passages (emphasis mine) show below, the forger tried to claim several kinds of authorities—political, religious, philological, historical—as guarantors of the *Diary*'s truthfulness, ending with its official enshrinement in an imperial library.

> [The shepherds'] master, *whose name was Eupraxides*, recognized the characters, and *presented the books to Rutilius Rufus, who was at that time governor of the island.* Since Rufus, when the books had been presented to him, *thought they contained certain mysteries*, he, along with Eupraxides himself, *carried them to Nero.* Nero, having received the tablets and having noticed that they were written in the Phoenician alphabet, *ordered his Phoenician philologists to come and decipher whatever was written.* When this had been done, since *he realized that these were the records of an ancient man who had been at Troy*, he had them *translated into Greek*; thus *a more accurate text of the Trojan War* was made known to all. Then he bestowed gifts and Roman citizenship upon Eupraxides, and sent him home.
>
> *The Greek Library, according to Nero's command, acquired this history that Dictys had written*, the contents of which the following text sets forth in order.[49]

The passage argues for the *Diary*'s truth and credibility, its *literary* authority, by narrating how it was duly passed up the chain of *legal* custody from the shepherds to the emperor, and transforms Nero's authority into architectural solidity by recording his insistence that the *Diary* be preserved in the "Greek Library."

The *Diary* has another, even stranger feature. In addition to its preface, which seems to be original to the text, it contains a "translator's foreword" that purports to be by a later Roman writer, giving us two adjacent, contradictory versions of the text's history. The foreword declares that "Dictys of Crete originally wrote his *Diary of the Trojan War* in the Phoenician alphabet, which Cadmus and Agenor had spread throughout Greece."[50] It omits the dramatic detail of the earthquake recounted in the preface, claiming instead that the tomb collapsed from sheer decrepitude, thereby implying more strongly that the text was authentically antique. The foreword notes that the "little box" containing the tablets was "skillfully enclosed in tin," evidently to protect the wooden tablets and container from rot and vermin, recalling precautions in accounts of "Numa's Books."

The foreword abridges the preface's elaborate account of the philological activity of Nero's scholars. However, it occasionally amplifies and improves details from the preface: when the shepherds broke open the box, instead of merely "expecting treasure," they "*brought to light, instead of gold or some other kind of wealth, books* written on linden tablets." Other details credit the shepherds with a richer psychology: "Their hopes thus frustrated, they took their find to . . . the owner of that place." Unlike the preface, the foreword does not implicitly expect shepherds to distinguish between familiar and unfamiliar alphabets; like real-life shepherds, these were illiterate.

The later writer had his own ways of solidifying the authority of the *Diary*. One was to emphasize his personal role in the transmission of the text, while affecting an exaggerated modesty. "When these little books had by chance come into my hands, I, as a student of true history, was seized with the desire of making a free translation into Latin. . . . I have preserved without abridgment the first five volumes which deal with the happenings of the War, but have reduced into one volume the others which are concerned with the Return of the Greeks. Thus, my Rufinus, I have sent them to you."[51] The first-person voice gives the texts an additional pretense of eyewitnessed authenticity, as does the writer's claim to be a connoisseur, "a student of *true* history." The "Quintus Aradius Rufinus" to whom the translation is dedicated and the translator Lucius Septimius are not readily identifiable.[52]

Unlike Pseudo-Dictys's *Diary*, "Dares the Phrygian's" *On the Destruction of Troy* may have originally been written in Latin.[53] It seems to have been composed in the fifth century CE. During the Latin Middle Ages, "Dares" was better known than "Dictys." Not coincidentally, Homer's *Iliad* mentioned a Trojan priest of Hephaestus named Dares, so the credibility of the imaginary Trojan antidote to Homer was supposedly guaranteed by Homer himself. But because medieval Latin readers had no access to the *Iliad*, the irony of the attribution would have been lost on them.

The sole paratext of the *Destruction of Troy* is compact and simplistic, and more aggressively polemical than the twin paratexts of Pseudo-Dictys. "*Cornelius Nepos* sends greetings to his Sallustius Crispus. While I was busily engaged in study at Athens, I found the history which Dares the Phrygian wrote about the Greeks and Trojans. As its title indicates, this history was *written in Dares' own hand*. I was very delighted to obtain it and immediately made *an exact translation* into Latin, *neither adding nor omitting anything, nor giving any personal touch*" (emphasis added). We may wonder what besides the title led "Cornelius" to believe the manuscript was an autograph (original) document. No matter: his preface reveals a forger far more ambitious, though less talented, than Pseudo-Dictys. The historical Cornelius Nepos (d. 25 BCE) was a Roman biographer and a friend of Cicero and Catullus. The addressee of Nepos's letter, Gaius Sallustius Crispus (d. 35 BCE), was an eminent historian, known in English as Sallust.

If he was aware of Pseudo-Dictys, the forger of Pseudo-Dares may have felt that "Dictys" did not contradict the *Iliad*'s account of the Trojan War explicitly enough: "Thus *my readers can know exactly what happened* according to this account *and judge for themselves* whether Dares the Phrygian or Homer wrote the more truthfully—*Dares, who lived and fought at the time the Greeks stormed Troy, or Homer, who was born long after the War was over. When the Athenians judged this matter, they found Homer insane for describing gods battling with mortals*. But so much for this. Let us now turn to what I have promised" (emphasis added). For medieval writers, especially the earlier ones, Pseudo-Dares's claim to be an eyewitness gave his *Destruction* impressive authority.[54]

"Dares" stands out from "Dictys" in another way: "Dares" lacks information about alphabets and writing materials that the other forger provided so abundantly. Aside from the famous names he drops, "Dares" assumes that a Greek text should appropriately be discovered in Athens (presumably in a private or public library, where it was well cared for). Compare this to "Dictys's" careful philological scene-setting. "Dares's" claim that the original manuscript was an autograph (or, as textual scholars say, a holograph), "written in Dares' own hand," is naively simple and not only because Dares was supposedly a Phrygian rather than a Greek. Whoever wrote the preface to Pseudo-Dictys made the same claim much more subtly: he simply assured his readers that the *corpus* of Dictys's writings was discovered in the grave with his *corpse*. Thus, by implication, the "hand" of the manuscript had to be the dead man's. (Although he created a real tomb, the best "Dictys" could do was imply that the corpse was too ancient to survive.)

WRITING AND HUMANITY: HANDS, CORPSES, CORPUSES

The connections between hands, corpses, and corpuses are so fundamental that they seem organic rather than symbolic. The inextricable link between the human hand, writing, and the book was not a refined literary metaphor nor was the discovery of a book in an antique grave. Forgers often claimed the "rediscovered" book had been found under the head or in the hand of the author's bodily remains, making identifying a dormant literary *corpus* equal to producing a decayed corpse almost a rebus for authorship. The reemergence of the book from a grave was a compact metaphor for the resurrection of its author's wisdom, an "unveiling" (revelation) or "uncovering" (discovery). One ambitious eighteenth-century forger even claimed to have discovered Homer's tomb, complete with his desk and writing implements.[55]

Heliodorus of Emesa's *Ethiopian Romance* was not a forgery but a forerunner of the European novel, composed in the third or fourth century CE. It took the organic relation of book to body to its absolute limit by telling the miraculous biography of a living woman instead of describing the corpse of a dead author-figure. Heliodorus's heroine, Chariclea, is the anomalously snow-white daughter born to the faithful Black wife of the Ethiopian king. Fearing her choleric husband will suspect adultery, the queen exchanges her baby for a Black infant, sending her own child away in the care of a servant. Chariclea's only Black feature is a birthmark shaped like the Greek letter *omicron*; this perfect *O* stamps her as a female Odysseus, destined for a long, circular journey of exile and return. At the vastly deferred happy ending of the romance, the adult Chariclea escapes execution, reunites with her parents, and marries the faithful lover who has shared her travails. Together they inherit the throne of Ethiopia, all thanks to her mother's foresight. The queen had sent her white baby away clothed in its own biography, which she inscribed on its swaddling clothes in "ink" made of her own blood and tears, and sealed by enclosing the king's own signet ring. Chariclea's faithful servant, who translates the document for the reader, remarks transparently that "on reading this . . . I perceived the hand of the gods and marvelled at the subtlety of their governance." Thus the "hands" of the gods, her mother, and her father authenticated Chariclea's status as heir to Ethiopia.[56] Although unknown to the Latin Middle Ages, throughout the sixteenth and seventeenth centuries, the *Ethiopian Romance* (also called *Aethiopica*) was a reliable best seller in Latin translation, in vernacular translations (French, Italian, and others), and in the original Greek. It became a model for early modern romances such as Cervantes's *Persiles y Sigismunda* (1617) and Henry Fielding's *Tom Jones* (1749).

MISCELLANIES

An immense storehouse of lore about the culture of antiquity is found in the *Deipnosophistai*, or *Dinnertime Philosophers* by Athenaeus of Naukratis (ca. 150–200). The work imagines a typically Greek symposium, an all-male party with dinner and drinking, but its most notable passage about writing concerns the best-known female poet of ancient times, Sappho of Lesbos (d. ca. 570 BCE). One of the learned diners relates that, in his *Sappho*, the comic dramatist Antiphanes (d. 334 BCE) portrayed the poet posing riddles to a male character. Sappho says, "It is a female creature that keeps its children safe beneath the folds / of its garment.[57] And though they are mute, they raise a resounding cry / through the sea-surge and the whole mainland / to whichever mortals they wish, and even those who are not there / can hear them, deaf though their perception is." Antiphanes showed Sappho mocking the man's inept guess rather mercilessly, says the speaker. Then Sappho offered the correct solution: "The female creature is a writing-tablet [*epistolé*],[58] / and the children she carries around inside herself are the letters [*grammata*]. / Even though they're mute, they speak to anyone they want / who's far away. And if someone else happens to be standing / nearby, he won't hear the man who's reading."[59] It is tempting to imagine what else "Sappho" said and did in this lost play. The real Sappho's own works are scantily preserved, but Antiphanes's comedy may have offered some clues about her personal and literary life. The fictional Sappho's sarcasm contrasts with the riddle's playfulness and even more with the earnest poignancy of the historical Sappho's surviving, fragmented lyrics. But the deepest significance of Athenaeus's anecdote lies in its claim that writing negates time and space.

The reference to silent reading in "Sappho's" riddle includes, but also reverses, the standard praise of writing as breaker of spatial and temporal barriers, showing how it can also insulate a reader, foreclosing bystanders' access to the message. Most modern reading is silent. But was unvoiced reading common in Sappho's time? Conceivably. There is abundant evidence that in Athenaeus's day silent reading was, if not usual, not extraordinary either. Plutarch recounts that Alexander the Great, a contemporary of Antiphanes, read silently; a century after Sappho, Euripides and Aristophanes showed characters reading silently on stage. Saint Augustine's *Confessions* attest to his own silent reading, although earlier in life he had marveled that his mentor Saint Ambrose read voicelessly. There were halfway measures as well. Saint Cyril, a contemporary of Augustine, admonished women to move their lips without speaking when reading to themselves during public worship. Totally silent reading would have been more difficult in these centuries than later, since punctuation, capitalization, and even the separation of one word from another by blank spaces were not practiced until the age of Charlemagne (d. 814).[60] But reading silently was not unheard of.

To quote Jerrold Cooper, "The domains in which early writing was used were, in fact, invented along with writing itself." Inventories, accounts, registers, lists, and labels "have no oral counterparts" but instead "represent the extension of language into areas where spoken language cannot do the job."[61] Similarly, this chapter has explored how written culture opened possibilities for both analytical and representational thought, for philosophy and narrative, that would have been difficult or impossible to achieve in an oral culture. Thanks to writing, thought could be reproduced with less variation, as it leapt from country to country and century to century. But because it could be read silently, writing could also conceal thought, so that even "if someone else happens to be standing / nearby, he won't hear." Acrostics (to be read vertically as well as horizontally), allegories (or "alien speaking," *alieniloquium*), rebuses (visual riddles that isolate and recombine syllables to create spoken words), and cryptograms (written signs that are systematic substitutes for other written signs) were all made possible, or at least infinitely easier, by writing. Silent reading paradoxically extended the range and utility of writing while opening the possibility of secrecy and even, as the *Iliad* already implied, magical or murderous treachery. Much later, the metaphor of "reading the signs" of murder through detection would turn mundane or horrific details into writable narratives.

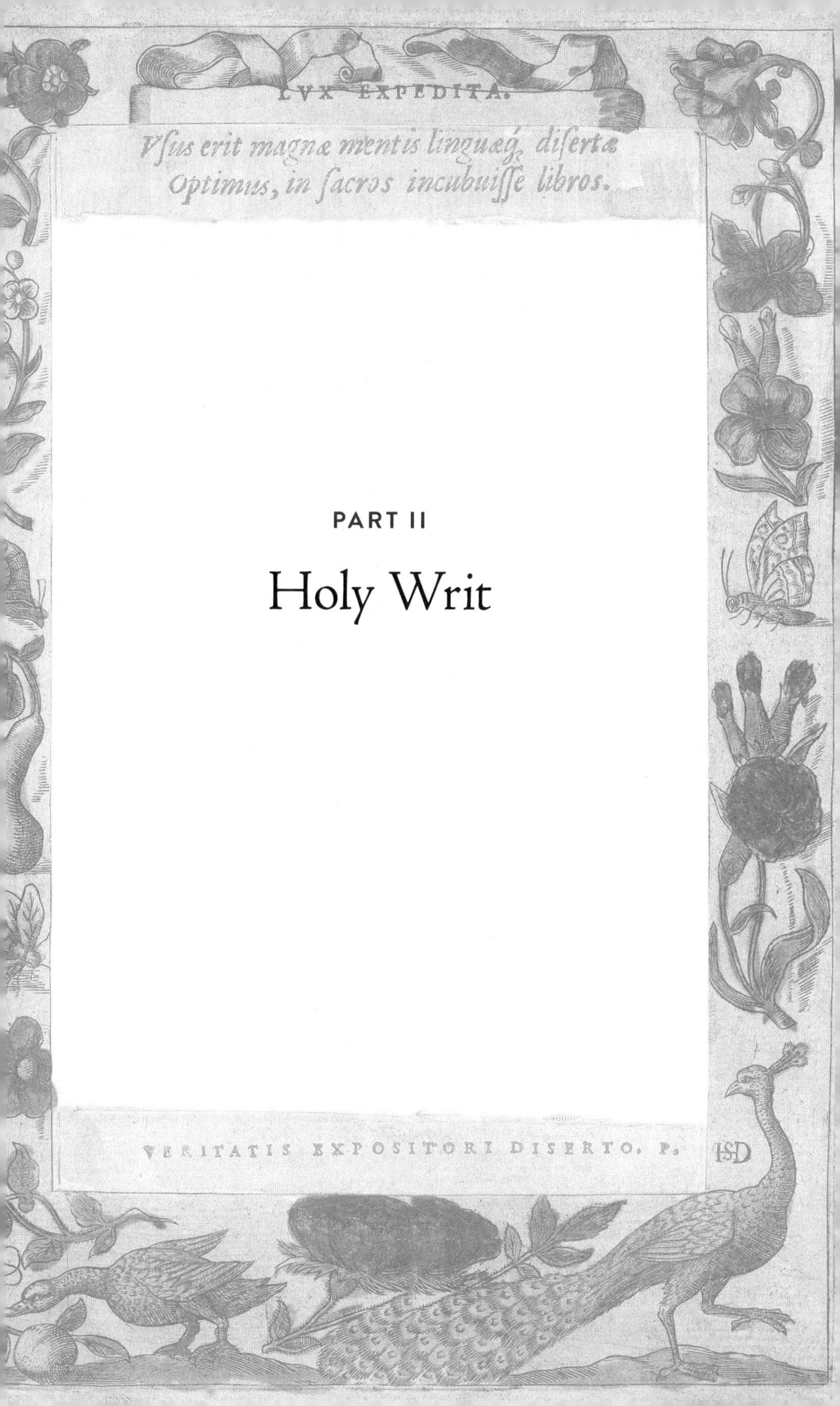

PART II

Holy Writ

CHAPTER FOUR

Writing and Scripture, 600 BCE–650 CE

Therefore the Book of the Wars of the Lord speaks of ". . . Waheb in Suphah, and the wadis: the Arnon with its tributary wadis. . . ."

—Numbers 21:14–15

"Scripture" is the conventional name for the holy books of any religion. In Latin, *scriptura* originally signified any handwriting, but in modern usage, it usually refers to the foundational books of Judaism, Christianity, and Islam. These religions' attitude towards scripture implies that writing is not merely culturally important but essential to humanity's nature and ultimate well-being. The three scriptures have differing histories, however, and are defined as scripture in divergent ways.[1] The Hebrew Bible and the New Testament are physical books—scrolls or codices—containing collections of writings, likewise known as "books," that originated in different eras and cultural contexts. Hebrew *sefer*, Greek *biblion*, and Latin *liber* are the terms used by Jewish and Christian Bibles for physical books; all three words referred broadly to "something written"; this appears to be the actual etymology of *sefer*. *Biblion* and *liber* originally designated the surface written upon. *Biblion* referred to papyrus; the name derived from Byblos, a Phoenician city from which that material was imported to Greece. *Liber* referred directly to tree bark, as we saw in chapter three. The Greek name for "the Bible," *ta biblia*, meaning "the books," became *biblia sacra*, "the holy books" in Latin and was eventually accepted by Western Christians as a singular noun referring to the entire collection.

The Qur'an is a single work (*al-Qur'an*). Devout Muslims consider it unitary (*al-Kitab, the* Book), while more secular-minded textual scholars consider it (like its Jewish and Christian analogues) a collection compiled over a period of decades. The Jewish and Christian Bibles are composed of many *books* subdivided fairly uniformly into *chapters* and *verses*, but the Qur'an is composed of 114 *suras*, resembling chapters, of widely varying length. Jewish and Christian scriptures are attributed to a number of writers, whereas Muhammad is the only human identified with the substance of the Qur'an (see plate 3).

Although considered timeless, scriptures, like the epics of Homer and other texts, were only written down after surviving for long periods as oral "literature." The word *scripture* itself indicates that the *mystique* of holy books, the complex of emotional responses to them, owes much to their writtenness. At a symbolic level, that mystical aura implied something akin to magic and reflected the relative permanence of writing, compared to oral communication. The Romans said, *Verba volant, scripta manent*: "Spoken words fly away, but writing remains." They were describing the durability of contracts between humans, but the same holds true of God's covenants with his people(s).

Monotheists' reverence for scripture because "it is written" reflects their ideas about God's own unchanging eternity. The present tense (*is*) in relation to the writtenness of scripture implies the same permanence as God's self-identification to Moses (Exodus 3:14). Although the Latin Bible translated God's words as, *Sum* qui *sum*, in Hebrew they mean less "I am *who* I am" than "I am *that* I am," just as the King James Version translates them.[2] "Am-ness" defines God: he is *semper idem*, always the same. Likewise, "is-ness" defines his scriptures: the devout treat them as unchanging, whether or not they believe this in historical, chronological terms. Jesus's rebuttal to anyone who contradicted him, including Satan, was *Scriptum est*, "It is written" (Matthew 4:4, 6, etc.). The Hebrew Bible, to which Jesus's words referred, is permeated by the assurance that the words on the page are all anyone needs to know. Yet both Jewish and Christian traditions overflow with commentaries on the Bible, suggesting that readers did not always find its words self-evident. The Gospel of John (written ca. 90–110 CE) even insisted that the Word of God had to be "made flesh" in his Son to be understood, that Jesus's existence and mission retrospectively clarified the sense of the Hebrew Bible. Nevertheless, despite repeated schisms and controversies, the major legacy of the Age of Scripture for both Jews and Christians is precisely this theological sharpening of the conviction that "It is written," that scriptures endure.

The three holy books are susceptible to two forms of treatment. Believers encounter them as scripture in a special sense, as God's words; although written down by humans, they are considered largely or completely unaffected by human intervention or filtering. The essence of scripture is *authority*, or confidence in its divine origin. Somewhat paradoxically, scriptural authority is felt most directly in the living, "oracular" pronouncements of prophets or preachers: "Verily I *say* unto you." Scholars, however, even the most religious, treat the three scriptures as *texts*, compiled and edited by humans. Acknowledging this written aspect imposes a concern with *accuracy*, confidence in the stable transmission of words over time. A lack of confidence in the accuracy of a scripture can diminish or destroy its authority.

Since the seventh century CE, there has only ever been one Qur'an, even though the history of its appearance in writing is uncertain. Yet Salman Rushdie's fictional *Satanic Verses* outraged many Muslims with (among other transgressions) its references to legends of a diabolically inspired, favorable mention of idolatry that temporarily destabilized the text of the Qur'an. Given their longer histories as texts, scholars of the Hebrew and Christian scriptures necessarily acknowledged the triage or canon-formation that distinguished sacred texts from all others. ("Sacred" derives from the Latin *sacrum*, meaning "set apart," and implies a taboo against change.) The holy books that appear in modern Bibles were not alone in claiming the status of scripture; they had numbers of unsuccessful competitors, books that groups of rabbis and/or priests disqualified on doctrinal grounds in early centuries. Many phases of editing were necessary before Genesis, Matthew, and other canonical scriptures reached stable textual form. Even so, what now appear as single, unitary "books" of the Bible were often created when scholars bound related texts into a single scripture. The Book of Genesis still shows the seams where two earlier versions of Adam and Eve's creation were conjoined. Similarly, the earliest kernel of the gospels was apparently not a narrative but a collection of Jesus's *logia*, or sayings, around which evangelists wove differing biographies over several decades.

How any scripture reached its current textual form is no more essential to my overall purpose than deciding whether it is truly "God's Word." Yet monotheists' claims about the authority—or cultural importance, in secular terms—of their scriptures are deeply entwined with the way those texts discuss writing, particularly their own formation and transmission. Put simply, the divine authority claimed for scripture constitutes the ultimate assertion that a particular writing demands and deserves our attention and wonder. Yet while scriptures claim to be both unique and complete (the Qur'an and the Book of Revelation assert this overtly), one plausible definition of a scripture is a book that spawns an almost infinite series of other books and writings. The suspicion can arise that the transformation of oral tradition into a written text both creates the need for a definitive, unambiguous "word" and acknowledges its impossibility.

The three religions are connected by respect for the Hebrew Bible as Scripture;[3] their common designation as "Abrahamic" recalls the earliest prophet or God-friend named there (Genesis 11:26). Yet each religion considers its foundational document "holy," "from God," "a book," and "scripture" in a different sense from the others. The Hebrew Bible is esteemed by Christians as well as Jews; yet Christians do not revere it unconditionally: they refer to it as the "Old Testament," considering it partly superseded by the "New Testament." Moreover, the Old Testament collection is not identical with the Hebrew Bible, nor do all Christians venerate identical "Old Testament" collections.

The scripturality of the Hebrew Bible arose from a long textual tradition. Its earliest books were supposedly written by Moses, sometime in the fourteenth, fifteenth, or sixteenth century BCE, lost during the Babylonian Captivity but then restored intact by the scribe Ezra in the sixth century BCE. Yet there is no evidence that any parts of the Hebrew Bible are that ancient. Although the Hebrew Bible is "from God," God's words form a relatively small part of its text and are framed or surrounded by the words of God's human interlocutors and messengers. The New Testament identifies even fewer words as God's own utterances; its overriding purpose is to describe the earthly life, mission, and ultimate significance of Jesus. The first three canonical *evangels*, or "gospels" ("good news" in Greek and Old English respectively), attributed to Matthew, Mark, and Luke, propose that Jesus "fulfills the prophecies" of the Old Testament in one sense or another. The fourth evangelist, John, suggests that Jesus was not simply an extraordinary man whose life and deeds were foretold centuries earlier by Old Testament prophets. Nor was he merely the "son of God," as in the other gospels, but God's own living *Logos* (Word or Utterance), His incarnate, salvific message to humanity, eternal and co-divine: "In the beginning was the Word, and the Word was with God, and the Word was God" (John 1:1).

Believers consider both Old and New Testaments *inspired*, that is, messages revealed by God to humans, who wrote them down. The writers are not authors but mere amanuenses, or scribes, who recorded God's *revelations*. Moses, God's human interlocutor and his co-protagonist in the books of Exodus, Leviticus, Numbers, and Deuteronomy, was traditionally considered to have written the Torah (Law) or Pentateuch, "five books of Moses"—the four just mentioned plus their prehistory in Genesis. Likewise, God was believed to inspire the four canonical evangelists, along with Saint Paul and other New Testament authors, revealing things that could not be known any other way. Whereas the Torah *presents* a revelation from God, the New Testament *tells about* a revelation. It describes Jesus's life *as* the revelation. This is clearest in the Gospel of John, which presents him as the incarnate *Logos* of God.

Things are very different with the Qur'an. It presents itself as not merely inspired by God but composed of his very words, exactly as spoken to Muhammad, who then recited them verbatim to his followers (*qur'an* means "recitation"). The "recitation" was written down—again verbatim—at a somewhat later date (pious traditions and scholarly estimates for the date of transcription range from sometime later in Muhammad's life to two or three centuries after his death). The Qur'an thus presents itself entirely as a revelation; although the Torah and the Gospels give their readers some information about Moses, Jesus, and the Evangelists, the Qur'an has no personal information about Muhammad. He appears, as it were, as nothing more than a mouthpiece: the reader of the Qur'an finds the Recitation and nothing more. Muhammad's life had to be reconstructed separately, from numerous *hadith*, or testimonies.

Unlike the Jewish and Christian scriptures, the Qur'an underwent no process of canon-formation. Rather than a book comprised of selected books, it is a single, unaccompanied book. Unlike the other two scriptures, it had no competitors. There are no Qur'anic apocrypha or pseudepigrapha, no equivalents of the unsuccessful books that clerics excluded from the Hebrew and Christian Bibles. The Muslim canon has only ever contained a single book, the Qur'an. Its canonicity is self-conferred by its claim to be God's exact words. It declares itself unique and inimitable, a judgment shared unwaveringly by observant Muslims. That non-Muslims may disagree is not only irrelevant or ridiculous to devout Muslims, but blasphemous. Imitating the Qur'an would be as futile and impious as creating an idol: adding to God's words would be as sinful as any other imitation of his activity. While Muslims accord respect to the Jewish and Christian scriptures, they define these differently and treat Moses and Jesus as mere participants in a long succession of prophets, beginning with Abraham and ending—forever—with Muhammad. Jesus was an exceptional man but not a divine "son of God," and neither the Old nor New Testament represents God's actual revelations: for Muslims they are defective, having been corrupted by their adherents.

Aside from defining Jesus as revelation incarnate, in other ways the New Testament differs more from the Torah and the Qur'an than they do from each other. The crucial factor here is, in the most fundamental sense, a matter of writing, and requires a brief explanation. The New Testament was written originally and entirely in the Greek alphabet, a writing system that has both consonants and vowels. Greek writing had evolved over the previous millennium and was refined by scholars and text-editors of Alexandria. By contrast, Hebrew and Arabic both employ consonantal writing systems, written without vowels, as explained in my "Complement" chapter. Both languages eventually developed supplementary graphic systems for indicating vowels, but this happened long after the Torah and the Qur'an originated. The writing systems that first recorded these scriptures were not only consonantal but also—in an objective, nonjudgmental sense—primitive: being new, they lacked the millennium of development behind New Testament Greek. Even aside from lacking vowels, neither Hebrew nor Arabic was a "complete" writing system when it first recorded the words of a scripture. For a long time thereafter, Hebrew and Muslim societies were transitional, moving slowly from exclusively oral transmission to literacy.

The process was very like the development that transformed the *Iliad* and *Odyssey* from oral bardic performances to "classical" written texts possessed of a nearly scriptural authority. Even now, those who seriously study the Torah or the Qur'an from within Judaic or Islamic culture approach their scriptures in a radically different fashion from readers of the New Testament. The Torah and the Qur'an were composed for restricted ethnic audiences, with no regard for linguistic translation or religious proselytization, whereas the story of Pentecost (Acts 1–2) proclaims that the gospel message was intended from the beginning to be translated and spread "to all the world."

Hence, in any language, the New Testament can be simply *read* aloud during worship, using the same performative techniques needed to present any other book. The oral reader need not be a cleric or even particularly studious. But to be presented properly during worship, the Qur'an and the Torah must be presented in the original language. Furthermore, they have to be *chanted*, or "cantillated," according to designated, quasi-musical performative techniques. These complex rules are best learned not by reading but by imitating a skilled practitioner. Codification of the techniques is not just a matter of aesthetics: they are necessary so that worshipers who do not see the Hebrew or Arabic text, or cannot read it, may follow along. In this and several other ways, the Torah and the Qur'an remain embedded in orality even for literate believers. Translation of the holy text into other languages is possible but, particularly in the case of the Qur'an, not considered authentic. As with performance of the Catholic liturgy, which was restricted to Latin before the Vatican II reforms of the 1960s, the ritual of worship is crystalized in a canonical language.

By contrast, the New Testament emerged from a cosmopolitan Greek-speaking society that enjoyed a high degree of literacy. Not only Christians but Jews as well encountered the Bible most frequently in Greek, the lingua franca of the Mediterranean world since the time of Alexander the Great. Many religious Jews were incapable of comprehending, much less reading, the Hebrew Bible in its original language and relied on Greek or Aramaic translations. Aramaic was the native language of Jesus and the early disciples, and he seems to have taught in that language, but all the New Testament texts were composed in Greek. Whether or not they knew Greek, Jesus and some of the disciples may have been capable of reading from the *Sefer Torah*, or Hebrew scroll, during worship, as Luke 4:16–20 seems to show Jesus doing, but it would not have been unusual if they could not, as John 7:15 seems to imply of Jesus. The monopoly of Greek over Jewish and Christian religion continued under Roman occupation of the Holy Land long after Jesus's time.

Although it is dangerous to exaggerate the differences, attaining literacy in Greek was less arduous than in Hebrew or Arabic. Apparently, the culture of seventh-century Mecca, like that described in Genesis, was entirely oral, and Muhammad's earliest followers lacked any direct experience with scrolls and codices. "The Qur'an came into existence in a seventh-century Western Arabian world that may have seen or heard of books, but, as far as we can tell, had neither produced nor read them. The same is likely true of the Israelites who stand behind—'composed' is still a step beyond them—the earliest books of the Bible."[4] Hebrew, Greek, Latin, and Persian book-cultures were entirely foreign to Muhammad's first audience; even poetry in Arabic seems not to have been transmitted in pre-Islamic writing. As a result, the Qur'an's notions of books and scripture differ radically from Jewish and Christian understandings. "God's revelations to Muhammad are self-identified as 'The Recitation' (*al-Qur'an*), and then, only somewhat later in the series of revelations, as 'The Book' (*al-Kitab*). This latter expression appears to refer not to the bound artifact, which would have been beyond the capabilities of Muhammad or his contemporaries to produce, but to the notion of a book revelation, to what we call 'Scripture.' What you are hearing from the Prophet's lips is *Scripture*, the Qur'an announces, with as little embarrassment or even awareness of the paradox as we ourselves have in repeating it."[5] "Scripture" was not a physical codex or scroll but "this Heavenly Book *sent* down but not yet *written* down by mortal hands."[6] It originated in "the eternal and uncreated Heavenly Book, the Qur'an's 'Well-Guarded Tablet' (28:20) on which God's Word is inscribed." This heavenly original was "the Mother of the Book." Muhammad did not see or read it but *heard* it as "the 'Recitation' pronounced privately but audibly by God—the Muslim theologians professed not to know how." It was then "repeated publicly, and identically, by Muhammad in the historical circumstances of Mecca and Medina between 610 and 632 CE. And finally, the Qur'an is Muhammad's 'recitation' committed, inerrantly . . . both to memory and to writing . . . during the Prophet's own lifetime, we are told."[7]

The Qur'an's scriptural claim to be immutably and inerrantly transmitted from its heavenly original to the book in the reader's hand conflicts with scholarly consensus about the transmission of texts through history, and with what is known or conjectured about the Qur'an's own textual history. However, the claim is invaluable confirmation of the mystique of writing in societies undergoing transition to literacy; the Qur'an's subsequent history, like that of the Jewish and Christian Bibles, leaves no doubt about the durability of that mystique even in highly literate societies. In its attitude to writing and books, Muhammad's oral society likely resembled the culture that initially produced the Torah. For such cultures, "scripture" as writing is rare, even unique, and thus possesses what F. E. Peters calls an "odd power or 'glamour'" (a word that implies magic). Citing the "handwriting on the wall" in the Book of Daniel (5:5–28), Peters notes that its words (*Mene mene tekel upharsin*) are "in translation, a nonsense jingle of Aramaic coinage." However, King Belshazzar and his Babylonian courtiers perceive the inscription as "pure magic," which can only be understood and interpreted by the wonder-working Hebrew prophet Daniel.[8] This "glamorous" view of writing as something produced and understood by the powerful but deadly to others reminds us of the mentality behind Homer's story about Bellerophon and the tablet of "murderous symbols." Indeed, the Babylonian king, who has defiled ceremonial vessels looted from the Jerusalem temple, dies the same night. In Homer the power behind the handwriting is mysterious but human, whereas in Daniel it is unmistakably divine. In both cases, to paraphrase Arthur C. Clarke, writing-technology is equivalent to magic.

The original emotional reaction to scripture as writing is wonder, astonishment at a technology perceived as supernatural. That wonder is strongest in contexts where literacy is rarest, particularly when entire societies, like those of ancient Israel and Mecca, are transitioning from oral to written culture, or from written back to oral, as in Homeric Greece. The charisma or "glamour" of scripture is later reinforced and carefully maintained through ritual, especially communal celebration, and through the continuous supplement of written commentary. But scripture itself contains robust textual evidence of the wonder originally provoked by writing. In what follows, I will concentrate on the Jewish and Christian scriptures, since they rarely claim—and even more rarely display—the immutability claimed for the Qur'an. It is true that medieval and later European writers treated the Bible with a respect they often denied to the Qur'an, but that is, again, a separate issue.

CHAPTER FIVE

The Jewish Scriptures

God from Heaven giving Laws to Men, gave not an oral, but a written Law, and it was from Him, that Letters were cloathed with Sounds.

—DANIEL DEFOE, 1726

Among religions "of the Book," that of Moses provided the most dramatic and enduring model of the original, "glamorous" association between divinity and writing. Despite the centrality accorded to writing in the Hebrew Bible, the first mentions of it do not come in the Book of Genesis, where we might expect them, but in its second book, Exodus, where Moses first appears. Moreover, even in Exodus, writing enters the scene very late in the text, most dramatically when God delivers the stone tablets containing the Decalogue, or Ten Commandments, to Moses. For many centuries, as Daniel Defoe confirmed in 1726, Jews and Christians accepted this episode as the moment when writing first began.[1] But this hoary commonplace misrepresents what the text of Exodus actually says. In fact, several incidents explicitly mention writing as existing *before* God delivers the stone tablets to Moses in chapter 31.

The earliest reference to the Mosaic commandments, in Exodus 20:1–14, asserts that God originally gave them in oral form only.[2] When Exodus first mentions Moses's ascent of Mount Sinai, God simply speaks the Ten Commandments to him, and adds several more detailed commands in the three following chapters (Exodus 21–23). Chapter 24 asserts that the prophet then descended to the Israelites and likewise recited the commandments orally. But after Moses's followers agreed to honor God's injunctions, the text continues, "Moses *then wrote down* all the commands of the Lord" (emphasis added). In this, the first reference to written commandments, Moses himself simply transcribes them—how and on what material, we are not told. Such a matter-of-fact presentation implies that writing is already a familiar human activity. After writing the commandments, Moses "took the record of the covenant and read it to the people" once more, and they reiterated their vocal agreement (see plate 4).

Although at this point God has not yet written anything, Moses's reading the written text aloud apparently bestows a more official and powerful mystique on the commands than merely reciting them from memory. After the reading, "Moses took the blood [of a sacrifice] and dashed it on the people and said: 'This is the blood of the covenant that the Lord now makes with you concerning all these commands'" (Exodus 24:4, 7–8). Together the blood and the writing dramatically confirm this moment as a binding contractual agreement between God and the Israelites.

Later in the same chapter, we get what seems to be another version entirely of the story behind the contract; this time, it is God's turn to write, and he tells Moses: "Come up to Me on the mountain and wait there, and *I will give you the stone tablets* with the teachings and commandments which *I have inscribed*" (Exodus 24:12; emphasis added). Then, God decrees how the Ark of the Covenant must be constructed, commanding Moses to "deposit in the Ark [the tablets of] the Pact which I will give you" (Exodus 25:16). At this point, preoccupation with the Ark makes Exodus lose sight of the tablets for six whole chapters. Then: "When [God] finished speaking with [Moses] on Mount Sinai, he gave Moses the two tablets of the Pact, *stone tablets inscribed with the finger of God*" (Exodus 31:18; emphasis added). Presumably, God's bestowal of stone tablets after Moses had already written down the commandments could be read as simply describing a permanent record of the covenant. But the mention of God's finger as the instrument of inscription seems intended as the definitive version of how the tablets of the law came to be written (figure 5.1).

However, in the form in which we now read the text of Exodus, these are only the *original* tablets. Three chapters later, Moses dis-

DE

LITERIS

INVENTIS

LIBRI SEX.

Ad Illustrissimum Principem

THOMAM HERBERTUM,

Pembrokiæ Comitem, &c.

Auctore GULIELMO NICOLS, A. M.

DEDITQUE DOMINUS MOSI DUAS TABULAS TESTIMONII LAPIDEAS SCRIPTAS DIGITO DEI.

J. Nutting Sculp.

LONDINI.

Apud Henricum Clementem Bibliopolam, ad Insigne Lunæ Falcatæ, in Cœmeterio D. Pauli.

M DCC XI.

FIGURE 5.1. "The Lord Gave Moses Two Tablets of Stone, Inscribed by the Finger of God." Title page of William Nicols's *De litteris inventis libri sex*. London: Henry Clement, 1711. Sheridan Libraries Special Collections, Johns Hopkins University, P2521 .N53 1711 C c. 1.

covers the Israelites worshiping the idol of the golden calf and smashes the "tablets inscribed with the finger of God." Thus we read that they are about to be replaced with facsimiles, also written by God. "The Lord said to Moses: 'Carve two tablets of stone like the first, and *I will inscribe upon the tablets* the words that were on the first tablets, which you shattered'" (Exodus 34:1). Moses follows God's instructions and makes the blank tablets; but now some further curious inconsistencies intervene. God does not deliver an exact repetition of the commandments familiar from Exodus 20:3–14 but a longer and different instruction. It contains no equivalents for commandments five through ten, which prescribed relations between people; instead, all commandments now concern human relations with God. They repeat some detailed commands from Exodus 23, including some that Moses had previously written down even before God delivered the original inscribed tablets (Exodus 24:4, 7, 12). Despite this expansion, the new tablets are referred to as "the Ten Commandments," as if identical to the original laws.

Between Exodus 20 and 34, the number and nature of the commands are in constant flux. The fact that we refer to the *Ten* Commandments appears to be sanctified by tradition rather than the biblical text. As if these differences were not sufficiently dramatic, God also now declines to inscribe the new tablets himself, instead telling Moses to take dictation.

> The Lord said to Moses: *Write down* these commandments, for in accordance with these commandments I make a covenant with you and with Israel.
>
> And he was there with the Lord forty days and forty nights; he ate no bread and drank no water; and *he wrote down on the tablets the terms of the covenant, the Ten Commandments*. (Exodus 34:10, 27–28; emphasis added)

Given the much lengthier commandments of Exodus 34, Moses might well have needed forty days and nights to carve them.

Thus, even in the same chapter, Exodus gives contradictory evidence: Who wrote down the commandments on the second set of tablets, God (Exodus 34:1) or Moses (34:28)? It is particularly significant that references to the tablets here begin with "the first tablets, which you shattered," and end with "the terms of the covenant, the Ten Commandments." These sound like editorial interpolations, and they appear to organize several preexisting versions into a single, more or less unified story. The reference to "the first tablets" in 34:1 is particularly crucial and was evidently intended to avoid confusion. It implies that there were at least two preexisting versions of Moses ascending the mountain and receiving the commandments on two tablets. But who inscribed them? In addition, the phrase "the Ten Commandments" in 34:28 fits no description of God's instructions except the first one, fourteen chapters earlier in Exodus 20:3–14. In between, we find several versions of God's commandments, all of them longer and more complex than the one in chapter 20. So chapter 34's belated reference to "the Ten Commandments" looks like a retrospective attempt to organize all of chapters 19 through 34 into a single narrative. It is only partly successful from a modern point of view, but there is no further attempt to consolidate the narrative: after the first three verses of chapter 35, references to the tablets and commandments cease.

In fact, historical scholarship considers the text of Exodus a composite document conjoining several earlier variants of the story about God's written contract with his people. The quantity of repetition, overlap, and even contradiction argues for intervention by several editors who (possibly over several centuries) streamlined several earlier accounts as far as possible into a single tale. Despite the remaining inconsistencies, two things are noteworthy. First, the necessity of *writing down* God's contract with the Israelites. Second, Exodus emphasizes the importance of the written text by relating how God commanded the Israelites to create the Ark of the Covenant to house the tablets *before* he gave the first set to Moses: "And deposit in the Ark [the tablets of] the Pact which I will give you." As Exodus now stands, the tablets' supreme importance requires that a suitable place be prepared for them in advance, which is described in the six-chapter hiatus between descriptions of the tablets that I mentioned above. Strangely, by the time the Ark is completed, not one but two sets of tablets will have been carved. Not until Exodus 40:17 do we learn that

Moses "took the Pact and placed it in the Ark." Nevertheless, despite their many contradictions, all the events serve to emphasize the tablets' scriptive—written, scriptural—authority.

In keeping with the reverence shown to the tablets, the Ark was no ordinary box (*arca* in Latin). It was a kind of portable mountain-top, a place for God's presence to appear; moreover, it would become an instrument for him to communicate further commands. "There I will meet with you, and I will impart to you—from above the cover, from between the two cherubim that are on top of the Ark of the Pact, all that I will command you concerning the Israelite people."[3] Despite the anathemas heaped on idols and "graven images" in Exodus and repeated throughout later centuries, the two golden cherubs on the Ark are there by God's command. Together, the Ark, the tablets, and the golden cherubim were meant to accommodate—and to signal—God's presence in the same way as the clouds and fire had previously done. Once the Ark was built, these two golden images would function as a visible indicator, showing the location of God's invisible presence, as if they were the twin poles of an electrical circuit (see plate 5).

By contrast, the rebellious Israelites blasphemously identified the golden calf *as* God's presence. At the end of chapter 31, the exasperated Israelites, still awaiting God's message, forced Moses's brother, Aaron, to create the idol. Afterwards, "they exclaimed: 'This is your god, O Israel, who brought you out of the land of Egypt'" (Exodus 32:4), showing their need for a visible, tangible God. In context, the Israelites sinned less by creating an image than by identifying it as "our god."

Thanks to the golden calf, the importance of the Ark as a portable venue for God's presence can also be understood another way. The calf had no connection to writing, whereas the cherub-images were consecrated by their framing of God's writing: they indicated the presence of the tablets inside the Ark just as they located God's invisible but audible presence above the Ark. For the priestly writers and editors of Exodus, writing made God and his words inseparable, whereas the idolaters identified him with an object, the concrete image of an animal.

The calf and the Ark are opposed in another way. Although God in his wrath threatened to destroy the idolaters, Moses convinced him not to. But on descending the mountain, Moses unleashed his own wrath. "Thereupon Moses turned and went down from the mountain bearing the two tablets of the Pact, tablets inscribed on both their surfaces: they were inscribed on the one side and on the other. The tablets were God's work, and the writing was God's writing incised upon the tablets. . . . As soon as Moses came near the camp and saw the calf and the dancing, he became enraged; and he hurled the tablets from his hands and shattered them at the foot of the mountain" (Exodus 32:15–16, 19; figure 5.2). Here again, the narrative juxtaposes God's writing with the golden idol.

Strangely, although God fabricated and inscribed them, the tablets suffer destruction, just like the golden calf: "Moses took the calf that they had made and burned it; he ground it to powder and strewed it upon the water and so made the Israelites drink it" (Exodus 32:20). Both the legitimate written form of God's presence and the image illicitly identified *as* divine presence are destroyed, creating a kind of vacuum of the divine. However, although the tablets are shattered, they will be recreated, whereas the golden calf is further debased: having been ingested, the powdered gold was necessarily mixed with excrement in the Israelites' bellies.

The passage implies that the crucial difference between idolatry and proper worship, as between the idol and the true God, is the presence of writing. Such an interpretation seems to be borne out in the rest of chapter 32, after Moses has broken the tablets, where another form of writing appears. After having the Levites slay three thousand errant Israelites, Moses gathers the remainder, berates them, and declares he will intercede for them with God. Both Moses and God then refer to something they call God's "written record" as the gauge of his favor toward humans. "Moses went back to the Lord and said, 'Alas, this people is guilty of a great sin in making themselves a god of gold. Now, if You will forgive their sin [well and good], but if not, erase me from the record which You have written!' But the Lord said to Moses, 'He who has sinned against Me, him only will I erase from My record'" (Exodus 32:31–33). The shattered

FIGURE 5.2. ". . . and Moses hurled the tablets from his hands and shattered them" (Exodus 32:19). Illustration by Gustave Doré, Doré's English Bible, 1866. Public domain, Wikimedia Commons.

tablets of the law symbolize the broken relation between God and the Israelite collective; but they also correlate to the "record"—not explicitly indicated as a metaphor—which God has written concerning his relations with individuals, a record he can erase or revise if he chooses. Moreover, whether it is literal or metaphoric, the record seems to have the same reality—or even to be identical with—something seeming to be God's ledger: "Go now, lead the people where I told you. See, My angel shall go before you. But when I make an accounting, I shall bring them to account for their sins" (Exodus 32:34). This reference to accounting seems to introduce the plague that God sends "upon the people, for what they did with the calf that Aaron made" (Exodus 32:35).

The question of God's presence is, then, central to the portrayal of the Ark. Before it was built, we are told, God habitually encountered Moses in the "Tent of Meeting." The people would see only a pillar of cloud at the entrance to the Tent of Meeting, but "the Lord would speak to Moses face to face, as one man speaks to another."[4] Yet this face-to-face familiarity seems contradicted later in the same chapter when Moses asks God to appear visibly to him so as to confirm the favor shown him thus far. To the request, "Oh, let me behold Your Presence!" God answers: "You cannot see My face, for man may not see Me and live." Instead, God proposes a compromise: "As My Presence passes by, I will put you in a cleft of the rock and shield you with My hand until I have passed by. Then I will take My hand away and you will see My back, but My face may not be seen" (Exodus 33:18–23). God's back and his writing are his only visible attributes, and both are mere traces of him: Moses can at least see God's back, but writing is all anyone else ever sees. By allowing Moses the special privilege, God shows him the sign of his leaving, of his *having been present*. But even this trace will be denied to everyone but Moses. For them, writing remains as the only residue of divine presence. Writers have understood for millennia that writing can serve as a substitute for the author's presence, but some have acknowledged that writing is thereby a sign of bodily absence. As the presence of an absence, writing is potentially an oxymoron. Yet the Ark, as Exodus represents it, contradicts this pessimistic idea—for Moses, at least—by directly and literally connecting the text of the commandments to the presence of God.

The peculiar mix of repetitions and contradictions in the Exodus story presumably resulted from the editors' obligation to satisfy several constituencies who possessed different narratives (and perhaps cultic traditions) concerning Moses, God, and the tablets. It may be that, when the differing accounts were edited into one—a process that probably happened more than once—the redactors were reluctant to eliminate many of the conflicting details. If so, they presumably considered it more important that different constituencies recognize elements of their own accounts than that the composite narrative cohere in perfect temporal sequence.

Another consequence of the editorial process is that in Exodus the Israelites seem to revere the tablets of the Decalogue before God has ever even mentioned them to Moses. Already in chapter 16, after God has distributed the first manna to the hungry Israelites, Moses commands his brother, Aaron, to place a jar of that substance "before the Lord, to be kept throughout the ages. As the Lord had commanded Moses, Aaron placed it before the Pact, to be kept" (Exodus 16:33–34). Although no tablets are mentioned here, subsequent chapters of Exodus identify the Pact as the tablets of the Decalogue; so the reference here hardly fits anything else. Yet God will not inscribe the tablets for another fifteen chapters, at the end of chapter 31.

There is one further major irony. Although the "Pact" of chapter 16 merely implies writtenness, writing is explicitly mentioned in chapter 17. After God calms the quarreling Israelites by giving them a military victory over the Amalekites, he commands Moses to "inscribe this in a document as a reminder, and read it aloud to Joshua" (Exodus 17:14). This is the first unambiguous mention of any act of writing in the Bible, seven chapters before the tablets of the commandments appear, and seventeen before the "finger of God" inscribes them.

So Defoe's—and many others' before him—idea that writing had a "divine original" is not biblical. Instead, it is an homage to the respect that the "religions of the book" developed for writing, perhaps because Exodus mentioned it in connection with "God's finger." According to the logic of monotheism, the tablets God inscribed *should* be the original of writing: people *should* write only because God wrote. But in the *text* of Exodus, God writes because humans wrote first, just as he does from a historical perspective on writing. The many contradictions about writing that remain in Exodus are probably not due to the priestly editors' carelessness or inattention but rather to the importance they accorded to writing as a durable record of actions and agreements, and more importantly, thanks to the Ark, as a means of access to God's presence, or at the very least, a sign of it.

IS IT A FORGERY?

There is one further aspect of this composite story that merits attention. For most of the past three millennia, the relation between God, Moses, and the writing of the Decalogue has been *sacred*. As noted earlier, the modern term comes from a Latin word—*sacrum*—meaning "set aside," "reserved," or "untouchable." Discussion of the Moses/God/Tablets story as a document invented by human beings rather than a revelation handed down by God has often been taboo for observant Jews and Christians. Whatever their private opinions on the matter, until rather recently, the majority have treated the entire story *as if they believed it*, and as if the discrepancies that a modern nonbeliever notices were unimportant or somehow intentional. (Whether in fact, any particular person believed this is, of course, unknowable.)

But if we regard the Exodus story in narratological terms—that is, *purely as a story*—it has features that are common to other texts that have been dismissed as forgeries, such as the ones discussed in chapter three. As with the books of "Numa," "Dares," and "Dictys," many such tales were obviously intended to make literary forgeries believable. In the case of Exodus 19–34, there may have been no cynical intention to deceive, even when the versions were combined to create a single narrative. Not all forgeries are deliberate, and many documents that have the literary characteristics of forgery were not demonstrably intended to deceive, as we saw in chapter three with Iamblichus's description of the *Corpus Hermeticum*. Still, it is certain that the Moses/God/Tablets story demonstrates an intention to *authorize*, to make official. Moreover, its effect is to solicit belief: to argue that this story is true—things happened in just this way—and that we should therefore accept (1) the rules for living given in the "Ten" Commandments, (2) the special relationship between God and Moses, and (3) the special relationship between God and a particular group of people.

We will see further examples of authorization strategies in future chapters.[5] For now, one paramount feature of the Moses/God/

Tablets story is notable: Moses's actions at crucial points in the narrative are unverifiable. On whose authority are we expected to believe the story? Forget for a moment the three thousand years of religious tradition: What does the text in its current form say about itself? In essence, it implies that *no one at the time proved that events did not happen* in the way Moses claimed. Thus, presumably, the ancient compilers of Exodus intended the acquiescence of Moses's followers to serve as a model, so that, when you or I read the text of Exodus, we should likewise not dispute the alleged facts of the story. However, some of those alleged facts open the story to suspicions that are considered entirely legitimate when provoked by other documents: such skepticism is conventional not only regarding other stories in books but also for assertions of proof in law courts and discussions of scientific experiments.

The most notable feature of the story is in fact Moses's lack of human company at crucial moments. From the religious viewpoint, he is solitary because he is "chosen," set apart by a deity to perform a task: he is sacred, untouchable. God commanded, "No one else shall come up with you, and no one else shall be seen anywhere on the mountain; neither shall the flocks and the herds graze at the foot of this mountain" (Exodus 34:1, 3). But we have only Moses's word, transmitted by the text, that God, not Moses, made the prohibition. From the narratological perspective, Moses becomes a sacred leader *because* he is solitary—the precautions above ensure that no human could credibly say, "I witnessed you at that time, and you did not experience what you claim." The final—and humanly most important—stage in the history of the tablets admits that Moses carved the tablets himself. He ascended the mountain carrying blank tablets he had created, saying that God would inscribe them on the mountaintop—where, for forty days, no one but Moses would even be allowed to approach. At the end of that time, Moses returned with two inscribed tablets, admitting that, in point of fact, God did not inscribe the commandments himself but dictated them to Moses.

In any other story about the origin of a historically important text—and we will examine several in this book—such a changed, "downsized" claim about authorship would make us suspect falsehood. So how do we know that even the least dramatic version of Moses's tablets is true? Other than the inscribed tablets, what evidence did Moses's followers have? Since no one was allowed to see Moses's encounter with God, Exodus had to provide a substitute. Thus it claims that Moses's own body displayed the aftereffects of God's presence, as if branded: "As Moses came down from the mountain bearing the two tablets of the Pact, Moses was not aware that the skin of his face was radiant, since he had spoken with him. Aaron and all the Israelites saw that the skin of Moses' face was radiant, and they shrank from coming near him" (Exodus 34:29–30). The Christian Middle Ages mistranslated these luminous rays as "horns," as illustrated in famous depictions, including Michelangelo's *Moses*. The rays or horns were presented as proof of Moses's encounter for "Aaron and all the Israelites." But for readers—both Moses's contemporaries and ours—the ultimate proof of God's presence is that "it is written."

Within the Exodus story, these proofs are part of a larger narrative about evidence that begins with the Israelites' attempts to leave Egypt. Until Pharaoh allows the people of Moses to emigrate, the plagues visited on the Egyptians demonstrate to them that God "means what he says." For readers, plagues demonstrate that Moses is telling the Egyptians—and Exodus is telling us—the truth about an agent named God. But the most important and crucial proofs in Exodus concern the Decalogue and the authority of writing. What proof do we have that any of this ever happened? That it was actually Moses who originally wrote it down? That Moses existed as a single, historical, flesh-and-blood person? Only that "it is written."

MOSES'S HUMAN SOURCES

If we assumed that a single man named Moses wrote the five books that tradition attributes to him, we would need to account for the fact that, on occasion, Moses quotes what can only be a written source. While recounting the Israelites' battles for possession of Canaan, the Book of Numbers mentions one encampment "beyond the Arnon [River], that is, in the wilderness. . . . Therefore the Book of the Wars of the Lord speaks of '. . . Waheb in Suphah, and the wadis: the Arnon with its tributary wadis. . . .'" (Numbers 21:14–15). Although this reference is an insignificant detail for moderns, it caused centuries of discussion among Jewish and Christian commentators: Was "the Book of the Wars of the Lord" a real, physical book or a metaphor? Was it another revelation made to Moses by God? Did its title actually refer to the book—Numbers—in which it occurred? To premodern audiences, it mattered greatly whether Moses's source was a human chronicler or God himself.[6] Most of them agreed, however, that if *The Wars of the Lord* had ever been a physical book, it was lost after Moses's time.

Traditional biblical interpretation identified the Book of Deuteronomy as another book lost in antiquity, despite being composed by Moses himself. Apparently it was only rediscovered when Josiah, the late seventh-century BCE king of Judah, decided to renovate the Temple of Solomon. For that purpose he sent the high priest Hilkiah to collect treasures stored within the temple. "As they took out the silver that had been brought to the House of the Lord, the priest Hilkiah found a scroll of the LORD's teaching given by Moses. Hilkiah spoke up to the scribe Shaphan and said, 'I have found a scroll of the Teaching in the House of the Lord.'" This incident is narrated in 2 Chronicles 34:14–15 (late 4th c. BCE); Hilkiah's discovery had already been told in 2 Kings 22:8 (late 7th to mid-6th c. BCE) but without the detail that the teaching was "given by Moses." Significantly, the Latin Vulgate, the official Roman Catholic version of the Bible before the twentieth century, translated this phrase as "per manum Moysi," *by the hand* of Moses, asserting the scroll's authenticity in no uncertain terms. Traditionally, the scroll that Hilkiah found was identified as the Book of Deuteronomy itself, and the story is still widely considered to reveal something about the sources and composition of Deuteronomy.

Traditional lore connected the vicissitudes of the Torah to the fortunes of Solomon's Temple, which Nebuchadnezzar's armies destroyed in 587 BCE when they conquered Jerusalem and took the Jews into exile (the Babylonian Captivity). According to later biblical sources, this atrocity eventually stimulated more researches in the archives, this time in those of Babylon. The Book of Ezra (Esdras) relates that the Persian "King Cyrus of Babylon" (d. 530 BCE), who conquered that city in 539, ordered that the Jerusalem Temple be rebuilt. By the time of Darius (d. 486 BCE), Jews had begun the rebuilding but were halted by the king's functionaries, who demanded documentary proof of Cyrus's authorization. "Thereupon, at the order of King Darius, they searched the archives where the treasures were stored in Babylon"—apparently to no avail. "But it was in the citadel of Ecbatana, in the province of Media, that a scroll was found in which the following was written: 'Memorandum: in the first year of King Cyrus, King Cyrus issued an order concerning the House of God in Jerusalem: "Let the house be rebuilt, a place for offering sacrifices.... And the gold and silver vessels of the House of God which Nebuchadnezzar had taken away from the Temple in Jerusalem and transported to Babylon shall be returned"'" (Ezra 6:1–5). The restoration of the temple thus had the highest earthly authority.

Even if genuine, Cyrus's command was apparently inadequate to reverse all the consequences of Nebuchadnezzar's conquest. According to another book attributed to Ezra, the Torah itself had been destroyed. Whether it actually happened or not, the erasure of the Torah symbolizes the idea that the Jews had become separated from their religion while captive in Babylon. The task of restoring orthodoxy fell to Ezra, since he was a scribe, and this entailed nothing less than completely reconstituting the Hebrew Bible. In the apocryphal Fourth Book of Ezra, God orders him to preach to the Jews, and explicitly compares his mission to that of Moses centuries earlier. Ezra accepts the charge, but asks God, "Who will warn those who will be born hereafter? For your Law has been burned, and so no one knows the things . . . done by you. . . . [S]end the Holy Spirit to me, and I will write everything that has happened in the world from the beginning, the things which were written in your Law, that men may be able to find the path."[7] God sends Ezra inspiration, not in the form of a book to be eaten, as was claimed by Ezekiel and John of the Apocalypse, but in a beverage: "a voice called me, saying, Ezra, open your mouth and drink." He was given "a full cup . . . of something like water, but its color was like fire, and when I had drunk it, my heart poured forth understanding, and wisdom increased in my breast."[8]

God commanded Ezra to recruit five scribes and "gave understanding to the five men, and by turns they wrote what was dictated, in characters which they did not know."[9] Presumably, the scribes' ignorance of the mysterious alphabet prevented the intrusion of human error, ensuring the purity of divine inspiration. Like Moses, Ezra and the scribes worked for forty days: they wrote "ninety-four books." Then God commanded Ezra to "make public the twenty-four books that you wrote first," that is, the entire Hebrew Bible. Both "the worthy and the unworthy" could be allowed to read them. However, Ezra was to "keep the seventy that were written last" for "the wise among your people. For in them is the spring of understanding, the fountain of wisdom, and the river of knowledge."[10] The identity of the seventy secret or "esoteric" books inspired centuries of speculation, as we shall see. Just as the returning Israelites rebuilt Solomon's Temple, so Ezra exactly duplicated the work of Moses by reconstituting God's Book: "It Is Rewritten," word for word. Interestingly, however, the seventy further books of interpretation were now necessary and reserved exclusively for the eyes of priests. Purely aside from questions about the Scriptures, these stories about Ezra reveal an ancient and common nervousness about the instability of writing and the extraordinary means necessary to stabilize it.

The Hebrew Bible refers to other books, such as the previously mentioned "Book of the Wars of the Lord" in a way that suggests they were its sources, but a modern scholar observes that "we know so little about these sources that we cannot be certain of the extent to which they were actual documents," even though many are referred to as annals or histories.[11] Such biblical references to writing, books, letters, archives, and libraries are too numerous to be useful here, though we shall see that scholars of later times nonetheless cataloged and commented on them.

OLD TESTAMENT APOCRYPHA AND PSEUDEPIGRAPHA

A relatively large number of ancient and medieval texts purport to be scripture and to provide important information about God, humanity, and the cosmos, yet many do not appear in the canon of the Hebrew Bible. These books, some of which now survive only as fragments, are known as apocrypha and pseudepigrapha. Simply put, the apocrypha are books that appeared in the pre-Christian Greek translation of the Hebrew Bible but were not accepted by the scholars who formed the canon of the "Old Testament," two or more centuries later. The term *apocrypha* derives from the idea that these books had been "hidden away," or "set aside," as being of unverifiable authorship. *Pseudepigrapha* is a more judgmental category and means "*falsely* titled or attributed." It refers to books which neither Jews nor early Christians accepted into their scriptures.[12]

In the long centuries before scholars aimed for objective historical criteria, commentary on biblical narratives habitually took the form of expansion and imitation, as in the Fourth Book of Ezra. Some apocrypha and pseudepigrapha were sequels and prequels to the Bible, like the multiple installments of *Star Wars* and other cinematic franchises. Other texts were creative imitations, alluding to or mimicking Bible narratives, much as the films *Clueless* and *Austenland* depend on familiarity with Jane Austen's novels. Given this fundamental bookishness of apocrypha and pseudepigrapha, the activity of writing is, unsurprisingly, one of their major themes. Attributing pseudepigraphic books to venerated figures of ancient Jewish legend was intended to make them *authoritative*, believable as history or doctrine. The predominating voice of the first-person singular "I" served the same purpose. However, excessive insistence on the venerable identity of the texts' authors backfired at times, making even ancient readers wary of trusting them because they seemed to "protest too much," to be "too good to be true."

The pseudepigrapha attributed to the patriarch Enoch provide an infamous example of readerly suspicion. According to Genesis 5:21–24, Enoch lived in the seventh generation after Adam and was the great-grandfather of Noah. As an alleged author, he had a number of advantages. The number seven seems to associate him with the heavens, since it was also the number of planets known to the ancients. Similar logic dictated his age: Genesis says he lived on Earth for 365 years, thus a "year" of solar years. Most importantly, he did not die at the end of that time but "walked with God," who "took" him, so that he "disappeared" and passed into eternity alive (see plate 6). This combination of antiquity, holiness, celestial numerology, and disappearance implied that Enoch's relationship with God was even more intimate than Moses's. Thus, Enoch became the most renowned putative author of pseudepigraphic texts. In fact, not one but three pseudepigraphic Books of Enoch are extant, composed at dates ranging from the second century BCE to the fifth or sixth century CE.

As their attribution to such an early patriarch implies, the Enoch-books are would-be supplements to Genesis, or even its alleged sources. They provide incidents missing from the "Books of Moses," allegedly revealed long before his time or edited out of Genesis by him. As Jacques Derrida observed many years ago, literary supplements have a tendency to overshadow the entities they document: thus, from four verses of Genesis, Enoch's ghostwriters produced over three hundred pages of his "revelations."

The book known as 1 Enoch or the Ethiopic Apocalypse of Enoch has survived only in Ge'ez, the language of ancient Ethiopia, although it must have been composed originally in Hebrew or Aramaic.[13] The New Testament Epistle of Jude (1:14–15) quoted it accurately, as did some early Christian writers. By the fourth century, however, major Christian theologians denounced it for contradicting the "facts" narrated in Genesis. Saint Augustine vehemently dismissed 1 Enoch for its "fables," particularly the assertion (consonant with Genesis 6:4) that angels copulated with women during Enoch's time, siring giants. This miscegenation implied that angels fell after Adam and Eve, as well as before, thus apparently contradicting scripture.[14] Clerical censorship eliminated Latin and Greek versions of the book almost completely by about 800, so they survived only in fragments until modern times, when more complete texts were discovered. *Beowulf* (composed ca. 700–1000 CE) makes one of the last favorable references to 1 Enoch.[15]

Although 1 Enoch initially presents itself as a vision, at moments it purports to narrate real physical experiences. References to writing are crucial to this "reality effect" and seem intended to vouch for the objective truth of narrated events. Already in chapter 12, Enoch is given the epithet "scribe of righteousness"; we are told that he preached to the fallen angels and at their behest wrote down a prayer asking God to forgive them. Chapter 14 opens self-reflexively by declaring

"This is the book" of Enoch; it also characterizes itself as "the words of righteousness and the chastisement of the eternal Watchers," that is, the angels God originally delegated to supervise humanity, who "fell" by copulating with women (14:1–4).[16] Thereafter, the text describes in detail just *how* Enoch was taken up into the heavens. In chapter 33, "Uriel, the holy angel who was with me," shows Enoch the stars and "the gates out of which they exit" at night. Enoch himself "wrote down all their exits for each [star]: according to their numbers, their names, their ranks, their seats, their periods, their months." But later "Enoch" emends this, saying that Uriel "wrote them down for me," along with "their names, their laws, and their companies" (33:1–4). This all-encompassing list now has both angelic and human authority. Yet "Enoch" does not divulge the list to his readers. He repeatedly tantalizes them by withholding the details of such revelations, suggesting that the mere assertion of unimpeachable authority is the ultimate point of 1 Enoch.

In chapter 65, Noah replaces Enoch as narrator, creating a sequel or supplement to Enoch's supplement to Genesis. The intrusion presents Noah as both eyewitness to the reality of the Flood and corroborator of Enoch's prophecies about it. The relation is carefully narrated and motivated: Noah has become aware that the world is doomed on account of its sins. So, somewhat like Gilgamesh, he travels to the ends of the Earth to consult his ancestor. Like Utanapishtim, Enoch has depressing news for his descendant: he explains that humans "have acquired the knowledge of all the secrets of the angels, all the oppressive deeds of the Satans, as well as all their most occult powers, all the powers of those who practice sorcery, all the powers of (those who mix) many colors, all the powers of those who make molten images" (65:6). Apparently, art, magic, and idolatry are intimately related.

There is so much more to know, says "Noah," that the archangel Michael subsequently "gave me instructions in all the secret things (contained) in *the book of my grandfather, Enoch,* and in the parables which were given to him; and he put them together for me in the words of *the book which is with me*" (68:1, emphasis added). What is the relationship between the manuscript Michael gave Noah and the book we are reading? It remains unexplained. Like Exodus, 1 Enoch is a composite text, blended from previous "revelations," one of which was evidently attributed to Noah. The ancient editor clearly valued the authority of Uriel, Michael, and Noah, as well as Enoch's, and so retained them all at the expense of narrative consistency.

Chapter 69 names the principal fallen angels and tells what forbidden knowledge they taught humans. Given the previous positive references to scribes, writing, and books, it is disconcerting that the chapter seems to condemn the very act of writing, as if writing were inherently hubristic or transgressive. "The fourth [fallen angel] is named Pineme'e . . . he caused the people to penetrate (the secret of) writing and (the use of) ink and paper [*sic*];[17] on account of this matter there are many who have erred from eternity to eternity to this very day. For human beings are not created for such purposes to take up their beliefs with pen and ink" (69:8–10). This sounds sinister enough, but "Enoch" condemns knowledge in general as a force for corruption: "For indeed human beings were not created but to be like angels, permanently to maintain pure and righteous lives. Death, which destroys everything, would have not touched them, had it not been through their knowledge, by which they shall perish; death is now eating us by means of this power" (69:11). So humans were meant to imitate the angels' purity, but knowledge and writing seem to corrupt purity. Thanks to the teachings of Pineme'e, writing seems to avenge or punish Adam and Eve's violation: death now eats humans because the primal couple ate the fruit of the Tree of Knowledge, just as God predicted (Genesis 2:16–17). To paraphrase Nietzsche, it seems that writing has made us "human, all too human."

In chapter 81, "Enoch" returns as narrator, informing us that "It is written," concerning all time, past, present, and future. Uriel tells him to "look at the tablet(s) of heaven; read what is written upon them, and understand (each element in them) one by one." Enoch dutifully "looked at the tablet(s) of heaven, and read all the writing (on them) and came to understand everything . . . all the deeds of humanity . . . upon the earth for all the generations of the world." That the "tablets of heaven" constitute a universal book written in the constellations was the basis of astrology and a constant theme from Babylon onward. But here the Book of the Heavens functions as a prototype of the book that Enoch himself is writing: the primal Book instructs him much as Uriel does, or as the book he is writing instructs his readers.

Throughout 1 Enoch, the claimed sources alternate between vocal and written instructions. But books carry more authority. If it were not already abundantly evident that books and writing are central to the message of 1 Enoch, chapter 82 would make it so. In effect, the book we are reading becomes fully self-reflexive, speaking of itself in the third person: "Now, Methuselah, my son, *I shall recount* all these things to you *and write* them down for you. *I have revealed* to you *and given* you the book concerning all these things. Preserve, my son, the book from your father's hands in order that you may pass it to the generations of the world. *I have given* wisdom to you, to your children, and to *those who shall become your children* in order that they may pass it (in turn) to their own children and to the generations that are discerning. All the wise ones shall give praise" (emphasis added).[18] Writing ensures permanence. In fact, the fluid time-frame of this announcement seems to encompass all time: the book speaks of itself before it was written, after its composition, and indeed as it makes its way through all of history. 1 Enoch barely seems to distinguish its present from its past or future. It seems, like God, to be eternal.

SCRIPTURE AND FALSIFICATION

As 1 Enoch winds down, it opens new territory, considering aspects of writing not previously introduced, including an opposition between true and false scripture. It accuses its future imitators of both forgery and plagiarism: "And now I know this mystery . . . ; [sinners] will speak evil words and lie, and they will invent fictitious stories and write out my Scriptures on the basis of their own words" (104:10–11). A reader of any era might ask how we know we are not already reading one of these falsifications. Yet despite this possibility, "Enoch" concludes: "Again, know another mystery!: that to the righteous and the wise shall be given the Scriptures of joy, for truth and great wisdom. So to them shall be given the Scriptures, and they shall believe them and be glad in them; and all the righteous ones who learn from them the ways of truth shall rejoice" (104:12–13). Are we now reading those very "Scriptures of joy?" Hard to tell: the author has momentarily slipped into a discussion of scripturality itself.

We might now expect 1 Enoch to end, but it seems incapable of doing so, proving Derrida's point about the tyranny of supplements. Four chapters later, we find "another book of Enoch—which he wrote for his son Methuselah and for those who will come after him." We are assured that "when sin passes away, . . . the names of the sinners shall be blotted out from the Book of Life and the Books of the Holy One." Among the punishments reserved for the wicked, "some of (these things) were written and sealed above in heaven so that the angels may read them (the things that are written) and know that which is about to befall the sinners," presumably by reading the heavenly tablets that keep being mentioned (e.g., in 106:19–107:1). As for the righteous, "Enoch" reassures us that "I have recounted in the books all their blessings," though he predictably does not specify either the books or the blessings, and finally declares, "Here ends the Revelation of the Secrets of Enoch" (108:1, 7, 10, 15).

The steady drumbeat of references to writing and books, to the tablets of heaven and God's record-keeping, and particularly to 1 Enoch itself, manifests a prideful sense of accomplishment: the text is so massively self-involved that the process of writing competes for attention with the awesome phenomena that the narrators "Enoch" and "Noah" are supposed to have witnessed. Far more than the biblical visions of Ezekiel, Isaiah, or John of the Apocalypse, 1 Enoch concerns itself with itself, obsessively writing its own autobiography as scripture. The reader or hearer of the book should ideally be overcome by a sense of wonder simply because "it is written." Later, in more book-saturated cultures, such awe at *scriptura* was no longer possible.

TWO SUPPLEMENTARY ENOCH-BOOKS

In addition to 1 Enoch, there are two related works, 2 Enoch in Old Slavonic and 3 Enoch in Hebrew. Their contents and manuscript histories vary, but they both depend on 1 Enoch. Like it, both dwell on the composition of "Enoch's book," as well as on angelic and divine record-keeping. But they presuppose a far more book-oriented culture than 1 Enoch, perhaps as late as the tenth century CE (dating is still uncertain). In 2 Enoch,[19] we read that "the Lord summoned one of his archangels, Vrevoil by name, who was swifter in wisdom than the other archangels, and who records all the Lord's deeds. And the Lord said to Vrevoil: 'bring out the books from my storehouses, and fetch a pen for speed-writing, and give it to Enoch and read him the books'" (22:10–11 [version J]). Vrevoil instructs Enoch for thirty days and nights, "and his mouth never stopped speaking. And, as for me, I did not rest, writing all the symbols." Following in the footsteps of Moses and Ezra—but claiming to foreshadow them—"Enoch" narrates, "I wrote everything accurately. And I wrote 366 books"—an appropriate, one-upping "leap year" number for a patriarch who lived to age 365. There is a direct relation between Vrevoil's revelations and the book we are reading, and the narrator, even more than his counterpart in 1 Enoch, prides himself on the care, accuracy, comprehensiveness, and sheer quantity of his writings (23:1–6 [version J]).

As in 1 Enoch, there is a relay of narrating voices. In chapter 33, God himself replaces the narrator, and commands Enoch to write down everything he has shown and told him but also "whatever *I* [God himself] have written in the books." Enoch must "give them the books in [his own] handwriting." He is commanded to distribute "the handwritings of your fathers," beginning with those of Adam and Seth, which "will not be destroyed until the final age." God has delegated two angels to preserve the texts, "so that they might not perish" in the flood that He will send "in your generation."[20]

Despite emphasizing his own prowess as a writer, "Enoch" tells his sons that "there have been many books since the beginning of creation, and there will be until the end of the age." (Indeed, apocrypha were attributed to other antediluvian people, beginning with Adam and Eve.) But the most significant development is in chapter 64, where Enoch's family, hearing God call him to heaven, ask his blessing. They exclaim that "you are the one whom the Lord chose in preference to all the people upon the earth, and he appointed you to be the one who makes a written record of all his creation, visible and invisible." This is a fine summary of "Enoch's" scattered boasts:

he has written the ultimate and total book, the exact representation of everything. But then the onlookers make what a modern editor calls "an astounding encomium, for which the early extravagances of the book scarcely prepare the readers," and which "could hardly please a Christian or a Jew." They tell Enoch that he is "the one who carried away the sin of mankind, and the helper of your own household."[21]

Despite his hyperbolic authorial boasts, "Enoch" does not expect to frighten others away from writing, at least not about themselves. Chapter 65 declares that God created time so that everyone "might keep count of his own life from the beginning unto death, and think of his sins, and . . . write his own achievements both evil and good. For no achievement is hidden in front of the Lord" (65:4–5). Apparently, each person's tally will be compared with God's (perhaps something like His ledger in Exodus 32:34). There is a sense here that the uncataloged and unwritten life—and not merely the unexamined one—is not worth living. This implies a prodigious archive of autobiographies by the end of time—to be read by whom? God, or another of his overworked angels?

Writing is so important to 2 Enoch that God leaves an explanatory note when He transports the prophet alive to heaven. As Enoch converses with the bystanders, a sudden darkness provides cover for angels, who "hurried and grasped Enoch, and carried him up to the highest heaven." When the darkness dissipates and light returns, "the people looked, but they could not figure out how Enoch had been taken away. And they glorified God. And they found a scroll on which was inscribed: THE INVISIBLE GOD."[22] Given the importance of writing to both God and Enoch, it is fitting that the deity explain Enoch's sudden disappearance in a note. As with Moses's sojourn on the mount, absence, invisibility, writing, and authority are causally linked.

3 Enoch[23] differs from both 1 and 2 Enoch. It belongs to a Talmudic tradition known as Merkabah mysticism, named after the heavenly chariot in the first chapter of Ezekiel. 3 Enoch refers frequently to the Hebrew Bible, valued because "it is written"—the phrase recurs dozens of times, thrice in chapter 5 alone (vv. 8, 10, 14). Merkabah was an esoteric set of doctrines, meant to be carefully concealed from

noninitiates who might defile or misuse them. Common features of Merkabah mysticism are ascent to God's presence (as with Enoch and Elijah), apocalyptic prophecies, and an extensive angelology. The chariot itself is composed of angels, and Enoch not only ascends to heaven but is transformed into an angel named Metatron.

The primary narrator-figure of 3 Enoch is not Enoch/Metatron but one Rabbi Ishmael (early second century CE), whom Metatron instructs—or rather initiates. Ishmael learns that God loved Metatron "more than all the denizens of the heights," meaning the angels, and so "wrote with his finger, as with a pen of flame, upon the crown which was on [Metatron's] head, the letters by which heaven and earth were created . . . the letters by which all the necessities of the world and all the orders of creation were created. Each letter flashed time after time like lightnings, . . . like torches, . . . like flames, . . . like the rising of the sun, moon, and stars" (13:1–2). To create, God does not merely speak, as in Genesis: instead, He writes with an alphabet of light, as if foreshadowing neon or light-emitting diodes. Writing seems to have become a spectacle in itself.

In another way that recalls writing, angels serve as intermediaries between God and humanity (in fact, "angel" comes from Greek *aggelos*, "messenger"). Two angels are in charge: one oversees "the books of the dead, for he records in the books of the dead, everyone whose day of death has come." The other "is in charge of the books of the living" and "records in the books of the living, everyone whom the Holy One, blessed be he, is pleased to bring into life, by authority of the Omnipresent One" (v. 24). The two angels have almost identical names, separated by a single phoneme, much as the legendary Golem could be animated or deactivated by the single phoneme separating "truth" (*emet*) from "death" (*met*). In appearance, the two supervisory angels are indistinguishable, and each holds a pen of flame and a burning scroll. But in 3 Enoch everything is written on an enormous, cosmic scale: "The length of the scroll is 3000 myriads of parasangs; the height of the pen is 3000 parasangs; and the height of each single letter that they write is 365 parasangs" (v. 25). A parasang was five to six kilometers in length, while a myriad was the number ten thousand.

Satan is the recording angel for the nation of Israel, but the seraphim, who number four (each having sixteen faces), constantly erase his indictments.

> Why is their name called seraphim? Because they burn the tablets of Satan. Every day Satan sits with Samma'el, Prince of Rome, and with Dubbi'el, Prince of Persia, and they write down the sins of Israel on tablets and give them to the seraphim to bring them before the Holy One, blessed be he, so that he should destroy Israel from the world. But the seraphim know the secrets of the Holy One, blessed be he, that he does not desire that this nation of Israel should fall. What, then do the seraphim do? Every day they take the tablets from Satan's hand, and burn them in the blazing fire that stands opposite the high and exalted throne. (3 Enoch 26:12)

Thanks to the naivety of Satan and his colleagues, the incriminating tablets are kept from the presence of God "when he sits upon the throne of judgment and judges the whole world in truth" (v. 12). There is an apparent contradiction between God's judging the world "in truth" while being (or pretending to be) in the dark about the sins of Israel. In context, however, this paradox is outweighed by the symbolism of inscription and obliteration. 3 Enoch suggests that the ultimate realities, from creation until the future redemption of Israel, are brought into being by writing, whereas in Genesis 1:3, God simply speaks. Writing and the cosmos are all but identical for 3 Enoch.

To keep track of all this writing, God's heavenly court needs archives and an archivist. Metatron tells Ishmael that

> Above the seraphim is a prince, more exalted than all princes, more wonderful than all ministers. Radweri'el YHWH is his name, and he is in charge of the archives. He takes out the scroll box in which the book of records is kept. . . . He breaks the seals of the scroll box, opens it, takes out the scrolls, and places them in the hand of the Holy One, blessed be he. The Holy One receives them from his hand and places them before the scribes so that they might read them out to the Great Law Court which

> is in the height of the heaven of 'Arabot, in the presence of the heavenly household. (27:1–2)

This law court is composed of Watchers, the angels who supervise humans, and they deal with the aforementioned "books of the living" and "books of the dead." Like Satan "the Accuser" in the Book of Job, the Watchers behave like lawyers and prosecutors: "they take up and debate every single matter and they close each case that comes before the Holy One. . . . Some of them decide the cases, some of them issue the verdicts in the great court . . . some raise the questions . . . some complete the cases . . . and some of them carry out the sentences" (28:7–9). Law, justice, writing: all are inextricably linked in a perfect, spectacularly wearisome divine jurisprudence accounting for every human action.

This entwinement of writing and justice is so important to 3 Enoch that in one of its versions, we are told that

> Metatron sits for three hours every day and assembles all the souls of the dead that have died in their mothers' wombs, and of the babes that have died at their mothers' breasts, and of the schoolchildren that have died while studying the five books of the Torah. He brings them beneath the throne of glory, and sits them down around him in classes, in companies, and in groups, and teaches them Torah, and wisdom, and haggadah, and tradition, and he completes for them their study of the scroll of the law, as it is written, "to whom shall one teach knowledge, / whom shall one instruct in the tradition? / Them that are weaned from the milk, / them that are taken from the breasts." (48:12, version C, quoting Isaiah 28:9)

Even more than the other two Enoch-books, 3 Enoch marries a pre-occupation with completeness to a reverence for writing. The book's lengthy, redundant-seeming enumerations, tedious to a modern reader, imply that whatever is not written down will be forgotten in the most radical sense: not merely unremembered but somehow erased from being, as if it had never existed. As in magical formulas and Greco-Roman curse-tablets, an unenumerated detail potentially menaces the completion of the whole, and this concern seems to include the souls of infants. Lacking a knowledge of God and his law, these uneducated, not-yet Jewish children died in a state of incompletion. Metatron's tutelage restored, or rather created, the Jewishness of their souls.

The unfinished, uneducated souls of dead children are for 3 Enoch as worrisome as unbaptized children were for premodern Christians. Education, however, was not among the tragic losses Christian children suffered; they lacked a single brief ceremony, but its consequences were eternal. Few theologians were prepared to assert that unbaptized children would suffer actual punishment, since babies were incapable of sinning voluntarily. But their lack of baptism, which left them uncleansed of original sin, condemned them to Limbo. The incomplete, not-yet-Christian souls of unbaptized children were deprived of the "beatific vision," the eternal awareness of being in God's presence. Nothing could compensate for their lack of baptism. By contrast, dead Jewish children could make up for lost time, thanks to Metatron. His own status as a human transformed into an angel allowed him to shape the souls of children into the souls of Jews, by teaching them written lore. This otherworldly schooling is perhaps the ultimate indicator of the assumption, in all three Enoch-books, that writing is in some fundamental sense equivalent to existence, its sine qua non.

Other Old Testament pseudepigrapha have a generally lighter emphasis on the connection between existence and writing, but like the Enoch-books, they contain dozens of explicit references to books and inscription. One of their favorite topics is antediluvian writing and how it survived the Flood, that venerable theme already observed in Plato and Berossos. Antediluvian tablets are mentioned some twenty times in the Book of Jubilees, the Hebrew text of which predates 100 BCE.[24]

Jubilees also continues the countervailing emphasis on writing as fomenting the temptations of forbidden knowledge. It contains a legend that became a favorite of medieval writers, particularly in Byzantium. Jubilees contended that the patriarch Cainan was a grandson of Noah's son Shem (rather than of Adam's son Seth, as in Genesis 5:9), and thus lived after the Flood.[25] In Jubilees, Cainan was a writer who discovered a dangerous antediluvian book and transcribed it onto a tablet or stele.

> And [Cainan's] father taught him writing. And he went forth in order that he might seek a place where he could build a city. And he found a writing which the ancestors engraved on stone. And he read what was in it. And he transcribed it. And he sinned because of what was in it, since there was in it a teaching of the Watchers by which they used to observe the omens of the sun and moon and stars within all the signs of heaven. And he copied it down, but he did not tell about it because he feared to tell Noah about it lest he be angry with him about it. (8:2–4)

The story, interesting in itself, is even more intriguing because of its intertextual relations. In the first place, Jubilees claims to be a record of human history revealed to Moses during his forty days on Mount Sinai (Exodus 34:28). Second, it emphasizes the "Watchers," familiar from 1 Enoch 6–14, who teach transgressive knowledge to humans and sire giants with the daughters of men as in Genesis 6:4. The Watchers' transgression of sexual taboo and the forbidden knowledge they impart are forms of rebellion against God. In Jubilees Cainan sinned not only by using the divinatory knowledge of the Watchers but also by copying their book. Intriguingly, scholars have long suspected that the first thirty-six chapters of 1 Enoch derive from a preexisting "Book of the Watchers," so there could be some connection to the legend of Cainan.

A similarly intriguing antediluvian book appears in the *Life of Adam and Eve*, written around 100 CE.[26] Probably composed in Hebrew, it survived into the Middle Ages only in Greek and Latin translations. The Greek preface claims incongruously that God revealed it to Moses "when he received the tablets of the law from the hand of the Lord, after he had been taught by the archangel Michael."[27] The reason for this division of labor between God and Michael is unclear. Does it indicate that God hands over writing directly to humans but speaks to them only through angels? Are angels and writing somehow equivalent intermediaries?

The Latin translation relates the lives of the first couple in great detail, and, like 1 Enoch and Jubilees, strongly emphasizes Noah's Flood. Six days after Adam's death, Eve felt her own end approaching and convoked her sixty (!) children. The archangel Michael had previously revealed that to punish her and Adam's sin, "the Lord will bring over your race the wrath of his judgment, first by water and then by fire; by these two the Lord will judge the whole human race." The connection between Adam and Eve's sin and the Flood was not made in Genesis, and subsequent tradition traced the catastrophe to the sins of later antediluvians or even the giants, when "all flesh corrupted its way" (Genesis 6:12).

Eve, however, was concerned not only with sin but also with remembrance and continuity. Thus she adjured her children: "Make now tablets of stone and other tablets of clay and write in them all my life and your father's. . . . If [God] should judge our race by water, the tablets of earth will dissolve and the tablets of stone will remain; but if . . . by fire the tablets of stone will break up and those of clay will be thoroughly baked."[28] In 2 Enoch 33:11–12, the patriarchs' writings, from Adam through Enoch, were said to be preserved under the constant supervision of two angels, "so that they might not perish in the impending flood which I will create in your [i.e., Enoch's] generation." Berossos had entrusted such an archive to the agency of gods and humans. But Eve enlists the forces of physics rather than living supernatural guardians: the contrary forces of fire and water require contrary media, clay and stone. Her concern with history is personal—her and Adam's biographies—but also universal. Eve's eldest son, Seth, undertakes his mother's request: "Then Seth made the tablets."

A number of medieval manuscripts intimate that this abrupt ending was a jolt to Christian sensibilities.[29] They add a long appendix that explicitly connects the *Life* to the entire course of history, declaring that Seth added "what he had heard from [his parents] and his eyes had seen." Seth consecrated the tablets by putting them "in the oratory where [Adam] used to pray to the Lord."[30] We are told that Seth's tablets had a curious reception history after the Flood: they "were seen by many persons but were read by no one." Does this suggest that literacy was rare or that the letters were undecipherable?

Eventually, "the wise Solomon . . . saw the writings, and was entreating the Lord, [when] an angel of the Lord appeared to him," and promised that "you shall be wise in writing so that you might know and understand all that is contained on the stones." The angel also declared that "I am he who held the hand of Seth, so that he wrote with his finger onto stone," underscoring Seth's holiness by enabling him to imitate the tablets of the Decalogue, "written by the finger of God" (figure 5.1).

Thanks to the tablets, Solomon discovered "where the place of prayer was where Adam and Eve used to worship the Lord God." This knowledge was important because "it is fitting for you to build the temple of the Lord, the house of prayer, at that place." Once again, writing and continuity are thoroughly entwined: because Solomon's Temple occupies the space originally consecrated by Adam and Eve, it connects Jerusalem indirectly to the earthly paradise. Nor are loose ends allowed to mar the history of the text itself. Once Solomon completes the temple, he "calls forth" the documents of Seth, "written without the knowledge of words by the finger of Seth, his hand being held by the angel." Presumably, the documents will be preserved in Solomon's Temple. And although Seth wrote the tablets, "on the stones themselves was found what Enoch, the seventh from Adam, prophesied before the Flood, speaking of the coming of Christ, 'Behold the Lord will come in his holiness to pronounce judgment on all, to convict the impious of all their works.'"

This is indeed the prophecy of 1 Enoch 1:9, traditionally seen as foretelling the Flood. Moreover, the author of the Christian appendix recalled that the Epistle of Jude (vv. 14–16) interpreted "Enoch's" prophecy as foretelling the rise of "false" Christians, Christ's Second Coming, the Last Judgment, and the end of the world. The circuit of time is closed by a succession of texts linking Adam and Eve, Seth, Enoch, Noah, and Christ. The *Life of Adam and Eve* declares that all history "is written" since the beginning of time and thus compatible with or identical to the *Life*'s own version. Thanks to Eve's concern for remembrance and her ingenuity at preserving her memoirs, Solomon obtained the aboriginal knowledge that Moses and later writers had edited out of—or was it into?—the texts of the Bible.

The pseudepigrapha contain numbers of Jewish and Christian supplements to the Hebrew Bible. They transmit curiously detailed, literalistic lore of writing and books, such as a declaration that the builders of the Tower of Babel wrote their names on the bricks they used; this explains why they said, "Let us build a city, and a tower . . . to make a name for ourselves" (Genesis 11:4).[31] As expected, favorite topics include the tablets of the law, the Ark of the Covenant, and Moses. One story tells that when Moses descended Mount Sinai, the writing on his tablets disappeared as soon as he spotted the golden calf.[32] The detail obviously symbolizes the Israelites' abandoning God for idol-worship, but by breaking *erased* tablets Moses avoids sacrilegiously smashing God's own handwriting. The same text relates that twelve stones, each inscribed with the name of an Israelite tribe, miraculously appeared from heaven; God commanded the people to put them in the Ark of the Covenant along with the tablets of the law, "and they are there to this day."[33] The prophet Jeremiah is said to have caused the Ark to be swallowed up by a rock to protect it from capture: "In the rock with his finger he set as a seal the name of God, and the impression was like a carving made with iron, and a cloud covered the name, and no one knows the place nor is able to read the name to this day."[34] Like God and Seth, Jeremiah apparently inscribed the rock with the mere pressure of his finger. Thus sealed, the Ark was to remain inviolable until God was ready to bring it forth once again.

In a sense, the apocrypha and pseudepigrapha are parodies of scripture: as supplements to the Hebrew Bible, they consume, replace, and obscure it. Moreover, their main subject is not the biblical texts, but themselves. Their pretensions to their own scripturality are emphasized by their efforts to elicit wonder. By so obsessively focusing their rhetoric of wonder on themselves, they ironically leave the Hebrew Bible unrivaled as Scripture, since its revelations concern so much more than the process of its own composition. Their textual narcissism may be one reason, if not the main or explicit one, why the Jewish and Christian arbiters of authenticity remained unconvinced these books should be admitted to the biblical canon. In a monotheistic tradition, holy books ought logically to reveal the prerogatives of God and the duties of humanity rather than preening about their own composition. Nonetheless, biblical apocrypha and pseudepigrapha emphasize themes of writing so lavishly that they define scripturality more thoroughly than the Hebrew Bible had done. Whether or not God is a mystery, writing itself exerts a mysterious charisma, from Exodus onwards. This charisma was to become a constant in Jewish writings, most notably in the grand traditions of the Talmud and the Kabbalah, the latter of which was to exert profound influence over Christian writers, as well, particularly in and after the European Renaissance.

CHAPTER SIX

The Christian Scriptures

> And this they said tempting him, that they might accuse him. But Jesus bowing himself down, wrote with his finger on the ground. When therefore they continued asking him, he lifted up himself, and said to them: He that is without sin among you, let him first cast a stone at her. And again stooping down, he wrote on the ground.
>
> —John 8:6–8 (Douai-Reims Version)

The Christian New Testament competes credibly for the title of most successful scripture ever. But what, exactly, is the New Testament? Most Christians speak confidently of it, as a known quantity. However, as with the "Old Testament," the canon of the New Testament was unstable for a long while; its collection of twenty-seven books was not fixed for over three centuries. Just as the Hebrew Bible and the Old Testament are not identical, so, until nearly 400 CE, there were many opinions as to what books composed the New Testament. The four gospels that became canonical (Matthew, Mark, Luke, and John) were attributed to disciples of Jesus, yet knowledgeable scholars agree they are all pseudepigraphic, along with a number of other New Testament books. In fact, none of the four claims explicitly to be authored by an eyewitness, yet Christians attributed them to Jesus's disciples by the early second century; the gospel of Mark, considered the oldest, cannot have been written before 70 CE. Many other would-be gospels existed, including several attributed to other companions of Jesus. Not until about 180 CE did the four successful gospels begin to be considered canonical (see plate 7).

Early Christianity was in flux, divided into two main currents, known as "proto-orthodoxy" and Christian gnosticism. The labels can be somewhat deceptive. Proto-orthodox Christianity was the forerunner of both Catholic and Eastern Orthodox confessions, which did not part company definitively until the Great Schism of 1054, over controversies including papal supremacy and the nature of the trinity. Before and after 1054, Eastern and Western Christianity both considered themselves "orthodox" (meaning "right thinking") and "catholic" (i.e., "universal"). Gnostic Christianity and proto-orthodoxy disagreed far more radically than Catholicism and Eastern Orthodoxy would later do, particularly concerning the nature of Jesus and his message. "For gnostics, a person is saved not by having faith in Christ or by doing good works. Rather, a person is saved by knowing the truth." *Gnosis*, as knowledge or enlightenment, is liberation from ignorance about humanity's true nature rather than redemption from guilt and sin. Gnostic truth is "knowledge of where we came from, how we got here, and how we can return to our heavenly home. According to most gnostics, this world is *not* our home. We are trapped here in these bodies of flesh, and we need to learn how to escape."[1] Whereas orthodoxy "implicitly affirms bodily experience as the central fact of human life," many gnostics dismissed it as "a distraction from spiritual reality—indeed, as an illusion."[2] The necessary knowledge is a secret hidden from most Christians, even though "it is Christ himself who brings this secret knowledge from above."[3] The crucial human faculty for gnosis is intellect rather than will: gnosticism was an initiation into ultimate truth rather than a declaration of orthodox "faith" or the will to believe "rightly."

Gnostic religion was fundamentally elitist: it assumed that because salvation came only through knowledge, not everyone was eligible. Proto-orthodoxy, like the orthodoxies that developed from it, taught that all people could be saved, so long as they adhered to a set of rituals and professed certain beliefs. In practice, this meant voluntary submission to clerical authority. The rigid division between priesthood and laity was solidified when the emperor Constantine convoked bishops to the Council of Nicea in 325, which established the definition of orthodox faith (in the Nicene Creed) and set principles of ecclesiastical organization. Such inflexible, legally defined hierarchical divisions did not exist among gnostics. Gnostics often assigned priestly roles, such as leading aspects of worship, to initiates temporarily, by drawing lots. Such differences reflect opposing views held by orthodox and gnostic Christians concerning both human and divine nature, and the consequent duties of humans. Despite its egalitarianism (all may be saved), orthodoxy presupposed an abyss between God and human beings: God was totally "other," absolutely unlike even the most righteous human—except for Christ, who was both fully divine and fully human. God could never actually be known in himself. Conversely, despite its elitism in human terms (only those "in the know" may be saved), gnosticism taught that knowledge of one's own human nature could and should lead to knowledge of God's being.

Like other Christians, gnostics wrote gospels and attributed them to authoritative figures of the past. But scripture served radically different roles in gnosticism and in orthodox Christianity. Neither confession assumed that everyone was personally able to read scripture—far from it. But the orthodox assumed that the doctrine most necessary for salvation was contained in scripture, whereas gnostics considered writing too blunt an instrument for imparting the necessary dogma. "Much of gnostic teaching on spiritual discipline remained, on principle, unwritten. . . . Gnostic teachers usually reserved their secret instruction, sharing it only verbally, to ensure each candidate's suitability to receive it."[4]

Gnostics distinguished between *exoteric* knowledge, which was basic and could be imparted to anyone, and *esoteric*, or secret, knowledge, which required initiation. "According to the gnostics, some of the disciples, following his instructions, kept secret Jesus's esoteric teaching: this they taught only in private, to certain persons who had proven themselves to be spiritually mature, and who therefore qualified for 'initiation into gnosis'—that is, into secret knowledge."[5] Gnostics defined Jesus's exoteric teachings as the Christianity familiar to us from the four gospels that became canonical; such teachings, imparted through parables, were suitable for the humblest intellects, but were not the ultimate humanly understandable truth.

Yet intellectual or spiritual elitism was not entirely absent from the canonical gospels. The two oldest, Mark and Matthew, relate that Jesus confided the "secret" or "mystery" of "the Kingdom of Heaven" to the twelve disciples, but imparted his teachings exclusively via parables to everyone else. Contrary to later teaching concerning the parables, Mark and Matthew assert that the little stories are meant to reveal the truth only to the initiated, while making it *less* accessible to everyone else. In Mark's gospel, Jesus is particularly ruthless, specifying that he tells parables so that "those outside" may "see but not perceive" and "hear but not understand; lest they should turn again [or "convert"], and be forgiven." Matthew is only slightly less brutal: Jesus states that he tells parables "*because* they," meaning everyone but the disciples, "seeing see not; and hearing, they hear not." Yet his parables are not intended to remedy the masses' ignorance; rather they ensure that "whosoever hath, to him shall be given . . . but whosoever hath not, from him shall be taken away even that he hath."[6] Evidently, even in these two oldest canonical gospels, "having" and "not having" are dependent on personal initiation.

Although all noninitiates were equally ignorant of gnostic truth, enlightenment was not identical for every initiate. The deepest secret of gnosticism was that knowledge of God was obtainable only through knowledge of oneself, which had to be obtained through an arduous process that Elaine Pagels compares to psychoanalysis. Somewhat like oral parables, written doctrines in themselves—even gnostic ones—were either inadequate or actively detrimental to spiritual enlightenment. "Whoever follows secondhand testimony—even testimony of the apostles and the Scriptures—could earn the rebuke Jesus delivered to his disciples when they quoted the prophets to him: 'You have ignored the one living in your presence and have spoken (only) of the dead.'"[7]

This reprimand implies that the notion of scriptures was practically oxymoronic for gnostics. If the point of scripture is to establish authority—"believe this because it is written"—then gnostics practically assumed the contrary, that the truth is not and cannot be written anywhere. Writing can only reveal that truth is elsewhere. In the absence of a living, speaking initiator, a "guru," or "analyst," the secret of gnosticism is simply that *there is a secret*.

Like orthodox Christianity, gnosticism depended to a great extent on interpreting Old Testament scriptures; for gnostics, the early chapters of Genesis were particularly important, but in surprising ways. As Bentley Layton summarizes, the details of gnostic myths about God, the cosmos, and humanity were "parallel to the system of Genesis but quite incompatible with it."[8] That incompatibility was the basis of gnostic enlightenment. Contrary to orthodox Jewish and Christian interpretations, in gnosticism the creator and the highest god are not identical; instead, the highest and original god is purely spiritual, while the creator is an inferior, later-born god. As in Genesis, the creator god fashions the material universe, but gnostics taught that he is both ignorant and evil. Known as Ialdabaoth, Saklas, or Samael (names signifying ignorance or mendacity), he is identified with the God of the Old Testament, particularly through his characteristic boasts: "For it said, 'It is I who am god, and no other god exists apart from me,'" and also declared, "I am a jealous god." *The Secret Book according to John* asserts that these boasts conclusively prove the mendacity of this god: "'And there is no other god apart from me.' In uttering this it signified to the angels staying with it that another god did exist. For if no other one existed, of whom would it be jealous?"[9] Thus, for all practical purposes, the gnostic creator god is identical to Satan. As a result, it has been observed, "gnostic scripture may seem both Christian and anti-Christian, both Jewish and anti-Jewish: the strength of this paradoxical ambiguity eventually made it the classic example of heretical scripture."[10] Variants of gnostic theology espoused by Cathar (or Albigensian) heretics earned them violent persecution by orthodox Christians in France and northern Italy as late as the fourteenth century.

In gnostic theology one stimulus for hostility to the creator god is, paradoxically, the awareness of similarities between Genesis and Plato's *Timaeus*. Orthodox Christians esteemed the *Timaeus* because, despite its pagan theology, it told about the nature of God and discussed the origin of the cosmos and humanity's creation, transgression, and punishment, and other familiar theological topics. But whereas orthodox Christians saw these similarities as recognizing the truth of Genesis, gnostics made the opposite deduction, seeing Genesis as a corrupt and mendacious distortion. Their simultaneous embrace and rejection of Genesis led gnostics to evolve a convoluted and bizarre mythology that more nearly resembled the *Timaeus.* "Gnostic myth is the literary creation of philosophical poets—an elaborate theological symbolic poem, and not the spontaneous product of a tribe or culture," as Genesis was. Layton remarks that "'philosophical myth' of this kind was generally fashionable in the second century [CE]." Gnosticism came to prominence "following a revival of interest in Plato's mythic tale of creation, the *Timaeus,* in the previous two centuries."[11] Interestingly, the gnostic, philosophical attitude, a product of literate men, resembles the attitude of Plato's Socrates toward the unreliability of writing.

As a result of this intimate contradiction, gnostic gospels that explain the nature of God, the cosmos, and humanity are so complex that they impart no essential knowledge except that the secret, the enlightenment—and thus the deliverance from the bondage of earthly ignorance—comes through a process so arduous that it is, for all practical purposes, impossible. It is no accident that Jorge Luis Borges, the twentieth century's greatest enthusiast of labyrinths and paradox, chose the thought of Basilides (ca. 135 CE) to illustrate the complexities of gnostic cosmology and theology. Basilides taught that the cosmos contained 365 concentric heavens; our visible heaven, the lowest, was created and ruled by the God of the Old Testament. This god is not only a profoundly ignorant tyrant, but "his fraction of divinity is near zero."[12] He is, as it were, a pallid, 364th-generation photocopy of the original god who lives and rules in the most distant heaven. To reach redemption, the "enlightened" soul, following the example of Christ, must pass through the lower 364 heavens, a feat which requires knowing the secret names of all their divine and angelic inhabitants and other arcane lore: as Basilides asserted, "Few people can know these things—only one in a thousand, and two in ten thousand."[13]

The disagreement between gnostics and the orthodox over the content and message of Christianity was thus all but absolute. Even the nomenclature *orthodox* and *gnostic* was a polemical creation of the "orthodox" themselves. Scholars agree that proto-orthodox theology was already "conceived of as being *what gnostic theology was not.*"[14] As happened with Old Testament pseudepigrapha, the "orthodox" were so successful in combating and suppressing gnostic literature that, until the twentieth century, gnosticism was known primarily through denunciations titled "Against Heresy" by self-styled orthodox writers of the second through fourth centuries. The etymology of heresy is "choice," which the orthodox construed as a perverse choice, not a choice between equally valid interpretations. The orthodox thereby implied that their interpretation preceded heresy, either chronologically or logically, as the "obvious" choice. In fact, the two theologies emerged and developed at roughly the same times and rates, as rival interpretations of Jesus's message and mission.

The most "perverse" choice distinguishing gnosticism from orthodoxy was its interpretation of the creator-god of Genesis: evil, not good; ignorant, not omniscient; incompetent, not omnipotent. Regarding the Old Testament god, gnostics thought that theodicy, "justifying the ways of God to man," was impossible, while the highest god was essentially unreachable. By contrast, orthodoxy portrayed God the Father as relatively easy to understand, despite (or even because of) his vindictive irascibility. The "orthodox" had two other powerful advantages: first, they offered salvation (as forgiveness of sin) to everyone rather than only to the "happy few"; second, they insisted that Jesus's simple, exoteric message was his only message. Orthodoxy argued for a literal interpretation of his salvific promises and a simple, obvious doctrinal explanation of his parables. His refrain that "it is written" expressed a fundamental respect for the Old Testament. Gnostic interpretations, by contrast, contradicted the literal sense of the Old Testament on important points, sometimes quite explictly. One treatise, *The Secret Book according to John*, assures its readers, "It is not as you have heard that Moses wrote; for in his First Book (i.e., Genesis) he said that [God] made [Adam] lie down; no, rather (it means) in his perceptions." In an even more blatant passage, "John" explains that Noah and his family "did not—as Moses said—hide in an ark; rather . . . not only Noah but many other people . . . hid within a luminous cloud."[15] Even the most complicated orthodox interpretations taught that, at some level, the text of the Bible was almost always literally true. But gnostic writings attacked the beliefs of the orthodox, from the most down-to-earth to the most marvelous, including doctrines later identified as quintessentially

Christian: a gnostic *Gospel of Philip* claimed that the eucharist, the virgin birth, and the resurrection were all metaphorical.[16]

The Jesus encountered in the gnostic gospels is an elitist even toward his disciples, and he tends to choose unexpected personages as his confidants. Moderns may not be overly surprised to encounter a *Gospel of Mary Magdalene*, given progressive gender sensibilities and her prominence in Dan Brown's novel *The Da Vinci Code* and its pseudo-historical sources. But the gnostic Magdalene "startles Peter by saying that she knows not only what Peter did not happen to hear, but what Jesus *chose* not to tell him: 'What is hidden from you I will tell you.'" Mary has learned this secret from a vision. For orthodox readers, visions could be a problematic source, but Jesus explained to Magdalene that visions are not seen by a person's soul or spirit (as Ezekiel and Revelation asserted) but rather "through the *mind*, or *consciousness*." Visions are thus fundamentally different from dreams or hallucinations. Moreover, Jesus teaches Mary a gnostic version of "what happens after death. He explains that the soul encounters 'seven powers of wrath,' which challenge it, saying, 'Whence do you come, killer of humans, and where are you going, conqueror of space?'" Best of all, "Jesus teaches the soul how to respond, so that it may overcome these hostile powers."[17] This "password protected" afterlife is far stranger than anything in the canonical gospels.

Scandalously, other gnostic gospels show Jesus choosing to confide salvific lore to the two disciples that orthodox tradition reviled: Doubting Thomas and Judas the Traitor. Rather than reflecting "what really happened" around 30 CE, canonical gospels' negative portrayal of these two disciples may express anti-gnostic propaganda. Elaine Pagels asserts that the message of John, the only canonical gospel to describe Jesus as divine, "was written in the heat of controversy, to defend certain views of Jesus and to oppose others"; she remarks that "research has helped to clarify not only what John's Gospel is *for*"—the traditional interpretation of it—"but what it is *against*." Pagels maintains that "what John opposed . . . includes what the Gospel of Thomas teaches—that God's light shines not only in Jesus, but, potentially at least, in everyone. Thomas's gospel encourages the hearer not so much to *believe in Jesus*, as John requires, as to *seek to know*

God through one's own, divinely given capacity, since all are created in the image of God."[18] John's insistence that Jesus is divine "far oversteps anything found in the gospels of Matthew, Mark, and Luke," all of whom depict him as merely "a man who receives special divine power, as God's 'anointed one.'"[19] Pagels observes that "John—and only John—presents a challenging and critical portrait" of Thomas. Indeed, John "invented the character we call *Doubting* Thomas," arguably to caricature writings like the *Gospel of Thomas* that displayed "a version of Jesus's teaching that [John] found faithless and false." To counter the idea that Jesus's own enlightenment was potentially available to everyone, because all were created in the image of God, Pagels thinks that John must have "decided to write his own gospel insisting that it is Jesus—and only Jesus—who embodies God's word and therefore speaks with divine authority."[20]

The smear campaign allegedly begun by John's gospel was so thorough that most Christians even misunderstand Thomas's name: "Thomas is not a proper name, but means 'twin' in Aramaic, the language that Jesus would have spoken." It was an "Aramaic nickname," and "the gospel itself explains that [Thomas's] given name was Judas (but, his admirers specify, 'not Iscariot')." Strangely, even the Gospel of John admits that "Doubting Thomas" was Jesus's "twin": both gospels, Thomas and John, "also translate Thomas into Greek, explaining to their Greek readers that this disciple is 'the one called "Didymus," the Greek term for 'twin.'"[21] In a second gnostic gospel attributed to Thomas, Jesus tells the disciple that "you are my twin and my true companion," and that "you will be called 'the one who knows himself.' For whoever has not known himself has known nothing, but whoever has known himself has simultaneously come to know the depth of all things."[22] "Know thyself," *γνῶθι σεαυτόν, nosce te ipsum*: this maxim of the Delphic oracle, attributed to everyone from Thales to Socrates and Plato, is another clue to the Greek philosophical tradition that informed gnostic Christianity.

The canonical New Testament portrays Judas Iscariot even more negatively than Thomas, yet Judas has long puzzled Christians. If the mission of Jesus Christ was to be sacrificed on the cross for the salvation of all humanity, doesn't Judas deserve some recognition—or even thanks—for making it possible? Why are the canonical gospels so merciless toward him? If the fall of Adam and Eve was a *felix culpa*, or "lucky" transgression, in that it made Jesus's atonement possible, shouldn't Judas's betrayal be seen the same way? The gnostic *Gospel of Judas* shows that such questions were asked in early Christianity, probably by 150 CE. Although the Judas gospel was not discovered until 1978, Jorge Luis Borges's 1944 story "Three Versions of Judas" fictionalizes such questioning, working from a logic that was always implicit in orthodox texts. Moreover, Borges must have understood Judas's connection to the riddles of early gnosticism, since he begins "Three Versions" by referring to Basilides, who "proclaimed that the cosmos was a reckless or maleficent improvisation by angels lacking in perfection."[23]

As the two eponymous gospels do for Thomas, the one attributed to Judas Iscariot proclaims that he was the favorite of Jesus, who initiated him alone into the secret of salvation. Judas does indeed betray Jesus, but only "at the sincere request of Jesus. Jesus says to Judas, 'You will exceed all of them [the other disciples]. For you will sacrifice the man that clothes me.'"[24] Bart D. Ehrman explains that "the deed that Judas performs for Jesus is a righteous act. . . . By handing Jesus over to the authorities, Judas allows Jesus to escape his own mortal flesh to return to his eternal home."[25] Jesus has devised a test for the disciples, which reveals that Judas alone "has correctly perceived Jesus' [divine] character," so "Jesus takes him aside, away from the ignorant others, to teach him 'the mysteries of the kingdom.' Judas alone will receive the secret knowledge necessary for salvation. And Jesus informs him that he will receive this salvation—even though he will grieve in the process," being rejected by the other disciples and replaced by Matthias. As disciple number thirteen, superior to the others, Judas "can attain salvation, while the other apostles continue to be concerned about 'their god,' that is, the creator god of the Old Testament."[26] Like the other Judas—Thomas or Didymus—Judas Iscariot will become a kind of "twin" of Jesus.

Among the complexities of gnostic thought is a sporadic softening of its negative attitude toward writing. Along with other gnostic texts, the *Gospel of Judas* exalts Seth, the third son of Adam, whom we remember as a primordial writer from the *Life of Adam and Eve*. In the gnostic cosmology that Jesus teaches Judas, we find a reference to "Seth, who is called Christ."[27] Although this identification may seem strange, we recall that Genesis glossed Seth's name as "another seed" to replace the murdered Abel. Later Christians would see this assertion as a kind of resurrection, thus as prefiguring that of Christ. In fact, there is an entire gnostic literature centered on Seth. *The Revelation of Adam* begins by claiming to be "the revelation [or apocalypse] that Adam taught to his son Seth in the seven hundredth year," and ends by repeating, "These are the revelations that Adam disclosed to his son Seth. And his son taught them to his seed."[28] Another pseudepigraphon emphasizes Seth's writings, claiming to be the "Report of Dositheus, (consisting) of the three tablets of Seth, father of the living and immovable race." The Report quotes the inscription on the three tablets (or stelae), and asserts that Dositheus "delivered it, just as it was written there, unto the elect."[29] A fourth-century orthodox writer on heresies says that sects of "Sethians" claimed there were seven books attributed to Seth.[30] Indeed, Layton writes that "in their mythic tales" the gnostics did not "refer to themselves, their ancestors or their spiritual prototypes as 'gnostics'" but rather as "offspring (seed, posterity, race) of Seth" or "offspring of the light."[31]

As with Gilgamesh's autobiographical tablet, the "revelations" written by gnostics are overshadowed for modern readers by their dramatic history of ancient disappearance and modern reemergence from burial in desert sands. Thirteen gnostic manuscripts, including both of the revelations attributed to Thomas, were found in 1945 in a cave near Nag Hammadi in Egypt. They had been buried in a large clay pot, almost certainly owing to an order issued by Athanasius, the bishop of Alexandria, in 367 CE, who provided a list of acceptable writings "that constitutes virtually all of our present 'New Testament'" and demanded that Egyptian monks destroy all "secret writings."[32] Such secret or gnostic revelations had been anathema for a long time; around 180 CE Irenaeus had denounced them in his *Against Heresies*, establishing Matthew, Mark, Luke, and John as the only valid gospels. Thus, the monks who decided against destroying the Nag Hammadi trove were connoisseurs of scripture—either sympathetic to aspects of gnostic mythology or opposed to biblioclasm on principle.

Other troves of biblical literature were discovered beneath desert sands in the twentieth century, including the more famous Dead Sea Scrolls, which provided some of the earliest surviving manuscripts of Hebrew Bible texts. Similar dramatic finds include the Cairo Geniza, or "graveyard" of worn-out manuscripts—a roomful of records, ranging from biblical texts to business records, that mentioned God, and thus could not be destroyed. At the opposite pole of attitudes toward writing, the massive garbage dump of discarded ancient papyruses at Oxyrhyncus in Egypt included everything from mundane letters to classical texts. Such discoveries are impressive as unexpected data banks, revelations about ideas and the material means that recorded them. But the gnostic rediscoveries are particularly relevant to this book, since they document a long-running polemical struggle to establish the truth of assertions about the ultimate questions. In one sense or another, biblical pseudepigrapha were always about the nature and value of writing because they attempted to define scripture. In this light, what makes gnostic scriptures especially compelling is their paradoxical combination of writtenness with a doctrinal hostility to writing.

NON-GNOSTIC CHRISTIAN PSEUDEPIGRAPHA

For first- and second-century Christians as well as Jews, Scripture meant the Pentateuch and the Jewish prophets. Given that orthodox Christian tradition revered the Hebrew Bible as Scripture foretelling the mission of Jesus, it is unsurprising that New Testament texts frequently show him citing or quoting it. He sometimes answers queries or challenges by referring textually to Scripture, declaring, "It is written." Indeed, the New Testament depicts Jesus as a scholar beginning at age twelve, astounding the rabbis by his exposition of the Scriptures, and arguing with them at other times. Yet despite this extreme demonstration of familiarity with biblical texts, he is only shown writing on one occasion—using his finger like the God of Moses, but on the ground rather than stone—and Scripture does not reveal *what* he wrote (figure 6.1).

FIGURE 6.1. Jesus writes on the ground as Pharisees ask about stoning the woman taken in adultery (John 8:6–8). Print. Anonymous Netherlandish, ca. 1530–1550. Aankoop uit het F. G. Waller-Fonds. Rijksmuseum.

Identifying the words Jesus wrote is no more possible than answering the famous question about the number of Lady Macbeth's children, yet the riddle has remained intriguing, including to modern amateurs on the internet. Clearly, the only possible answer is that there is something metascriptural about Jesus's writing. On this occasion, he had been teaching in the Temple when the Pharisees interrupted him to test whether he would approve their citing Scripture to justify stoning a woman taken in adultery, so part of the mystery about what Jesus wrote presumably concerns its relation to Scripture. Was it a reference to preexisting Scripture or an entirely new writing? Fittingly, the Gospel of John (8:1–9), the only one which refers to Jesus as "the Word" (*Logos*), is also the only one to record this incident. Yet Jesus's entire interaction with the scholars and the accused woman took place orally. So was John implying that Jesus's action was also self-referential, aimed at inscribing his answer as Scripture? In any event, the entire incident *became* Scripture.

Moreover, New Testament pseudepigrapha, whether orthodox or gnostic, almost never depict Jesus as a writer. However, two quite ancient pseudepigrapha do claim to be written by Jesus around the time of his death, and they were widely publicized over the centuries. The more ambitious of the two, titled "A Copy of a Letter written by Our Blessed Lord and Saviour Jesus Christ," was promulgated in the sixth century (figure 6.2). A preface claiming to be "Signed by the Angel Gabriel" dated "seventy-four Years after Our Saviour's Birth" says Jesus's letter was found ten years earlier "about eighteen miles from Iconiam," covered by "a great Stone, round and large, at the Foot of the Cross," on which "was engraven 'Blessed is he that shall turn me over.'" As with King Arthur's sword-in-the-stone, grown men were stymied by the immovability of the stone, whereas "a little Child, about six or seven Years of Age," was able to turn it over "without Assistance, to the Admiration of every person that was standing by." The text of Jesus's "letter from heaven" urged Christians to begin celebrating the Sabbath on Sunday rather than Saturday and promised great blessings to anyone who publicized the command. What better authority for such a significant ritual change than Our Lord Himself? The letter belongs to the ancient genre of the *Himmelsbriefe*, or letters from heaven, texts supposedly delivered by meteors; their otherworldly origin was presumably guaranteed by the stony vehicles that brought them to Earth.[33] The "Letter by Christ" remained popular for centuries, as a 1795 London broadsheet printing attests.[34]

The same broadsheet contains another famous pseudepigraphon, a third- or fourth-century forgery mentioned by Eusebius of Caesarea, containing an exchange of letters between King Abgar of Edessa and Jesus.[35] The fictional Abgar had heard of Jesus's miracles and requested that he visit Edessa to cure Abgar's serious illness. In his brief answer, Jesus declines the invitation, blaming his busy schedule: "all the Things for which I am sent must be fulfilled, and then I shall be taken up, and return to him that sent me." However, he promises that "after my Ascension, I will send one of my disciples, who shall cure thy Distemper, and give Life to thee, and to all that are with thee." Abgar's response to this proposed substitution is not recorded.

Other early Christian texts that aspired to canonicity resembled the teachings of the New Testament far more closely, depicting Jesus and his disciples in more familiar settings. They tend to recount episodes from Jesus's life before his ministry began at age thirty, or after his resurrection, detailing his interactions with Mary, Joseph, and the disciples. Two books recount Mary's birth and her experiences before the birth of her son, episodes absent from the canonical gospels. In another text, Doubting Thomas testifies to his conversion after touching the wound of the resurrected Christ. There are sometimes bizarre incidents: in one of the Thomas gospels and one other text, the boy Jesus uncharacteristically kills a playfellow who had offended him, presumably just to demonstrate his divine power.[36]

These pseudepigrapha emphasize writing but, unlike Old Testament pseudepigrapha, not obsessively so. An exception is the *Gospel of Nicodemus*, also known as *Acts of Pontius Pilate*; it includes a narrative about the two sons of Simeon, the old man who had been promised that he would not die until he saw the Messiah (Luke 2:25–35); his sons died, went to hell, and were resurrected after Christ's crucifixion. At the behest of Jewish authorities, each separately wrote down everything he witnessed there. Miraculously, "their writings were found the same, neither more nor less by a letter."[37] Identical

A Copy of a Letter written by Our Bleſſed Lord and Saviour JESUS CHRIST.

AND found eighteen Miles from Iconiam, ſixty-three Years after our Bleſſed Saviour's Crucifixion.

Tranſmitted from the Holy City by a converted Jew.

Faithfully Tranſlated from the Original Hebrew Copy, now in the Poſſeſſion of the Lady Cuba's Family in Meſopotamia.

This Letter was written by JESUS CHRIST, and found under a great Stone, round and large, at the Foot of the Croſs Upon the Stone was engraven *Bleſſed is he that ſhall turn me over.* All People that ſaw it, pray to God earneſtly, and deſired, That he would make this Writing known unto them; and that they might not attempt in vain to turn it over. In the mean time there came out a little Child, about ſix or ſeven Years of Age, and turned it over without Aſſiſtance, to the Admiration of every Perſon that was ſtanding by. It was carried to the City of Iconium, and there publiſhed by a Perſon belonging to the Lady Cuba.

On the Letter was written.

The Commandments of Jeſus Chriſt.

Signed by the Angel *Gabriel*, ſeventy-four Years after Our Saviour's Birth.

To which is added,

King *Agbarus*'s Letter to Our Saviour, and our Saviour's Anſwer. Alſo, His Cures and Miracles. Likewiſe, *Lentulus*'s Epiſtle to the Senate of *Rome*, containing, *A Deſcription of the Perſon of Jeſus Chriſt.*

A LETTER of JESUS CHRIST.

WHoſoever worketh on the Sabbath-Day ſhall be curſed. I command you to go to Church, and keep the Lord's Day holy, without doing any Manner of Work. You ſhall not idly ſpend your Time in bedecking yourſelf with Superfluities of coſtly Apparel, and vain Dreſſes; for I have ordained a Day of Reſt: I will have that Day kept holy, that your Sins be forgiven you. You ſhall not break my Commandmets, but obſerve and keep them; write them in your Hearts, and ſteadfaſtly obſerve, That th was written with my own Hand, and ſpoken with my own Mouth. You ſhall not only go unto the Church yourſelf, but alſo ſend your Men-Seervants and your Maid-Servants; and obſerve my Words, and learn my Commandments. You ſhall finiſh your Labour every Saturday in the Afternoon, by Six o'clock; at which Hour the Preparation for the Sabbath begins. I adviſe you to faſt five Fridays in every Year, beginning with Good-Friday, and continuing the four Fridays immediately following, in Remembrance of the five bloody Wounds which I received for all Mankind. You ſhall diligently and peaceably labour in your reſpective Callings, wherein it hath pleaſed God to call you. You ſhall love one another with brotherly Love: and cauſe them that are baptized to come to Church and receive the Sacraments, Baptiſm and the Lor d's Supper, and to be made Members of the Church in ſo doing. I will give you a long Life, and many Bleſſings; your Land ſhall flouriſh, and your Cattle bring forth in Abundance and I will give unto you many Bleſſings and Comforts in the greateſt Temptations; and he that doth to the contrary ſhall be unprofitable. I will alſo ſend a Hardneſs of Heart upon them, till I ſee them, but eſpecially upon the Impenient and Unbelievers. He that hath given to the Poor, ſhall not be unprofitable. Remember to keep holy the Sabbath-Day; for the Seventh Day I have taken to reſt myſelf. And he that hath a Copy of this my own Letter written with my own Hand, and ſpoken with my own Mouth, and keepeth it, without publiſhing it to others, ſhall not proſper; but he that publiſheth it to others ſhall be bleſſed of me; and though his Sins be in Number as the Stars of the Sky, and he believes in this, he ſhall be pardoned; and if he believes not in this Writing, and this Commandment, I will ſend my own Plagues upon him, and conſume both him and his Children, and his Cattle. And whoſoever ſhall have a Copy of this Letter, written with my Hand, and keep in their Houſes, nothing ſhall hurt him, neither Lightning, Peſtilence, nor Thunder, ſhall do them any Hurt. And if a Woman be with Child, and in Labour, and a Copy of this Letter be about her, and ſhe firmly puts her Truſt in me, ſhe ſhall ſafely be delivered of her Birth.

You ſhall not have any Tidings of me, but by the Holy Scripture, until the Day of Judgment ——— All Goodneſs, Happineſs, and Proſperity ſhall be in the Houſe where a Copy of this my Letter ſhall be found.

CHRIST's Cures and Miracles.

HE cleanſed a Leper, by touching him. He healed the Centurion's Servant afflicted with the Palſy, Peter's Mother-in-law of a Fever. Several *poſſeſſed of Devils. A moſt violent Tempeſt ſtilled by him. A Man ſick of the Palſy cured. Raiſed a Man from the Dead. Cured two blind Men. A dumb Man, who was poſſeſſed of a Devil. Fed above five thouſand with five Loaves and two Fiſhes. Walked on the Sea. All Diſeaſes in Genaſaret with five Loaves Touch of his Garment. Cured a Woman of a Devil, and Multitudes, both Lame, Blind, Dumb, Maimed, &c. He fed above four thouſand with ſeven Loaves, and a few little Fiſhes.*

King AGBARUS's Letter to CHRIST.

I Have heard of Thee, and of the Cures wrought by Thee, without Herbs or Medicines: For it is reported, Thou reſtoreſt unto Sight the Blind, makeſt the Lame to walk, cleanſeſt the Lepers, raiſeſt the Dead, and healeſt thoſe that are tormented with Diſeaſes of long Continuance.——Having heard all this of Thee, I was fully perſuaded to believe one of theſe two Things, either that Thou art very GOD, and cameſt down from Heaven to do ſuch Miracles; or elſe, that Thou art the SON of GOD, and performeſt them: Wherefore I have now ſent theſe Lines, intreating thee to come hither, and cure my Diſeaſes. Beſides, having heard that the Jews murmur againſt Thee, and contrive to do Thee Miſchief, I invite Thee to my City, which is but little indeed, but exceeding beautiful, and ſufficient to entertain us both.

Our SAVIOUR's Anſwer.

BLeſſed art Thou, Agbarus, for believing in me, whom thou haſt not ſeen; for it is written, that they who have ſeen me, ſhould not believe, and they who have not ſeen, ſhould believe and be ſaved: But as to the Matter thou haſt wrote about, theſe are to acquaint thee, that all the Things for which I am ſent muſt be fulfilled, and then I ſhall be taken up, and return to him that ſent me But after my Aſcenſion, I will ſend one of my Diſciples, who ſhall cure thy Diſtemper, and give Life to thee, and to all them that are with thee.

LENTULUS's Epiſtle to the Senate of ROME.

THERE appeared in theſe our Days a Man of great Virtue, called Jeſus Chriſt; who by the People is called a Prophet: but his own Diſciples call him the Son of God. He raiſeth the Dead, and cureth all manner of Diſeaſes, A Man of Stature ſomewhat tall, and comely, with a reverend Countenance ſuch as the Beholders may both fear and love. His Hair is the Colour of a cheſnut full ripe, and is plain almoſt down to his ears, but from thence downwards it is ſomething curled, but more of the oriental colour, waving about his ſhoulders; in the middle of his Head is a ſeam or parting like the Nazarites. His forehead very plain and ſmooth. His Face without either Wrinkle or Spot, beautified with a comely red; his noſe and mouth ſo formed that nothing can be reprehended; his beard thick, the Colour of the hair of his head; his eyes grey, clear, and quick. In reproving he is ſevere, in counſelling courteous; he is of a fair-ſpoken, pleaſant, and grave Speech; never ſeen by any to laugh, but often ſeen by many to weep. In proportion to his Body he is well ſhapen and ſtrait; and both his hands and Arms are very delectable. In ſpeaking he is very temperate, modeſt and wiſe. A Man, for his ſingular Beauty far exceeding all the Sons of Men.

FIGURE 6.2. "A Copy of a Letter Written by Our Blessed Lord and Saviour Jesus Christ," including "King Abgarus's Letter to Our Saviour, and Our Saviour's Answer," and "Lentulus's Epistle to the Senate of Rome, Containing a Description of the Person of Jesus Christ." [London: 1775?] Printed broadsheet. Johns Hopkins University Libraries, Bibliotheca Fictiva Collection, no. 69.

transcriptions demonstrate truthfulness. And among *Nicodemus*'s several references to books and writing is the declaration, reminiscent of Pseudo-Dictys, that "[Pontius] Pilate himself wrote all the things that were done and said . . . and placed all the words among the public records of his praetorium," presumably adding a supplementary Roman guarantee of their authenticity.[38]

There was even an attempt to insert an "eyewitness" testimony to the existence and mission of Jesus into the appropriate book of the Jewish historian Flavius Josephus's *Jewish Antiquities*. The forger declared that "He was the Messiah" but neglected to explain how and why Josephus, a convinced Pharisee, would have known or agreed that Jesus was the Messiah or have devoted so few words to this momentous news.[39] "Newly discovered" gospels continued to attract forgers and readers long after the fourth century. Recently, a scrap of parchment bearing only the words, "Jesus said 'my wife'" was the focus of a brief controversy. Although the parchment was genuinely ancient, forensic tests soon revealed the ink was modern and the text a forgery. Given the vogue of *The Da Vinci Code* and the pseudo-historical sources it drew upon, the forgery was predictable, since these recent narratives depict Jesus as married and fathering a line of illustrious royal descendants.[40] "Jesus said 'my wife'" is a lapidary critique of his and his male disciples' celibacy in the gospels and an implicit vindication of female disciples such as Mary Magdalene. Like much else in Scripture, the gospels' male exclusivity seems less appropriate to some recent readers than it did to traditional Christians.

CHAPTER SEVEN

Cultural Clashes and the Defense of Uniqueness

The majority of these authors have misrepresented the facts of our primitive history, because they have not read our sacred books; but all concur in testifying to our antiquity.

—FLAVIUS JOSEPHUS, *Against Apion the Grammarian*

What happens when a scripture is published? Publication is here understood in its original sense of *made public*. How does a scripture's exposure to cultures beyond its original ethnic or cultural audience change the kinds of commentary it excites? What happens when it is no longer sacred—set *aside*—for its initiates but set *beside* texts from other cultures, compared to them and, frequently, found ordinary and human rather than exceptional and divine, a mere text, not a scripture?

The Hebrew Bible's division between Jews and Gentiles had a literary and cultural dimension as well as a religious one. Like other texts that trace their own origins to the beginning of time, some Jewish pseudepigrapha expressed elements of cultural polemics. Some declared that pagan wise men—such as Pythagoras, Socrates, and Plato—cribbed elements of their philosophy from Moses. Homer, Hesiod, and the legendary poets Orpheus and Linus were said to have "taken information from our books."[1] Another text from the second or third century BCE claimed the mythical primeval Greek poet Musaeus was the teacher of Orpheus rather than his pupil and was in fact none other than Moses. Moses/Musaeus was supposedly a great counselor of the pharaoh Chenephres, and the inventor of boats, stone-cutting tools, philosophy, and the "sacred letters" (hieroglyphics). Thus, Moses/Musaeus was the true inventor of writing, and the Egyptians called him "Hermes" to celebrate his interpretation (*hermeneia*) of those sacred letters.[2] Such free associations still appealed to some Christian scholars in the seventeenth century.[3]

Polemical Jewish attitudes were initially provoked by Greek cultural hegemony. Koinê, or Hellenistic Greek, became the lingua franca of the Mediterranean world following the conquests of Alexander the Great (d. 323 BCE). Translation into Koinê spread the Hebrew Bible beyond its original Jewish audience during the third century BCE. Many Jews of this era, including writers of the New Testament, knew the "Old Testament" primarily or only in Greek translation rather than Hebrew.

A curious legend accounts for the principal Greek translation being titled the Septuagint, or "Version of the Seventy." The tale evidently purposed smoothing over the frequently antagonistic encounters between Greek and Hebrew cultures. Sometime between 150 and 100 BCE, a writer signing himself "Aristeas" wrote a long letter providing an uplifting account of the Septuagint translation, connecting it to the Library of Alexandria.[4] According to Pseudo-Aristeas, the Egyptian king Ptolemy II (d. 247 BCE) commissioned his librarian, Demetrius of Phalerum, to collect "if possible, all the books in the world." Pseudo-Aristeas claims he was present when the king asked Demetrius how many books were in the royal library. The librarian answered over two hundred thousand but undertook to increase the number quickly to half a million. As if citing a random example of desirable acquisitions, Demetrius added that "information has reached me that the law books of the Jews are worth translation and inclusion in your royal library."[5] He went on to propose, "If you approve, O King, a letter shall be written to the high priest at Jerusalem, asking him to dispatch men of the most exemplary lives and mature experience, skilled in matters pertaining to their Law, six in number from each tribe, in order that, after the examination of the text agreed by the majority, and the achievement of accuracy in the translation, we may produce an outstanding version in a manner worthy both of the contents and of your purpose."[6]

For an Egyptian pagan, Demetrius shows himself suspiciously well-informed about the structure of Jewish society; the high priest of Jerusalem naturally jumps at his proposition, sending the requested number of elders to Alexandria. Ptolemy questions the Jews to determine the extent of their wisdom, and they predictably pass with flying colors. Recounting the King's esteem for the Jews and their culture, Pseudo-Aristeas loses himself in wonderstruck descriptions of the erudite proceedings, the sumptuous gifts exchanged, and the sagacity of the Jewish delegates. He goes on so long (ten pages) that he apologizes for the digression, but swears he is merely quoting the royal written record. Finally, the Jewish elders are sent to a luxurious retreat, where they undertake the translation. "The outcome was such that in seventy-two days the business of translation was completed . . . and they commanded that a curse should be laid, as was their custom, on anyone who should alter the version by any addition or change to any part of the written text, or any deletion."[7] Seventy-two translators, seventy-two days: the six-by-twelve numerology evokes both God's creation of the cosmos and the number of Jewish tribes. Hence, the translation is traditionally called the "Septuagint," rounding the numerology to fit the Latin word for seventy.

As usual with such legends, later writers embellished Pseudo-Aristeas's account to make it even more wondrous. By the time it reached Saint Augustine (d. 430), the legend of the Septuagint specified that each of the seventy-two translators had done his work in isolation from the others, yet there was no variation in their wording. "There was an agreement in their words so wonderful, stupendous, and plainly divine, that when they had sat at this work, each one apart (for so it pleased Ptolemy to test their fidelity), they differed from each other in no word which had the same meaning and force, or in the order of the words; but, as if the translators had been one, so what all had translated was one, because in very deed the one Spirit had been in them all." There could be only one reason for such miraculous unanimity, of course: "they received so wonderful a gift of God, in order that the authority of these Scriptures might be commended not as human but divine . . . for the benefit of the nations who should at some time believe, as we now see them doing."[8] Scripturality has never been more rigorously defined, and as we saw in chapter six, at least one would-be Christian gospel also claimed such a "word-perfect" miracle, though on a modest scale.

Augustine conceded that Jews of his time criticized the Septuagint as unfaithful to the Hebrew "in many places," while praising the fidelity of Jerome's new Latin translation, the Vulgate. The Bishop of Hippo remained unimpressed: "the churches of Christ judge that no one [translator] should be preferred to the authority of so many men, chosen for this very great work by Eleazar, who was then high priest." Just as Moses and the prophets had been inspired by God, so "the same Spirit . . . was also in the seventy men when they translated."[9] The question of translators' fidelity to the biblical text was destined to remain an issue, but Jerome's Vulgate eclipsed other Latin versions for a millennium, until its authority was contested in the fifteenth and sixteenth centuries. Translations into the modern languages raise many of the same problems and further complicate the question of scriptural authority.

FROM TRANSLATION TO ADAPTATION: FLAVIUS JOSEPHUS'S DEFENSE OF THE BIBLE

One writer who spread the legend of the Septuagint was Flavius Josephus (d. ca. 100 CE), a Greek-speaking Jew from Jerusalem. Moreover, in his *Jewish Antiquities*, Josephus took the Septuagint as his inspiration for paraphrasing, rather than translating, the Hebrew Bible in Greek. Josephus defended Jewish culture and religion far more aggressively than "Aristaeus." In its early books, his *Antiquities* paraphrased the Pentateuch's account of ancient history, following the tradition of midrash, a Rabbinic method of commenting on the Bible by retelling it. More specifically, Josephus's book can be called aggadah, which explains biblical narrative by "filling in the blanks," adding or extrapolating incidents and details. In his *Legends of the Jews*, Louis Ginzburg defined aggadah as "characterized first by being derived from Holy Scriptures and then as being of the nature of a story," but according to a modern scholar, "not all Aggadah is intended to clarify Scripture. . . . [I]t may include anecdotes and folktales completely unrelated to the Bible."[10]

This is the case with Josephus; like apocrypha and pseudepigrapha, his *Jewish Antiquities* included lore not presented in the Bible. But Greek-speaking Christians treasured its anecdotes as a supplement to Scripture, and it was translated into Latin around 600 CE. For the rest of the Middle Ages and long after, Josephus's book enjoyed "a reputation second only to that of the Bible."[11] But Josephus did not draw his history only from Jewish culture; the *Jewish Antiquities* had the added attraction of preserving passages from the lost books of pagan writers, affording precious glimpses into ancient Phoenician, Egyptian, and Mesopotamian lore about the history of writing. The most conspicuous example was Berossos's *Babyloniaka,* which until about 1600 remained known almost exclusively through Josephus's few quotations and paraphrases. Both Jews and Christians esteemed the *Jewish Antiquities* for emphasizing the extreme antiquity of Chaldean records and their seeming corroboration of events in Genesis, especially the story of Noah. But scholars have observed that, despite respecting Berossos as an astrologer, "classical antiquity did not value highly" his historiography, not thinking it "worth the effort to preserve." Eusebius preserved longer passages of Berossos's *Babyloniaka* than Josephus, but he derided its differences from Genesis, calling it ridiculous and mendacious. Jerome did not simply agree: he excised the Berossos material altogether from his translation of Eusebius. As a result, for centuries, Berossos's reputation as historian rested on the shoulders of Josephus.[12]

Josephus used pagan lore to correct a glaring absence from the Hebrew Bible—its failure to mention the origin or history of writing before the time of Moses—although it pinpoints the development of other essentials of civilization. Genesis attributes three supremely beneficial inventions—tents, music, and metalworking—to descendants of the arch-sinner Cain. Reading between the biblical lines, Josephus spelled out the negative implications of metalworking for early human society. "Jobêl [the biblical Jabal] . . . erected tents and devoted himself to a pastoral life; Jubal . . . studied music and invented harps and lutes; Jubêl [Tubal-cain], . . . surpassing all men in strength, distinguished himself in the art of war, procuring also thereby the means for satisfying the pleasures of the body, and [he] first invented the forging of metals."[13] This was not an ideal way to begin the history of civilization: Tubal-cain's bellicosity and sensuality spoil the positive-sounding achievements of his brothers. Indeed, Josephus declares that something went quite wrong in this generation. "Within Adam's lifetime, the descendants of Cain went to depths of depravity, and, inheriting and imitating one another's vices, each ended worse than the last. . . . [I]f anyone was too timid for slaughter, he would display other forms of mad recklessness by [his] insolence and greed" (*Jewish Antiquities* 1.66; hereafter JA). The Bible eventually offsets this negative episode by expounding worthy accomplishments of later individuals, but Josephus apparently felt he needed to show some positive developments much sooner than the Bible itself did.

So to offset the inventions of Cain's descendants, Josephus inserted a tale about Adam's third son, Seth, which has no biblical precedent. Josephus relates that Seth "cultivated virtue" and "left descendants who imitated his ways." Seth's progeny made major contributions to civilization by inventing astronomy and, apparently, writing. "They also discovered the science of the heavenly bodies and their orderly array. Moreover, to prevent their discoveries from being lost to mankind and perishing before they became known . . . they erected two pillars, one of brick and the other of stone, and inscribed these discoveries on both; so that, if the pillar of brick disappeared in the deluge, that of stone would remain to teach men what was graven thereon, and to inform them that they had also erected one of brick" (JA 1.69–70). The Sethians erected their pillars because Adam (rather than Michael or Eve, as in the *Life of Adam and Eve*) foretold "a destruction of the universe, at one time by a violent fire and at another by a mighty deluge of water." Perhaps out of respect for the biblical account, Josephus declines to specify whether Adam based his prediction on his sheer human intelligence or a specific revelation from God.

As we shall see in later chapters, Josephus's tale of the Sethians was to have a long afterlife among Jewish and, particularly, Christian accounts of world history. More radically than Berossos, Josephus's imitators would intimate that writing is *monumental*, fixed in place. By contrast, even if baked into terra cotta, Berossos's clay tablets might have been scattered by the turbulence of the Flood due to their small size. And if, as often happened in Babylonian times, the tablets were not baked, they would have dissolved and become indistinguishable from the surrounding earth, whereas the heavy, durable materials imagined by Jews and Christians would not.

Logically, the *Jewish Antiquities* needed a story such as this. Given the Bible's emphasis on genealogy, the individuals it names as inventors of tents, music, and metalworking are tainted by their descent from Cain. Cain himself invented cities (Genesis 4:17), after killing his brother Abel, the archetypal shepherd, so Adam's entire remaining progeny gave civilization a thoroughly bad name. Seth and his posterity had to redress the balance in favor of a providential guidance of history. But why did Josephus choose—and possibly invent—this particular story? The *Book of Enoch* and *Jubilees* had both already attributed the discovery of astronomy to the patriarch Enoch (though neither work had a sterling reputation among the orthodox).[14]

Conjectures vary about the model for Josephus's inventive Sethians, but I suspect it was mainly stimulated by his favorite pagan source for primeval history. Without openly referring to Berossos, he seems to be concocting a concise Jewish riposte to Berossos's claims about the remote origins of human civilization.[15] As we have seen, Berossos's chronology dwarfed the timescale of Genesis, and in his encyclopedia, Josephus's contemporary Pliny the Elder quoted Berossos's claim that Babylonian written records were hundreds of thousands of years old. As much or more than writing, astronomy was considered the Babylonian discovery par excellence; both biblical and classical authors used the term *Chaldaeus* as a synonym for astronomer or astrologer. Berossos himself was a renowned astrologer, according to Pliny, Vitruvius, and other Roman writers. So Josephus implied that his own ancestors' accomplishments antedated Babylonian science, even while he denounced the Babylonian chronology of civilization as grossly inflated.

Like the *Life of Adam and Eve*, Josephus emphasized the importance of writing for preserving human culture rather than exclusively as a means for transmitting God's commands. He no more named an inventor of writing than Genesis and Exodus had done, but writings antedating the Sethian stelae are conspicuously absent from his account. The columns could even be interpreted as the origin of historiography: a "colophon" on each pillar recounted its own creation and its relation to the other copy. As a final touch, Josephus defended his claim by invoking archaeology, assuring his readers that the stone stele was undamaged by Noah's Flood and still standing in the "land of Seiris" (JA 1.68–71).

Far from telling an impartial history, Josephus's *Antiquities* stress the history of writing so as to emphasize the superiority of Jewish culture, as the legend of Enmerkar had done for Sumeria. Josephus expected his readers to be familiar with Plato's *Timaeus*, and quoted the Egyptian priest's rebuke to Solon that "in the Greek world, everything will be found to be modern, and dating, so to speak from yesterday or the day before."[16] Writing is the true foundation of culture, but "the land of Greece . . . has experienced countless catastrophes, which have obliterated the memory of the past." Recalling the old priest's contrast between unreliable Greek myths and meticulous Egyptian archives, Josephus declared that "the tradition of keeping chronicles of antiquity is found rather among the non-Hellenic races than with the Greeks."[17] Thanks to the Sethians, and not to Plato's Egyptians or Berossos's Chaldeans, non-Greek historiography was based on an unbroken chain of written documents that stretched to the beginning of the world. Jewish culture was anterior and therefore superior to all others.

No one knew better than Josephus that superiority and hegemony were not the same, since he witnessed the Romans' devastating conquest of Judea in 66–73 CE. A leader of the Jewish rebellion, he was captured by the Romans under circumstances that impugned his bravery, and he then became a Roman citizen. This fact tarnished his reputation among Jews until very recently, but he was no mere traitor. Like Berossos, he wrote in Greek to defend his native culture against the scorn of Greeks and Greek-speaking Romans. These Gentiles derided the Jews as a people without history, Jewish culture as barbaric, and the Hebrew Bible as a collection of lies and fairy-tales. Like Berossos's, Josephus's cultural patriotism was serious and urgent. So he integrated Plato's tongue-in-cheek anecdote into a deadly serious denial of Greek sophistication.

Josephus's claims about the Jewish origin of civilization caused scandals among his contemporaries. Soon after publishing the *Antiquities*, he had to defend them against Greek-educated critics. His rebuttal, *Against Apion the Grammarian*, reveals that the *Antiquities* outraged Gentile readers by claiming the Jewish race was "of extreme antiquity," having a history that "embraces a period of five thousand years," and that Jews were characterized by "the purity of [their] original stock." To these claims Apion and his fellow detractors apparently retorted that since the Jews "had not been thought worthy of mention by the best Greek writers," their civilization must be "comparatively modern."

The task of *Against Apion* was to explain the historical causes of Greek ignorance about Jewish history, and demonstrate that, contrary to Greeks' arrogant claims, their civilization was the "comparatively modern" one. As proof, Josephus invoked "the writers who, in the estimation of the Greeks [themselves], are the most trustworthy authorities on antiquity as a whole." By contrast, the Greeks were rank beginners as historians: "the most recent . . . of their attainments is care in historical composition." Even worse, Greek historians never agreed about their own culture; Josephus sarcastically declined to discuss "what discrepancies there are between Hellanicus and Acusilaus . . . how often Acusilaus corrects Hesiod, how the mendacity of Hellanicus . . . is exposed by Ephorus, that of Ephorus by Timaeus, that of Timaeus by later writers, and that of Herodotus by everybody."[18] This remark was ironic and unfair in the extreme: we have seen that Herodotus stressed the antiquity and sophistication of barbarian cultures half a millennium before Josephus.

Yet later Greek writers deprecated barbarians as "babblers" whose grotesque languages betrayed their lack of proper civilization. As Herodotus had done, Josephus turned Greek prejudice on its head by shifting the cultural emphasis from speech to writing: he declared that "[Greeks] were late in learning the alphabet and found the lesson difficult. . . . [T]hose [Greeks] who would assign the earliest date to its use pride themselves on having learnt it from the Phoenicians and Cadmus. Even of that date no record, preserved either in temples or on public monuments, could now be produced."[19] Josephus ignored Berossos's outlandish chronology and interpreted his assertions about the primeval history of writing as if they corroborated the antiquity and historical accuracy of the Hebrew Bible. For Josephus, these and other "barbarian" cultures demonstrated that Greeks were not culturally superior but boastful, undocumented, upstart latecomers. Josephus's distortion of Berossos guaranteed that, for Jews and Christians, proof that "it is written," not current political hegemony, would remain the measure of absolute historical and cultural importance.

Berossos's own cultural patriotism was so pronounced that Josephus cannot have failed to notice it. Judging from his claims about ethnic antiquity and historiographic accuracy, it looks as if Josephus's tale of the antediluvian stelae was not merely a weapon against the Greeks but also an implicit riposte to Berossos's elaborate stories of the demigod Oannes and the ancient king Xisouthros. Both authors trace the same cultural trajectory: the presence of writing and the other arts at the beginning of time, measures taken to preserve them from the flood, and afterwards a kind of renaissance when the antediluvian writings were recovered. One might therefore imagine that Josephus would recount a story in which God explicitly warned Noah to rescue knowledge as well as life from the Deluge. But he one-upped Berossos by attributing foreknowledge of the Flood to Adam—the first human—rather than to humans in the tenth generation (figure 7.1).

That Josephus was tacitly contesting Babylonian claims about writing and civilization is most evident in his account of exactly how Seth preserved written records from the flood. He streamlined the history of inscription and erasure that Plato and Berossos had consigned to long, dilatory narratives; by reducing every element of the myth to essentials, he gave compressed and elegant expression to the fundamental realities of culture—memory and forgetting, writing and obliteration. The survival of the Sethians' stone stele was a compact analogue of Berossos's claim that Xisouthros had rescued "all the tablets" from water by burying them in the citadel of the sun. Josephus's two stelae, differentiated by their materials, isolate and resist both of the archetypal enemies of writing, fire as well as water. Both Berossos and Plato had referred to bulky archives on clay or papyrus, which were inherently portable, thus leaving records susceptible to dispersion and erasure. By contrast, Josephus's two media are compact and immovable; together they ensure the text is indelible. Rather than Plato's countless cycles of fires, floods, and recoveries, Josephus exploited Genesis's limited chronology to imply that a single great flood or fire could suffice to destroy all civilization. By the same token, a single stele could renew it.

To fortify the Jewish claim, Josephus invoked authorities who were neither Jewish nor Greek, though they wrote in Greek. Even ethnic Greeks, said Josephus, admit that "the Egyptians, the Chaldaeans, and the Phoenicians . . . possess a very ancient and permanent record of the past." As if quoting Plato, he claimed that "all these nations inhabit countries which are least exposed to the ravages of the atmosphere, and they have been very careful to let none of the events in their history be forgotten, but always to have them enshrined in official records written by their greatest sages." *Against Apion* claims inaccurately—and possibly in bad faith—that these "neutral" authors corroborated the historical accuracy of the Pentateuch. In particular, said Josephus, Berossos's account of the Flood was practically identical to the one in Genesis. Unlike modern scholars, Josephus simply assumed that Xisouthros was another name for Noah.

> I will now proceed to the allusions made to us in the records and literature of the Chaldaeans: on various points these are in close agreement with our own scriptures. My witness here is Berossos, a Chaldaean by birth, but familiar in learned circles through his publication for Greek readers of works on Chaldaean astronomy and philosophy. This author, following the most ancient records, has, like Moses, described the flood and the destruction of mankind thereby, and told of the ark in which Noah, the founder of our race, was saved when it landed on the heights of the mountains of Armenia. Then he enumerates Noah's descendants, appending dates, and so comes down to Nabopolassar, king of Babylon and Chaldaea.[20]

Genesis 11:31 said Abraham came from "Ur of the Chaldees," so Josephus claimed that the Chaldeans "are the original ancestors of our race, and this blood-relationship accounts for the mention which is made of the Jews in their annals."[21]

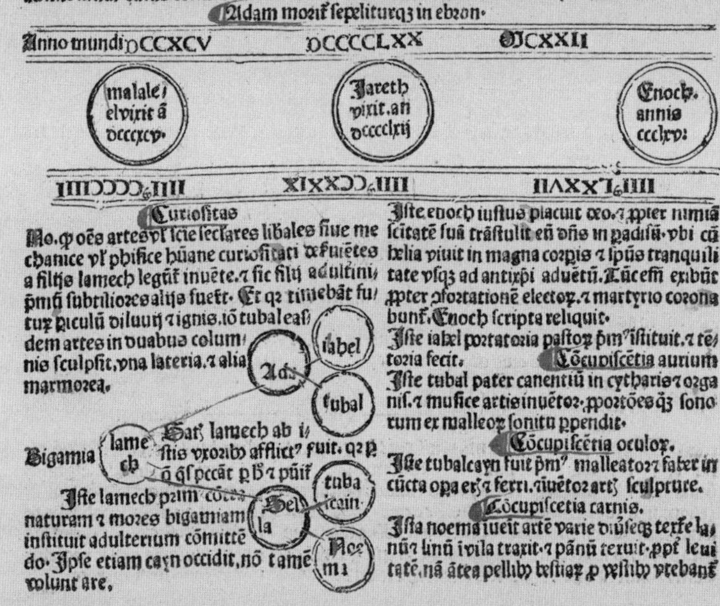

Prima etas. Adam

Mundi etates sin̄ ordinarie accipiūt'. s'm etates hoīs. ⁊ a diuersis diuersimode assignan t' cō nib3 mlt3 astruūt. q̄re sic vel sic puenienti⁹ inchoent vel terminēt. de q̄b3 nihil ad p̄ns: s3 q̄ntum nūc sufficit ē aduūtēdū q̄r. vi. sūt etates mūdi. q̄r p̄ma incipit a mūdi creatōe durat vsq3 ad dilu uiū. ⁊ h3 s'm hebraicā vitatē annos. Mdclvi. s'm. lxx. interp̄tes sicut Isidorus ponit ⁊ alij q̄3 plures q̄s hic sequimi h3 annos. ij⁹ cclij. ⁊ sic vrn̄t in annis. dlxxvi. q̄s hebrei min⁹ h̄nt in hac etate. ⁊ iuxta hūc cōputū matusale mortu⁹ ē ante diluuiū. eodē tn̄ anno q̄ diluuiū fuit. Se cūda etas incipit a diluuio ⁊ durat vsq3 ad natiuitatē abrahe. ⁊ h̄nit s'm hebreos ānos .ccxcij s'm. lxx. dccccxlij. ⁊ sic vrn̄t in annis. dcl. quos iter hebrei min⁹ h̄nt. Que aūt sit ratio tante diū sitatis inuenire nō potui. Tertia etas incipit a natiuitate abrahe ⁊ durat vsq3 ad initiū re gni dauid. h̄ns s'm hebreos annos. dcccc. xli. s'm. lxx. dcccc. xl. Quarta incipit a principio rgni dauid. ⁊ durat vsq3 ad transmigrationem babylonis. ⁊ habet annos. cccclxxiij. s'm hebreos. s3 s'm septuagita. cccclxxv. Quita etas incipit a trāsmigratōe babylonis. sc3 q̄n irl'm destru cta fuit. ⁊ tēplū in ea incēsū. ⁊ durauit vsq3 ad bn̄dictā natiuitatē xp̄i. h̄ns ānos s'm hūc mo dū p̄etactū quē hic sequim̄. dxc. ⁊ sicut p̄dictū ē grādis est altricatio de supputatōe annoꝝ hu ius etatis. diuersi diuersimode cōputāt. s3 nō mun⁹ mltū differūt inter se. Eligat quiuis q̄d sibi placuerit. q̄ nobis visum fuit posuim⁹. Sexta etas incipit a xp̄i natiuitate. ⁊ durat vsq3 ad finē mūdi. cuius terminū solus deus nouit. ⁊ hec dicit' senectus siue hora vltima.

Adam morit' sepeliturq3 in ebron.

Anno mundi DCCXCV — DCCCCLXX — MCXXII

malale / el vixit ā / dccc xcv.

Iareth / vixit. an / dcccclxij

Enoch. / annis / ccclxv.

IIIICCCC. IIII — XIXXCC. IIII — IIVXXL. IIII

Curiositas

Ho. q̄ oēs artes vl' scie seclares libales siue me chanice vl' p̄hisice hūane curiositati deseruiētes a filijs lamech legūt' inuēte. ⁊ sic filij adultini, p̄miū subtiliores alijs fuerit. Et q̄r timebāt fu tur̄ p̄iculū diluuij ⁊ ignis. iō tubal eas dem artes in duabus colum nis sculpsit. vna lateria. ⁊ alia marmorea.

Iste enoch iustus placuit deo. ⁊ p̄pter nimiā scitatē suā trāstulit eū dn̄s in p̄adisū. vbi cū helia viuit in magna corp̄is ⁊ ip̄us tranquili tate vsq3 ad antixp̄i aduētū. Tūc esti exibūt p̄pter p̄fortationē electoꝝ. ⁊ martyrio corona bunt. Enoch scripta reliquit.

Bigamia

Sat3 lamech ab i stis vxorib3 afflict⁹ fuit. q̄r p̄ ō q̄ speccāt p̄ b⁹ ⁊ p̄niī

Iste lamech p̄rim contra naturam ⁊ mores bigamiam instituit adulterium cōmittē do. Ipse etiam cayn occidit. nō tamē volunt are.

Iste iabel portatoria pastoꝝ p̄m⁹ istituit. ⁊ te toria fecit.

Cōcupiscētia aurium

Iste tubal pater canentiū in cytharis ⁊ orga nis. ⁊ musice artis inuētor. p̄portōes q̄3 sono rum ex malleoꝝ sonitu p̄pendit.

Cōcupiscētia oculoꝝ.

Iste tubalcayn fuit p̄m⁹ malleator ⁊ faber in cūcta opa eris ⁊ ferri. inuētor art3 sculpture.

Cōcupiscetia carnis.

Ista noema inueit artē varie diuiseq3 texte la nū ⁊ linū i vila traxit. ⁊ pānū texuit. p̄p̄t leui tatē. nā ātea pellib3 bestiaꝝ p̄ vestib3 vtebant'

FIGURE 7.1. "The First Age [of the World]." The Patriarchs from Adam to Noah, and a diagram of the Ark, showing "the habitation of the wild animals," "the habitation of men and birds," "the habitation of tame animals," the "dispensaries" of fruits and herbs, and the "stercoraria" (sewage-hold or bilge). At lower left, "Tubal [*sic*] sculpted those arts on two columns, one of brick and the other of marble." Werner Rolewinck, *Fasciculus temporum*. Venice: Erhardt Ratdolt, 1484. Sheridan Libraries Special Collections, Johns Hopkins University, Incun. 1484 .R6 c. 1.

Nota s'm doctores debitã penã h' tpe mũdo inflictã. Quia eñi luxuria abũdauit que corpa polluit. iõ p aquã terra lota 7 mũdata fuit. In fine aũt mũdi q2 cupiditas abũdabit. p ignẽ exuret. Aur ei 7 argẽtũ igne purgari solẽt

Archa noe hũit in lõgitudie. ccc. cubitos. in latitudine .l. cubitos. 7 ĩ altitudie. xxx. 7 ĩ sũmitate. i. cubitũ geñ. vi.

MCCLxxvij. Mccccliiij. ij° ccxlij.

mathusale vixit ã. dcc. cclxix.

lamech. vixit ã. dcclxxvij.

Tempus diluuij.

Hic matusalẽ senissim⁹ quo ad multitudinem annoꝝ oĩm q̃s scriptura ꝯmemorat. Cũ eñi. d. ãnos hret: dixit ei dñs. Edifica domũ si vis. qñ adhuc. d. ãnis viues. rñdit. Propter tantillũ tpis non edificabo domũ. 7 sub arboribus 7 circa vepres dormiuit vt prius consueuerat.

Iste Lamech cũ iã senuissꝫ 7 caligassẽt ocli ei⁹ a puero ducebat. qui putãs ꝙ ferã vidissꝫ idicauit ei vt sagittaret. 7 sic ipm caym transfixit. Vñ ob h' pueꝝ verberib⁹ adeo afflixit. vt etiã moreret

Hoc signũ federꝭ qđ do inter me 7 vos. 7 ad oẽs aĩam geñ. ix.

Arcus pluuialis siue iris duos colores principaliter hꝫ. qui duo iudicia representãt. Aque⁹ diluuiũ pterritũ figurat ne ãpli⁹ timeat. Igne⁹ futuꝝ iudiciũ significat p ignẽ. vt certitudinaliter expectetur.

Ionichus filius noe

Hic incipit Secunda etas mundi durans vsq ad abraham. Et habet annos. dcccclxij.

ãno vite noe. dc. facta est diluuium

Noe vixit. ãnis. dcccccl

Noe vir iustus grãm inuenit corã domino. Cũ eñi eẽt ãnoꝝ. d. genuit sem cham 7 iaphet. Arcã de mandato dñi edificare cepit eãq3 in. c. annis pfecit.

Centisimo ergo ãno iã arca ꝯpleta iteꝝ dñs apparuit ei. mãdãs vt cũ vxore sua 7 filijs eoꝝq3 vxorib⁹ archã intraret. 7 cũ aĩalib⁹. 7 statim diluuiũ inũdauit. Stetitq3 aq̃ sup altissimos mõtes cubitꝭ. xv. geñ. vij. Et nota ꝙ eodẽ die dñico in mayo in q̃ ingressus fuit. ãno reuoluto cũ vniũsis que ibi erãt egressus est. P⁹ diluuiũ accidit ipi noe illa famosa ebrietas. cui⁹ occasiõe ipe filijs. s. sem 7 iaphet p honore pẽno 7 honesta verecũdia bñdixit. filio vo chã p irrisiõe 7 irreuerẽtia maledixit. Et hic s'm augu⁹. fit pria mẽtio de suitute. 7 p oppositũ de nobilitate. Nec putãdũ ẽ ꝙ oẽs de chã descẽdẽtes fuerint ignobiles 7 ipotẽtes cũ ipi cepẽt prĩo esse potẽtes sup terrã. vt ptꝫ de nemroth 7 regib⁹ chanaã 7 affroꝝ 7c. Nec oẽs de sem 7 iaphet fuẽt vtuo si siue nobiles aut potẽtes. cũ pene oẽs in idolatrie crimẽ cecidert 7 ab alijs oppressi sepe fuẽt sꝫ hec maledictio 7 benedictio vicia 7 virtutes respicit. pter q̃ vel quas homo veraciter di- x

Berossos made it possible for Josephus to advance an even broader claim, based, again, on archaeology: "This flood and the ark are mentioned by all who have written histories of the barbarians. Among these is Berosus [*sic*] the Chaldaean, who in his description of the events of the flood writes somewhere as follows: 'It is said, moreover, that a portion of the vessel still survives in Armenia on the mountain of the Cordyaeans, and that persons carry off pieces of the bitumen, which they use as talismans.'"[22] Although Josephus intended to outdo his Chaldean model, he carefully avoided making Berossos an explicit target of his polemics. Instead, he included Berossos in a host of "neutral" authorities.

To strengthen the case for Noah's historicity, Josephus delightedly heaped up references to other "barbarian" historians: "These matters are also mentioned by Hieronymus the Egyptian, author of the ancient history of Phoenicia, by Mnaseas and by many others. Nicolas of Damascus in his ninety-sixth book relates the story as follows: 'There is above the country of Minyas in Armenia a great mountain called Baris, where, as the story goes, many refugees found safety at the time of the flood, and one man, transported upon an ark, grounded on the summit, and relics of the timber were for long preserved; this might well be the same man of whom Moses, the Jewish legislator, wrote.'"[23] Josephus used evidence selectively: here he ignores the fact that, besides this Noah-figure, Nicolas referred to other people who survived the Flood by simply climbing the mountain. What mattered was Nicolas's explicit mention of Moses as author of the Flood-story, and the putative agreement between Berossos, Nicholas, and other "neutral" authors.

Throughout the *Antiquities* and *Against Apion*, Josephus emphasizes the conclusiveness of written testimony and deprecates oral myth as unverifiable. But his use of written sources hardly met modern standards. Like other ancient historians, Josephus implied he was quoting directly from authors whom he actually knew only second- or thirdhand. Berossos and others of Josephus's sources were probably already inaccessible in their original form to the first readers of the *Antiquities*. So the *Antiquities* quoted Berossos and other writers by repeating excerpts from their works transmitted by authors in the intervening centuries.[24]

Josephus is an important witness to the history of lost libraries and erased literature. His exploitation of Berossos was not only indirect but a lucky accident for him and us, since the *Babyloniaka* survived exclusively through such quotations and paraphrases. Many other works Josephus quoted have been well and truly lost, as a tour through various scholarly anthologies titled "Fragments of the Greek Historians" confirms. To these authors, some of whose names are now known mainly because he mentioned them, Josephus added vaguer references to archives and libraries that his Greek and Roman readers probably could not hope to consult, blocked by barriers both geographic and linguistic. Nor is it even certain that all these collections existed as Josephus remembered or imagined them.

Even if he had read Berossos's history of civilization firsthand, he would likely have avoided quoting it too faithfully. Most obviously, it starred pagan Babylonia. Second, the Apkallu Oannes cut a very strange figure for a cultural hero. As we saw in chapter one, two later writers who excerpted Berossos extensively, Eusebius of Caesarea and Eusebius's own excerpter George Syncellus (fourth and ninth centuries, respectively), describe Oannes as a "frightening monster" or a "silly beast." Worse, both writers ridicule Berossos's lack of agreement with the Book of Genesis, whereas Josephus claimed the Chaldean historian corroborated the biblical account in detail.[25] In his Latin translation of Eusebius, Jerome gave up on the entire Berossian mess and left it out, thus validating Genesis's chronology by default and leaving Berossos's account unknown in the West for another millennium.

Although he prized evidence gleaned from non-Jews, Josephus was not content to base his claims on it or on destroyed libraries in places or times remote from his own. He proudly enlists Jewish tradition in the Bible's defense: "Our forefathers took no less, not to say even greater, care than the nations I have mentioned in the keeping of their records—a task which they assigned to their chief priests and prophets." He concludes that "down to our own times, these records have been, and if I may venture to say so, will continue to be, preserved with scrupulous accuracy."[26] Once again, Josephus compares the putative written sources of the Hebrew Bible to the kinds of priestly archives that Plato's *Timaeus* imagined for the Egyptian archives at Sais.

Throughout, Josephus exploits the concept of written sources to argue that the Book of Genesis recounts a history both factual and believable. Thus he blurs the distinction between the Bible as history, as culture, and as dogma. Or rather, these modern distinctions meant little or nothing to Josephus and his original audience. He presents Moses as an exemplary Greek or Roman historiographer, an "intellectual" intent on chronological accuracy and philosophical veracity, rather than a dogmatic self-styled confidant of the Almighty. By making the Bible seem more Greco-Roman, Josephus's touches occasionally disrupt his fidelity to the biblical account, as we saw in the antediluvian history of writing.

Both Berossos and Josephus differed from Greek historians by claiming that there was an ultimate book, the book from the beginning of time, and each averred that his own culture had composed it. Josephus's curious story makes a number of crucial assumptions, none of which have any importance to the Bible as religious dogma: like pagan historians, Josephus is overtly describing technological and political advances, not ostensible moral or theological developments. Moreover, in contrast to both Berossos and the story of the Ten Commandments, Josephus's primordial book of the Sethians is not a gift bestowed by the deity. Its composition and preservation from erasure were an essentially human affair.

Yet although human discovery and invention explain the origins of culture in a fashion more congenial to classical Greek and Roman culture, Josephus did not eliminate God's role in Jewish superiority. He could not drastically change "Moses's version" of history, yet he clearly believed it had potential credibility problems. He cautioned non-Jewish readers against scoffing at the enormous life spans the Bible attributes to antediluvian patriarchs and attempted to make them more credible to pagans by giving partially naturalistic explanations. He declared that God allowed the Sethians extreme longevity so as to give them adequate time to make their momentous discoveries. "In the first place, they were beloved of God and [were] the creatures of God Himself; their diet too was more conducive to longevity: *it was then natural that they should live so long*. Again, alike for their merits and to promote the utility of their discoveries in astronomy and geometry,

God would accord them a longer life; for they could have predicted nothing with certainty, had they not lived for six hundred years."[27]

To assure his readers that he was not recounting fables, Josephus enlisted his familiar shelf-load of barbarian historical sources, adding a few Greeks for good measure. "My words are attested by all historians of antiquity, whether Greeks or barbarians: Manetho the annalist of the Egyptians, Berosus the compiler of the Chaldaean traditions; Mochus, Hestiaeus, along with the Egyptian Hieronymus, authors of Phoenician histories, concur in my statements; while Hesiod, Hecataeus, Hellanicus, Acusilaus, as well as Ephorus and Nicolas [of Damascus], report that the ancients lived for a thousand years."[28] Why was Josephus so intent on piling up authorities for this statement? The obvious answer seems to be that the outlandish life spans the Bible attributed to Adam and his descendants were problematic. However, Josephus seems to have reasoned that if he could make them "scientifically" plausible, they would bolster his claim that Adam's family could have discovered and developed such complicated sciences over a mere couple of generations.

With Josephus, we seem to come full circle in the Jewish history of attitudes to writing. Exodus combines several accounts of how the Bible began when someone wrote down the Lord's commandments. The mere fact that "It is written" made writing wondrous and authoritative. But precisely because writing appeared so marvelous, it reproduced itself continuously: the apocryphal and pseudepigraphic supplements to the Pentateuch expanded in number and complexity to the point that they must be culled, so that limited, authoritative canons might be agreed on.

When two or more ethnic or national traditions of writing come into contact, matters become even more urgent. Since the authority of a body of writings is so bound up with its origins—whether chronological or metaphysical—theoretically only one tradition can prevail. This is the point at which Pseudo-Aristeas and Flavius Josephus enter the scene: writing no longer provokes wonder simply because "It is written." A culture's claims must be defended, its wondrousness explained; a criterion is needed. Skepticism and intercultural hostility are in the offing.

Flavius Josephus had defended the content of Scripture, its validity as history, against pagan critics who considered it merely a text and not a very impressive one at that. But for Christians as well as Jews, there was more at stake in cultural polemics than one-upmanship. For scriptures, historical precedence was a matter of absolute, ultimate religious truth. As we saw, some biblically derived texts claimed that the least objectionable pagan poets and philosophers were echoing revelations that God originally made to Moses; indeed, the legendary Musaeus might actually *be* Moses. Moving from mere similarity of authorial names to doctrinal compatibility, Plato's *Timaeus* was prized because, like Genesis, it described the creation of the cosmos by a benevolent deity. Jews and Christians could not reduce Plato's consonance with the most fundamental doctrine of their religion to simple coincidence: either Plato learned Moses's creation story from human traditions or the Holy Spirit inspired the philosopher to intuit the truth.

Pagans also saw such parallels but in reverse: Saint Augustine took particular offense at accusations of plagiarism, "levelled by readers and admirers of Plato, who had the nerve to say that our Lord Jesus Christ learnt all his ideas—which they cannot but marvel at and proclaim—from the works of Plato." However, said Augustine, his mentor Saint Ambrose (d. 397) "discovered that Plato went to Egypt . . . at the time of Jeremiah," making it more likely "that Plato had been introduced to our literature by Jeremiah" than that Christ cribbed his doctrines from Plato. "The usefulness of history" taught Augustine that "the literature of the Hebrew race . . . was not preceded even by Pythagoras, from whose followers [pagans] claim that Plato learnt his theology." From the Egyptians, Greeks had learned a culture that was originally Jewish and only derivatively Egyptian. A study of chronology was sufficient to refute the "quite crazy idea" that Christ was a Platonist; it was "much easier to believe that the pagans took everything that is good and true in their writings from our [Jewish and Christian] literature." Augustine denied harboring arbitrary sectarian allegiance because, he averred, "what has already gone into the past and cannot be undone must be considered part of the history of time, whose creator and controller is God."[29] If the One God created the world and time, and if Jews and Christians worshiped Him, then clearly their view of history was the correct one.

GOD'S WRITING

Christian apologists, both gnostic and not, fundamentally altered the concept of scripture and went beyond Josephus, reading the Hebrew Bible selectively according to a new definition of history. The utterances of Hebrew prophets allegedly foretold the coming of Jesus as the promised Messiah, indeed as the Son of God. Christians proclaimed that some events in the "Old Testament" were true in a new and radical sense: not only as historical fact but more importantly as prophecy about Christ and Christianity. Events such as Noah's Flood and the journey to the Promised Land were not only "literally" true as history but were also God's figurative way of foretelling the eternal salvation of humankind by His Son. More generally, the cosmos itself should be read as a scripture. Paul's Epistle to the Romans proclaimed that "what can be known about God is plain" to both Jews and Gentiles "because God has shown it to them. Ever since the creation of the world, his invisible nature, namely, his eternal power and deity, has been clearly perceived in the things that have been made" by Him. Thus, pagans' lack of access to divine revelation through the Bible was irrelevant, and "they are without excuse."[30] The cosmos around us is another Bible, a scripture written in the language of things rather than words, and is an open book. Later Christians would turn the Book of the World into a *topos*, or commonplace, citing both Paul's assertion in Romans and striking parallels to it in chapters 11 through 13 of the second-century BCE Book of Wisdom.

AUGUSTINE ON HUMAN WRITING

Augustine gave compelling form to the semiotic assumptions behind his guidelines for interpreting Scripture. His exhaustive presentation of human language made *On Christian Doctrine* a fundamental text throughout the Latin Middle Ages. Augustine remarked that "spoken words cease to exist as soon as they come into contact with the air, and their existence is no more lasting than that of their sound; hence the invention, in the form of letters, of signs of words. In this way, words are presented to the eyes, not in themselves, but by certain signs peculiar to them."[31] Later Christian writers would imagine letters as fetters, binding thoughts to prevent their escape into oblivion. But for Augustine, human signification extended beyond writing to wordless communications. Idolaters worshiped demons nonverbally, expressing veneration through objects—amulets and "characters," or marks resembling writing.[32] Conversely, they worshiped statues as gods, treating signs as the demonic deities they signified, just as the Hebrews had done with the golden calf.

Augustine distinguished human veneration between "use" and "enjoyment": only God merited being *enjoyed* on his own account; everything else was to be *used* as a sign of God, enjoyed exclusively in reference to God.[33] Enjoyment of any other object or person for its own sake was idolatry—worshiping a creature rather than its creator, revering a sign rather than its referent.

Since he was trained in Roman rhetoric and oratory, Augustine also confronted the question of literary style in Scripture, asking the loaded question whether it was simply "wise, or eloquent as well." He concluded that the Bible's style "transcends that of others, not in grandiloquence, but in substance."[34] This praise masked his concern over "those who despise our prophets as unlearned and unacquainted with eloquence,"[35] and was shared by his contemporary Jerome, who defended the Bible's "low" or "humble" style when translating it into Latin. Jerome dreamed that God rejected him for loving his Roman books excessively and rebuked him for following Cicero the stylist rather than Christ.[36] Yet scriptural style was not always simple or straightforward; when Scripture commended something reprehensible, Augustine argued, it should be interpreted allegorically—figuratively rather than literally.[37] If an assertion *should* not be true literally (because the Bible can only espouse the good), it *could* be true only as *alieniloquium*, or "alien speech," as allegory about something else entirely. In a less radical act of creative misreading, the Israelites' crossing of the Red Sea, their travails in the desert, and their entry into the Promised Land could be seen to prophesy the baptism, life struggles, and ultimate redemption of Christians.

After Augustine, the question of scriptural style was rarely unconnected either to its "salvific" contents or to a meditation on the history and glamor, or charisma, of writing. The question of style would return to haunt the humanist scholars of the Renaissance, who revived the stylistic excellence and philosophical profundity of pagan literature. An extreme case was Giovanni Pico della Mirandola (d. 1494), an enthusiast of classical and Hermetic literature who defended biblical simplicity in his *Heptaplus* even as he argued for its underlying "sevenfold" complexity. Whatever one's conception of writing, at any level from the alphabet to the utmost niceties of style, or to ultimate truth-value as doctrine, the Bible could only be the best form of writing. And since tradition declares that older is better, the Bible could only be the oldest writing. Subsequent chapters will show the extent to which monotheists pushed this idea: even if Scripture did not achieve its full form until Moses, or even later, as with the gospels, it should have—indeed, must have—existed throughout time.

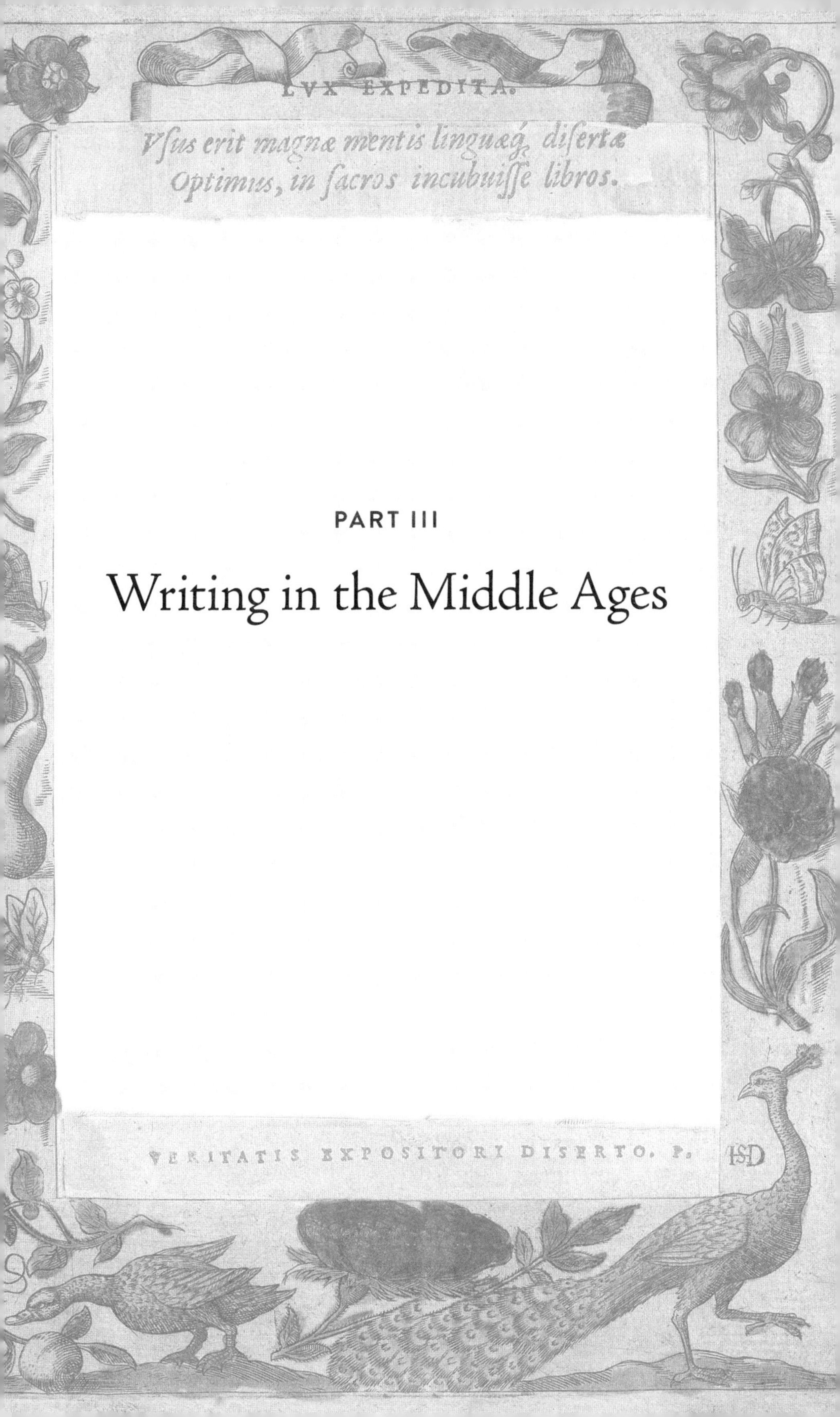

PART III

Writing in the Middle Ages

CHAPTER EIGHT

An Age of Paradoxical Optimism, 650–1350

> The letters of the Hebrews started with the Law transmitted by Moses. Those of the Syrians and Chaldeans began with Abraham, so that they agree in the number of characters and in their sounds with the Hebrew letters and differ only in their shapes. Queen Isis, daughter of Inachus, devised the Egyptian letters when she came from Greece into Egypt, and passed them on to the Egyptians. Among the Egyptians, however, the priests used some letters and the common people used others. The priestly letters are known as *hierós* (sacred), the common letters *pándemos* (common).
>
> —ISIDORE OF SEVILLE, *Etymologies*

Two eminent historians of scholarship write, "Although few ages are so dark that they are not penetrated by a few shafts of light, the period from roughly 550 to 750 CE was one of almost unrelieved gloom for the Latin classics on the continent [of Europe]; they virtually ceased being copied."[1] The eclipse of these foundational texts affected every area of intellectual endeavor; some have still not been recovered. As for vernaculars, or everyday languages, they were rarely recorded; to be literate was to read and write Latin. Only around 1000 CE were vernacular texts like *Beowulf* and *The Song of Roland* committed to writing, after being transmitted orally for centuries.

Much of the lore discussed in previous chapters failed to influence medieval Western Europe. Knowledge of Greek was reduced by the seventh century to a few peripheral scholarly holdouts. Herodotus, Diodorus, and even Homer had to await Latin translation in the fifteenth century to influence Western writers. Aristotle, virtually unknown in the West until the 1100s, was mostly translated from

Muslim translations not from Greek originals, while a fourth-century Latin translation of *Timaeus* was the only work of Plato known in detail during a thousand years. In addition to the Greek classics, most Byzantine authors remained closed books, owing not only to Catholic ignorance of Greek but also, after the mid-eleventh century, to theological and organizational hostility between the Orthodox and Roman churches. Disputes over the Roman pope's claims of universal authority, plus theological disagreements over the legitimacy of images in Christian worship (the iconoclasm controversies) and whether the Holy Spirit emanated only from the Father or from the Son as well (the *filioque* controversy), kept the two churches apart. The many crusades against Muslim "infidels" after 1095 brought little reconciliation between Eastern and Western Christians. The nadir was the disastrous Fourth Crusade (1202–1204), when Latin crusaders sacked Christian Constantinople itself.

Pessimism about the state of learning would therefore seem an appropriate attitude for medieval writers, and indeed it was not unknown. Rodulfus Glaber, an eleventh-century French monk, wrote that for two centuries, "since the time of the priest Bede in Britain and Paul in Italy, there has been no one anxious to leave any written record for posterity. . . . This was certainly true of the many events which occurred with unusual frequency about the millennium of the Incarnation of Christ Our Saviour."[2] Glaber, whose name means "Bald Ralph," convinced some nineteenth-century historians that medieval Christians expected the world to end around the year 1000. This sounds like something medieval people *ought* to have believed, but modern historians maintain that Ralph's pessimism reveals his unfamiliarity with works of history available in his time.[3] Ironically, his laments are also boasts, intended to show himself as the shining exception to his age's ignorance.

In fact, hopelessness as cosmic as Ralph's was rarely expressed before about 1350; more characteristic was optimism or complacency toward the benefits of writing, and a tranquil resignation about the limited lifetimes of books other than the Bible. From the mid-seventh to the mid-fourteenth century, writers projected confidence about the powers of the scribal art and the durability of *litteratura*, an attitude that now seems more unrealistic than Ralph's pessimism. The cultural losses were far more extensive than medieval authors imagined, and their histories of writing make one wonder whether ignorance caused their bliss.

Not that medieval optimists were unaware of the written word's vulnerability to warfare, earthquakes, floods, and catastrophic fires, or to the everyday ravages of mold, mice, bookworms, and neglect. But Latin Christians do not seem to have discovered biblioclasm as a civilizational problem until around 1350. Some medieval optimism was probably an effect of the mystique surrounding *scriptura* that we saw in previous chapters. Although confidence in the invariability of the Bible was not absolute, it probably encouraged belief in the durability of written culture as a whole. Once Saint Jerome's Latin translation of the Bible (the Vulgate, completed in 405 CE) came into general use, Western readers could imagine or assume that divine providence preserved the scriptural texts intact. Conversely, the devaluation of pagan religion by the Bible and by Christian polemicists encouraged indifference, and occasional enthusiasm, toward the erasure of ancient literature and philosophy.

But Christian writers' negative attitude toward pagan culture should not be overestimated. Christian scribes preserved much of "heathen" literature, mythology and religion, philosophy and science. Moreover, traditional stories about writing required no apologies. There was no a priori connection between writing and polytheism, thus no obvious conflict with monotheistic religion. King Cadmus might well have existed and introduced the Phoenician alphabet into Greece, and Augustine wrote that "we were not wrong to learn the alphabet just because they say that the god Mercury was its patron."[4] As we saw earlier, Christian scholars tended to domesticate such figures, considering them either alternate identities of biblical personages or their literary heirs.

TWO TRADITIONS

From the seventh to the fourteenth century, discussions of writing show two main lines of influence, one derived from Pliny the Elder and the other from Flavius Josephus. The Plinian strain is a more mundane history of writing; although it contains mythic elements, its predominant outlook is technological, its narratives factual-sounding, secular, concerned with human history. By contrast, the Josephan strain shows a more scriptural mentality, never straying very far from biblical and religious themes. But some writers show both Plinian and Josephan filiations. This dual heritage shows most clearly in the figure traditionally considered the last of the classical writers or the first of the medievals, the hinge between the two cultures.

ISIDORE OF SEVILLE

Ironically, the cultural devastations of the fifth and sixth centuries were followed by an optimistic vision of the history of writing, in the *Etymologies*, or *Origins*, of the Iberian bishop Isidore of Seville, who died in 636. Isidore's encyclopedia was the grandest since Pliny's *Natural History* six hundred years earlier. The *Etymologies* became a cornerstone of European thought and survives in over a thousand manuscript copies. Isidore quoted many earlier works; true to the pattern we have seen, his excerpts are sometimes all that remains of an ancient text. Isidore's book reflects the premodern belief that *nomina sunt consequentia rerum*, that every word's original form—its etymology—reveals the deepest nature of the thing it signifies.[5] But medieval etymologies are often based on phonetic association rather than history and can strike modern linguists as far-fetched or even ridiculous: "Letters (*littera*) are so called as if the term were *legitera*, because they provide a road (*iter*) for those who are reading (*legere*) or because they are repeated (*iterare*) in reading."[6]

Isidore bequeathed to medieval Europe the ancient, mundane assessment of letters, but his reverence for the wondrous power of writing already animates the optimistic third chapter of his *Etymologies*. "Letters are the tokens of things, the signs of words, and they *have so much force that the utterances of those who are absent speak to us without a voice*, for they *present words through the eyes, not through the ears*. The use of letters was invented for the sake of remembering things, *which are bound by letters lest they slip away into oblivion*."[7] For Isidore, as for Saint Augustine, letters are fetters, preventing memory's escape, whereas "spoken words cease to exist as soon as they come into contact with the air, and their existence is no more lasting than that of their sound."[8] Both men agreed that writing is synesthetic: language enters our mind almost magically, through the eyes, bypassing its natural receptors, the ears. Speaking without a voice, letters paradoxically enable communication with people whom death or distance have silenced.

Isidore's encomium contrasts starkly with Socrates's denunciation of writing as the destroyer of memory or its debilitating prosthesis. Isidore has far more in common with the Sumerian legend of Enmerkar's forgetful messenger, from twenty centuries earlier. But Isidore went further, declaring that letters are our only defense, not just against routine forgetting but against generalized information overload: "With so great a variety of information, not everything could be learned by hearing, nor retained in the memory"[9] (figure 8.1).

ISIDORE ON WRITING MATERIALS

Like Pliny's *Natural History*, the *Etymologies* discuss the history and geography of writing materials. Echoing Pliny, Isidore traces the Latin etymology of books to the tender inner bark of trees. "*Liber* is the inner membrane of bark, which clings to the wood. With regard to this, Vergil [says] thus: 'The bark (*liber*) clings to the high elm.' Whence what we write on is called a book (*liber*) because before the use of papyrus sheets or parchment, scrolls were made—that is joined together—from the inner bark of trees [*de libris arborum*]. Whence those who write are called copyists (*librarius*) after the bark of trees."[10] More fancifully, he says that *liber* is so called because it is freed (*liberatus*) from the cortex, or bark.[11]

Isidore mentioned palm-leaves as very ancient writing surfaces and gave a notable example: "Among the pagans, Dares the Phrygian was first to publish a history, on Greeks and Trojans, which they say he wrote on palm leaves."[12] Pseudo-Dares omitted this "fact" about his chosen stationery, but Pliny had contended that palm leaves were a primeval writing surface. Other etymologies, says Isidore, connect books to trees: *codex* derives from *caudex*, or the trunk of a tree or vine, "because it contains in itself a multitude of books, as it were of branches," while "a book (*liber*) is [comprised] of one scroll." Leaves, or

FIGURE 8.1. A saintly hermit in his study. "Evagrius was learned in the holy divine oracles and in the eternal dogmas of the law. From which he gave his holy precepts for a salubrious life, invincible arms against demons." Jean le Clerc, *Oraculum anachoreticum* (Paris: Jean le Clerc, ca. 1606). Sheridan Libraries Special Collections, Johns Hopkins University, Women of the Book Collection, Bib#9252715.

folia, "are so called from their likeness to the leaves of trees." He added that "before the use of papyrus sheets or parchment, the contents of letters were written on shingles hewn from wood."[13]

Although trees dominated the vocabulary of books and writing, reeds came a close second; Isidore mentions the ancient use of papyrus and the *calamus,* or reed-pen (though he says it comes "from a tree"). Quill-pens were rarely mentioned before Isidore, but he knew that the word *pinna* (or *penna*) still meant "feather."[14] His reverence for writing shines through his remarks on feather-pens, providing a brilliant example of medieval Christians' penchant for allegorical interpretation. *Scriptura* as both writing and Holy Writ inspires Isidore to claim that "the tip of a quill is split into two, while its unity is preserved in the integrity of its body, I believe for the sake of a mystery, in order that by the two tips may be signified the Old and New Testament, from which is pressed out the sacrament of the Word poured forth in the blood of the Passion." The words formed by the flowing ink "signify" Christ the Word (*Logos*) by analogy with the letters transcribing that other divine word, the Bible.[15] By extension, any book and any act of writing partakes of this scriptural mystery.

For Isidore the alphabet itself was soaked in allegories about the ultimate realities. The letter *Y* was Pythagoras's symbol of the necessary choice between the paths of vice and virtue, while Christ claimed to be Alpha and Omega; Isidore declares that *T* "shows the figure of the cross of the Lord," based on Ezekiel 9:4. *Theta* (θ) "has a spear through the middle, that is, a sign of death"; it is the first letter of *thánatos* (death), and "judges used to put this . . . letter down against the names of those whom they were sentencing to execution."[16] Such determined allegorization, foreign to both antiquity and the Renaissance, looks far-fetched to moderns but demonstrates Isidore's and his era's profound emotional connection to the art of writing.

Medieval Christians usually presumed that Moses was the oldest author, if not the actual inventor of writing. Yet this notion was more flexible in Isidore's time than among some later writers. He asserted that Greek and Latin owed their alphabets to the Hebrew, which began when Moses wrote down the Law; though he also claimed that Abraham had invented the "Syrian and Chaldean" letters much earlier, an idea that gained further influence after 1500.[17] But his view of cultural history was largely unpolemical, and he showed great respect for pagan writers. Although Moses was the first author of history *apud nos*, "among us Christians," Dares the Phrygian and Herodotus were the first Gentile historians, with Dares, as supposed eyewitness of the Trojan War, being older.[18]

Isidore might have transmitted Plato's myth of Solon and the Egyptian archives but did not; perhaps he disliked its implication that Egyptian records existed several thousand years before God created the world. Instead, he asserted that the Egyptians learned letters from Isis, whom he identified with Io, the Greek woman whom Jupiter's jealous wife, Juno, turned into a cow; Io's cloven hoofprint resembled a monogram of her Latin name (a capital O divided by an *I*).[19] Even the most patriotic of ancient Greeks would have been surprised to learn that Io/Isis brought letters "when she came from Greece into Egypt, and passed them on to the Egyptians." Unlike many authors, Isidore did not describe hieroglyphs as picture-writing. Since he thought Egyptian writing came from Greece, he evidently considered it alphabetic: "Among the Egyptians, however, the priests used some letters and the common people used others. The priestly letters are known as *hierós* (sacred), the common letters *pándemos* (common)."[20]

In fact, Isidore assumes all writing is alphabetic. He attributes the principle of representing discrete sounds by individual letters to the Phoenicians, on the authority of the Roman poet Lucan (d. 65 CE): "If the report is trustworthy, the Phoenicians were the first to dare to indicate by rudimentary shapes a sound meant to endure."[21] Isidore adds that "chapter headings are written with Phoenician scarlet, since it is from the Phoenicians that the letters had their origin." He also relates that the Greeks expanded their repurposed Phoenician alphabet from seventeen [*sic*] to twenty-four letters.[22] Because he drew his history from disparate Greco-Roman and Judeo-Christian sources, his account has inconsistencies and apparent confusions. He mentions Cadmus as inventor of Greek letters but declares that they derive from Hebrew, "the mother of all languages and letters."[23] For Isidore, all his sources were "authorities" (*auctores* or *auctoritates*), worthy of belief simply because they had long been considered so. To have contradicted one of them or opposed it resolutely against another was to lack respect for tradition.

RECYCLING: THE PALIMPSEST

Throughout much of the Middle Ages, writing materials were often in short supply. This was particularly true of parchment, the specially treated animal skins that supplanted papyrus as the preferred surface for writing books and other important documents. Parchment was an extremely expensive commodity; transcription of a single manuscript might require the skins of dozens or even hundreds of animals. But parchment lent itself better to the wholesale revision or substitution of texts. When fresh skins were unavailable, secondhand parchment was the only satisfactory substitute, so numerous manuscripts were "recycled": they were disbound, their pages were washed or abraded to erase as much of the original text as possible, and they were then overwritten with a different text (see plate 8).

A clamorous example of recycling was announced in 1998, when Christie's of New York auctioned off a remarkable medieval Byzantine manuscript. The price tag, over two million dollars, was far from being its most wonderful feature. "Barely visible underneath thirteenth-century Christian prayers were the erased words of an ancient legend and a mathematical genius: Archimedes of Syracuse. Incomplete, damaged, and overwritten as it was, this book was the earliest Archimedes manuscript in existence. It was the only one that contained *Floating Bodies*—perhaps his most famous treatise—in the original Greek, and the only versions of two other extraordinary texts—the revolutionary *Method* and the playful *Stomachion*."[24] Thanks to the exhibition mounted by the Walters Art Museum of Baltimore in late 2011, the Archimedes Codex is now the best-known example of a *palimpsest*, a word that means "rubbed again" in Greek but can be loosely translated as "recycled" or "prepared to receive writing again." Archimedes's works were copied onto this parchment in the ninth or tenth century, but the pages were recycled in the early thirteenth.

Today, this sounds like sacrilege, but the medieval insistence on the unique importance of the Christian religion meant that the decision to create a palimpsest was often made at the expense of ancient pagan culture. Recyclers evidently lost less sleep than we might if we had to make the same decisions, given our stronger preoccupation with cultural losses. Christian counterparts of Caliph Omar reasoned as he supposedly had about the books in the Library of Alexandria: pagan texts might deserve to be overwritten when they contradicted the teachings of Christianity, but they could also be sacrificed if their contents were considered irrelevant or unnecessary to readers' eternal salvation. Our era considers the text of the prayer book trivial in comparison to the Archimedian volume that was sacrificed to receive it, but thirteenth-century monks apparently held the opposite conviction. Luckily, their scale of values was usually applied to individual manuscripts rather than whole libraries, or we would doubtless have even fewer texts from pagan antiquity—and more medieval prayer books—than we now do.

There were relatively few readers in medieval society, and they were well aware that books were rare and precious, produced with enormous labor. Some works were extant in only a few copies, often scattered widely across Europe, so no one who recycled a manuscript could be certain it was not the only surviving copy of the "expendable" text. Yet the entire millennium between the fifth and the fifteenth centuries had no truly universal scale of values about books. Choosing which texts to erase was necessarily a localized decision, one that reflected the value placed on various kinds of written texts in individual monasteries and other Christian milieus at particular times. Which of the existing manuscripts in a library had the correct physical dimensions to be recycled for the new text and, more importantly, which ones were most expendable from a cultural or religious point of view?

These were the issues that determined the fate of the tenth-century codex containing Archimedes's three works. Portions of it were discarded during the process of recycling, and several other codices were recycled along with it to create the prayer book. Some of those codices contained works that are still unidentified.[25] Thanks to information contained in the colophon, or sign-off statement of the scribe, we know that his name was John (Ioannes) Myronas, and he says that he completed the transformation of Archimedes into a prayer book on April 14, 1229, the day before Easter Sunday.[26] To make the texts of Archimedes completely legible is now impossible: the mutilation of 1229 has been compounded by the ravages of time and the elements and, most importantly, by owners who ignorantly mishandled it and others who actively falsified it—not in the Middle Ages, but between 1907 and 1998. However, much of the text has been recovered, thanks to the massive combined efforts of manuscript conservators, paleographers, historians of mathematics, and physical scientists using the most sophisticated space-age imaging technology.

Let there be no misunderstanding about the nature of a palimpsest: creating it only appears to be vandalism or is at worst a partial vandalism. In the case of the Archimedes Codex, William Noel observes that, despite the manuscript's poor state of preservation, "the scribe was the unwitting savior of Archimedes, and not his nemesis. The Palimpsest was the creation of religion, not its victim," and the remote monastery where it was stored for much of its life "is better characterized as a safe house for Archimedes than as a tomb."[27] Although it is a melancholy object today, the battered manuscript protected the tenth-century scribe's shadowy transcription of Archimedes from complete destruction.

"LIFE RELATIONS" TO BOOKS AND WRITING

Few periods in history have been as saturated with reverence for writing as the Middle Ages. At every level, from the solid materiality of pens, inks, and parchment to the mystical significance of individual letters, medieval writers were thoroughly engrossed by their relationships with writing, books, and libraries. The Venerable Bede (d. 735) gave a strange but revealing example in his *History of the English Church and People.*[28] Relating that Ireland enjoys a gentler climate than Britain, he remarked that it nonetheless lacks serpents: indeed, snakes that stow away on English ships die as soon as they smell the air of Ireland. Bede says he witnessed people cured of snakebite through a curious remedy: the pages of books brought over from Ireland were scraped, the scrapings (*rasura*) drunk in water by snakebite victims. The venom immediately lost its force and the swelling abated. This sounds more like a compliment to Ireland than to writing, but scraping was the method employed to erase parchment for reuse as palimpsests (the etymology of *erase* is "to scrape away"). Bede's snakebite victims were drinking writing. Apparently any Irish book would do, since Bede does not mention the Bible specifically; but this was a peculiarly monastic medicinal magic, since only clerics had access to books. Even much later, water into which holy objects were dipped sometimes served for remedies and exorcisms, but Bede's book-water may be unique.

Scribes in Christian monasteries revered their tools,[29] but sometimes their nearly lifelong association with writing inspired calmer emotions, resembling affection more than Isidore's and Bede's dramatic veneration. A particularly whimsical example is in the "Riddle of Verona," written around 800 CE in a Latin dialect that was evolving into medieval Italian. The text reads: "He urged on the oxen / plowed white fields / held a white plow / sowed a black seed."[30] The oxen are the scribe's fingers, the white fields his parchment, the white plow his goose-quill pen, and the black seed his writing. Perhaps novice scribes learned reverence for their profession as well as humor from the riddle.

A century and a half earlier, Isidore was already aware of the metaphor. He recorded that a Roman author compared writing on wax-covered tablets to plowing: "It was established that [Roman scribes] would write on wax tablets with [styli made from] bones, as Atta indicates in his *Satura*, saying (12): 'Let us turn the plowshare and plow in the wax with a point of bone.'"[31] (Appropriately in the context of erasure, nothing is known of Atta except this verse and that he died in 77 BCE.) The eminent medievalist E. R. Curtius found no evidence before Isidore's quotation that ancient Romans used "plowing" to describe incising letters on wax tablets with styli. But he observed that writing on papyrus and parchment was compared to "the dressing of a field" as early as Plato, and that *exarare*, "to plow up," was so common a synonym for writing that it "seems no longer to be felt as a figurative expression but simply means 'to write,' 'to compose.'" Isidore seems to have thought the metaphor was logical or intuitive. "A verse (*versus*, also meaning 'furrow') is commonly so-called because the ancients would write in the same way that land is plowed: they would first draw their stylus from left to right, and then 'turn back' (*convertere*) on the line below, and then back again to the right—whence still today country people call furrows *versus*."[32] Isidore describes the archaic writing technique called *boustrophedon*, or "ox-turning," although he shows no awareness that true *boustrophedon* writing also mirror-reversed the individual letters of alternate lines to reflect the lines' right-to-left direction.

Fate, a concept fundamental to Near Eastern and European cultures, was often expressed through metaphors of writing, as we saw with Ovid. The premises of astrology, an art that dates to ancient Mesopotamia, posit that our lives are preordained by, or "written in," the stars. Christianity inherited such thinking in the form of divine predestination, an idea made necessary by God's omnipotence but problematic because it implicitly or explicitly denied human free will. (Theologians regularly drew a distinction between God's foreknowledge of a person's actions and his determination of those actions, but the distinction failed to satisfy those who insisted on God's absolute power to save "whomever he willed.") In its most extreme form, fate has sometimes been compared to the relationship between literary authors and the personages they create. Luigi Pirandello's *Six Characters in Search of an Author* (1921) is the most compelling modern commentary on the old relationship; the incomplete, unfinishable play puts the unfortunate characters in permanent limbo, sparking their anguished complaints to the personage of "the Author."

As Curtius documented, such long-lived metaphors explore writers' "life-relation" to writing. But medieval writers also reveled "literally" in the sheer materiality and beauty of their tools.

> In the . . . twelfth century we find a pleasant poet who who had a particular fondness for the scribal art: Baudri de Bourgueil. Ausonius [d. c. 395] had written a poem to his writing paper, but Baudri outdoes him, for he dedicates numerous poems to . . . his wax-tablet notebooks. A particularly fine one . . . has eight wooden tablets, and thus provides fourteen writing surfaces [not counting the two unwaxed surfaces]. . . . To save the eyes they are coated with green wax. The craftsman (*tabularius*) who made them was an artist. If only a needlewoman would make a bag to keep the little wooden book in![33]

But the ultimate reverence for writing was made clear in an anecdote Curtius copies from an early life of Saint Francis of Assisi: "Thomas of Celano tells us that the saint picked up every written piece of parchment which he found on the ground, even if it were from a pagan book. When a disciple asked him why, he replied, 'My son, letters are what the most glorious name of God is composed of.'"[34] Centuries later, the author of *Don Quixote* claimed that his inability to avoid reading scraps of paper in the street led him to the "original manuscript" of his hero's exploits.[35]

Medieval angels and demons continued keeping literal-sounding registers of individuals' sins or merits, like the records described in the apocrypha. The stenographer-spirits were often represented quite seriously in sculpture and book-illustration, but there were humorous examples as well. A well-known misogynistic quip told of a devil who grew desperate while trying to record the rapid-fire gossip of some women: when his parchment began filling up, he tried stretching it with his teeth.

A DECLINE IN REVERENCE?

In the later Middle Ages, metaphors expressing fascination with the activity, implements, and social utility of human writing appear less frequently. The decline was perhaps due to increased numbers of writers: as monasteries and cities grew larger, record-keeping of all sorts became more frequent, and the gradual replacement of parchment by paper in the late Middle Ages reduced the cost of writing. After about 1000, writing and literacy spread ever more rapidly beyond Latin to the vernaculars. Poetry flourished in Anglo-Saxon, German, Spanish, Old French, and Provençal. Merchants kept commercial and accounting records; they and other laypeople corresponded at a distance by letters.

The vocabulary of writing shifted from being the *tenor* of metaphors and similes to being their *vehicle*: from "A book/pen/writing is (like) *X*" to "*X* is (like) a book/pen/writing." With increasing frequency, writers imagined lofty subjects, such as the nature of God, the world, and human life through similes and metaphors comparing them to writing. The later Christian Middle Ages went so far as to imagine the cosmos as a book, through the concept of "God's two books": the Bible, transcribed in ink by people, and the *liber mundi*, or Book of Creation written by God. Alan of Lille (d. c. 1203) famously declared, "Every creature in the world is like a book, a picture, and a mirror to us" (*Omnis mundi creatura / Quasi liber et pictura / Nobis est et speculum*). E. R. Curtius examined variations of the *liber mundi*, but he scarcely mentioned other crucial metaphors. Some medieval authors spoke of a Book of Doom, imitating the Apocalypse; for individuals, there was a Book of Experience and a Book of the Heart.

As vernacular writing increased in sophistication, so did recourse to these lofty metaphors. Dante (d. 1321) was singularly adept at finding metaphors to express life relations with writing and books. As Curtius observed, "The entire book imagery of the Middle Ages is brought together, intensified, broadened, and renewed by the boldest imagination in Dante's work."[36] Dante's poetic autobiography, the *Vita Nuova*, began by declaring, "In that part of the book of my memory before which little can be read, there is a rubric [chapter heading] that says '*Incipit vita nova* [here begins the new life].'"[37] "To compose poetry, then, is to copy the original text recorded in the book of memory. [T]he poet is both scribe and copyist."[38]

In the *Divine Comedy*, Dante personifies a book, Vergil's *Aeneid*, as the ghost of its author and makes "Virgilio" his guide through hell and Purgatory, on up to the threshold of the Earthly Paradise. There Dante meets the soul of Beatrice, "she who blesses." He inscribed her entry into his consciousness decades earlier under that first rubric in the book of his memory, and she escorts him through the spheres of the heavens to God's court in the Empyrean, where Dante experiences the godhead as the ultimate book: "In its depths I saw internalized, / bound with love in one volume, / what through the universe becomes unsewn quires."[39] The universe reflects God as if the booklike unity of his mind had been disbound and scattered to create the Book of the World. In a manner of speaking, Dante's cosmos is God-as-biblioclasm, an exploded or deconstructed view but one where all the pieces survive and point back to their original unity. To read the Book of the World, one has to "re-collect" it, "re-member" it, reconstitute it in one's own mind, hoping thereby to gain a glimpse, however imperfect, of God.

As Curtius reminds us, "For the Middle Ages, all discovery of truth was first reception of traditional authorities, then late—in the thirteenth century—rational reconciliation of authoritative texts," leaving no contradictions among them. No understanding was possible without writing and books: "A comprehension of the world was . . . an assimilation and retracing of given facts; the symbolic expression of this being reading. . . . For Dante and for the Middle Ages, the basic plan for any education of the intellect is, in general, the reading of books" rather than experience or experimentation with the physical world.[40]

As we saw, in the Latin Middle Ages, Plato's *Timaeus* was seen as agreeing with Genesis, confirming that the cosmos was created by a benevolent god rather than existing throughout eternity or coalescing from atoms crashing randomly in a void, as in some ancient philosophies. But for our purposes Plato's creation story is less relevant than his tale of Solon's visit with the Egyptian priests and their temple archives. Plato's fourth-century Latin translator used the immensely suggestive phrase *memoria literarum*. The Latin phrase implies that letters *contain* memory and, as Plato makes explicit, enable it to survive infinitely longer than a single lifetime, if only they can be protected from fire and water. The translation reinforces Plato's idea by sliding into prose-poetry: *ut necesse sit* novo *initio vitae* novo*que populo* novam *condi memoria literarum*—"so that [after floods and wildfires] it becomes necessary to establish *anew* the memory [held by] letters for a *new* beginning of life and a *new* people." The repetition of *new* is especially poignant since, in Latin, *novum* could also suggest strangeness, outlandishness, or baselessness. The Latin implies that the cultureless people who survived cataclysms were strangers to their ancestors, and thus to themselves as well. The danger, observes the old Egyptian priest, is not that a people may be utterly destroyed, for a few individuals (a *semen exiguum*, or tiny seed, in the translation) always survive. The real tragedy is that such people have amnesia, surviving in ignorance of their origins and of the great achievements of their ancestors.

But, while it was widely read, Plato's myth of the Egyptian archives was far less influential throughout the Middle Ages than the Sethian stelae, or "columns," of the *Jewish Antiquities*. Josephus's tale of the archetypal, antediluvian codex of civilization was repeated endlessly thanks to its poetic concision and mythic suggestiveness. It presented the fundamental themes of writing and obliteration—discovery, loss, and rediscovery, human enterprise and divine providence—in a form that evoked profound feelings of wonder. It became a template for Western Europeans' meditations on these themes and only began losing its hold on their imagination around the mid-eighteenth century.[41]

Few writers transmitted the myth as they found it, yet they rarely did violence to it, for it posed the problem of cultural continuity with stark simplicity. Both Greek and Latin writers found intriguing resonances between Josephus's stelae and the Bible's description of Seth as "another seed" born to replace Abel (*semen aliud pro Abel*, Genesis 4:25). Isidore and others glossed Seth's name as *resurrectio*, "as if he triggered the resurrection of [Abel] from the dead." A few suggested that the soul of Abel had literally transmigrated into Seth.[42] Following the Christian habit of reading "Old Testament" events as foreshadowing Christianity, Hrabanus Maurus (d. 856) noted that Seth prefigured the Resurrection of Christ, while Isidore had already recorded that the heretical sect of the Sethians asserted that Seth "was the Christ," an assertion we saw earlier in the *Gospel of Judas*.[43] So it was no coincidence that Seth's progeny implemented the "resurrection" of culture after the Flood.

An extraordinary number of medieval authors relied on the *Jewish Antiquities* as a supplement to the "histories of Moses" in Genesis. Josephus's original Greek text was not printed until 1544, so, as with the *Timaeus*, Westerners relied on a Latin translation of the *Jewish Antiquities*, commissioned about 600 CE by the educator-monk Cassiodorus. The *Antiquities* and *Against Apion* were among the most-read books of the Latin Middle Ages, with at least 118 manuscripts surviving from before the fourteenth century. The translation's popularity continued into the Renaissance and far beyond: first printed in 1470, it had fifteen editions before 1525.[44]

Most medieval Latin writers construed the Sethian story as recounting the invention of writing, whether or not they mentioned the Sethians' astronomy. But Latin authors often supplanted the Sethians by more familiar biblical figures. The contents of the stelae, the details of their composition, and even their number also varied widely, though the message remained the same: civilization is ultimately a book, a written text. The story clearly argued for the primacy of biblical history, but exactly who created the "columns" and what knowledge they recorded were subject to the vagaries of memory and the personal preferences of Josephus's followers.

Medieval writers were particularly impressed by Josephus's reference to the precise location of the surviving pillar, which led many to assume he had personally examined it in situ. Thus, for numerous scholars, his antediluvian book became an authoritative reference to the world's oldest text, not only for writing and science but for history as well. It gave readers a sense of direct connection to the beginning of human history, as well as to its ancient near-obliteration and its future end. From the age of Adam through Noah's Flood to the fiery conclusion of the world, the Sethian columns seemed to embrace all human time.

Previous to Isidore, references to books written before the Flood were not abundant in Latin, so Josephus's tale became the de facto archetype. But the desert monk John Cassian (d. ca. 435) did record a similar story. Deliberately or not, Cassian countered Josephus's virtuous Sethians by defining the antediluvian book as a magical *grimoire* (a magician's manual) and, perhaps under the influence of Enochbooks, implied that writing was abominable rather than laudable by connecting it to Noah's rebellious son Ham. Cassian wrote that

> as the ancient traditions testify, Ham, the son of Noah, who was instructed in these superstitions and sacrilegious and profane arts, knowing that he would be utterly unable to take a book about them into the Ark, which he was going to enter with his righteous father and his holy brothers, engraved these wicked arts and profane commentaries on plates of various kinds of metal, which could not be ruined by exposure to water, and on very hard stone. When the Flood was over, he sought for them with the same curiosity with which he had concealed them and handed them on—a seedbed of sacrilege and unending wickedness—to his descendants.[45]

Here, the knowledge preserved is not a revelation but, as Socrates said, a memorandum; its information must have been complex and technical if Ham feared forgetting it during his year in the Ark. His decision to hide the book outside the Ark is perhaps the oldest recorded instance of self-censorship; by concealing his book, Cassian's Ham prevents its erasure by humans rather than the elements. Being concerned only with the immediate danger of the Flood, Ham chooses materials that are vulnerable to fire. Cassian imagined the Flood preserving and concealing the documents rather than menacing them, as in Josephus.

Cassian's Ham casts writing in the same malignant role as did some of the apocrypha. But despite condemning Ham's writings, Cassian implicitly praised the acquisition of practical learning, through human effort as well as divine revelation. Going Josephus one better, he wrote that the newly created Adam "was at once immersed in the study of all natural things" and mastered "that true discipline of natural philosophy," which he "was able to grasp clearly and to pass on in unambiguous fashion to his descendants."

> For he had gazed upon the infancy of this world while it was as it were still tender and trembling and unformed, and by a divine inbreathing he was filled not only with a plenitude of wisdom, but also with the grace of prophecy. Thus, as the tenant of this as yet inchoate world, he could name every animal and not only discern the rage and poison of every sort of beast and serpent, but also distinguish the qualities of herbs and trees and the natures of stones and the changes of season that he had not yet experienced. He could realistically say: "The Lord has given me a true knowledge of the things that exist, so that I might know the arrangement of the earth and the powers of the elements . . . the courses of the years and the arrangement of the stars, the natures of animals . . . the power of spirits and the thoughts of human beings. . . . Whatever is hidden and open I have learned."[46]

Cassian left Adam room for curiosity and enterprise rather than attribute his natural knowledge entirely to divine inspiration.

Still, Cassian's presentation of writing is ambivalent; although he quotes Scripture incessantly, he implies that writing should not have been necessary for the maintenance of virtue: "Before the law, and even before the Flood, we know that all the holy ones observed the commandments of the law without having read the letter." The "natural and ingrafted law" left no necessity for "a law to be added from without, set down in writing."[47]

ISIDORE ON "THE FIRST BOOK"

Given his respect for writing, it is no surprise that Isidore was the first Latin author known to quote the *Jewish Antiquities* and that he lingered over the Sethian columns.[48] In his *Chronicon*, Isidore drastically simplified the anecdote, purged it of Josephus's patriotism, and repurposed it, elevating writing into a realm of optimistic myth and humanistic wonder. He wrote: "At the age of 190 years, Lamech begat Noah, who in his own five-hundredth year of life was commanded by divine oracle to build the Ark. In those times, as Josephus tells, since those men [*illi homines*] knew that they could perish by either fire or water, they wrote down their researches on two columns, made of brick and stone, so that the things they had so wisely discovered would not be erased from memory. Of those columns the stone one is said to have survived the Flood, and to still remain in Syria."[49] Isidore diverged from Josephus by emphasizing writing rather than astronomy, and generalizing the "sons of Seth" as "those men." He refers only vaguely to Adam's prophecy of fire and flood but emphasizes God's announcement to Noah of the imminent cataclysm. Isidore's changes imply that the stelae were created during the ninth or tenth generation of humankind, not the third. Cora Lutz observed that Isidore must have thought the columns "more appropriate" to the time of Noah than of the Sethians.[50] But, by referring to "those men's" "researches" (*studia sua*), Isidore emphasized general human initiative and discovery over God's primeval inspiration of Adam.

Isidore transformed Josephus's story, rendering it more optimistic than chauvinistic, more about human enterprise than Jewish tradition. What remained after Isidore's editing was irreducible: the "studies" of the antediluvian sages, the threat of obliteration by fire or water, the preservation of writing on an immobile, indestructible medium, and the survival of essential human knowledge as a parallel benefaction to God's salvation of Noah and his family. Isidore encouraged Christian writers to see the story as about the invention of writing, and his reference to "studies" led most of them to interpret the "columns" as a kind of encyclopedia.

The influence of Isidore's version of the stelae rivaled or surpassed that of Josephus's original throughout the Middle Ages. Until the fourteenth century, encyclopedists and chroniclers interpreted the stelae in Isidore's optimistic vein, as evidence of writing's ability to withstand erasure and preserve culture. They often included details from Josephus's version or others' and usually emphasized Adam's prophecy more than Isidore had, as evidence that divine providence actively guaranteed the survival of human culture. Heinz Schreckenberg found only twelve close quotations of Josephus's anecdote by the mid-sixteenth century, but derivative versions numbered in the hundreds.[51] Throughout the long history of the anecdote, the victory of the immovable stele over the Flood's irresistible force communicated the ultimate optimism about writing. The antediluvian columns even found a place in specialized early modern histories of writing, as late as Thomas Astle's in 1803. Even when historians of writing debunked it, the victory of the massive stone column over divinely ordained universal destruction remained a powerful symbol of the resistance of written knowledge to time and the elements.

Curiously, however, the legend of the stelae conceals a logical flaw, or perhaps a lack of imagination. Neither Josephus nor Isidore considered the possibility that Noah simply loaded scrolls or tablets into the Ark along with the animals. The force of textual authority was such that throughout the long medieval tradition of commentary on Noah's Flood, only one biblical pseudepigraphon considered this solution to the problem of erasure.[52] Cassian's story of Ham went further and explicitly rejected the possibility of storage in the Ark.

As Lutz demonstrated, Isidore's changes allowed even further metamorphoses of the story in medieval literature.[53] In the ninth century, Hrabanus Maurus produced a curious composite version of the columns in his commentary on Genesis. "Josephus, the historian of the Jews, relates it thus: 'Now Jubal,' he says, 'devoted himself to music, and excelled at the psaltery and the zither; and so that the things he had discovered should not be lost by humanity, or perish before they became known, since Adam had prophesied.'"[54] Whether deliberately or because he was quoting from memory, Hrabanus reconciled Josephus with Genesis. Whatever his intent, Hrabanus optimistically overlooked the fact that Jubal was a Cainite rather than a Sethian, a member of the sinful "race" that the Flood annihilated.[55]

Freculf of Lisieux, a chronicler who flourished about 830, requested Hrabanus's commentary from its author in an extant letter.[56] Collating his friend's version with Josephus's, Freculf credited Jubal and the sons of Seth as coauthors and imagined them writing the archetypal encyclopedia: "So then the sons of Seth with their kinsman Jubal wrote onto the columns the things they had discovered about the science of the heavens, *along with the other arts*." Freculf made explicit what Hrabanus and earlier commentators had implied, by asserting, "It is clear as day that *the invention of writing took place in the time of the first man*, who was still alive when his children and grandchildren studied wisdom so fervently."[57] As late as the seventeenth century, historians of writing took the idea to its logical conclusion and quoted Freculf as their authority that Adam himself had invented writing.

Byzantine writers followed a separate but similar paradigm. The encyclopedia *Suda* (or *Fortress*, ca. 1000) recorded that "arts and letters belong" to Adam as their inventor, but it also followed Josephus and attributed the invention to Seth. The Greek chronographer Joel (after 1204) retold the *Book of Jubilees'* story about Canaan, the great-grandson of Noah, explaining that he discovered "astronomical terminology, inscribed on a standing stone." But whereas *Jubilees* attributed the observations to the fallen Watcher-angels and condemned Canaan for transcribing them, Joel ascribed the stele to the Sethians and declared that Canaan wrote a commentary on it.[58] Thus, implied Joel, was born scholarship.

The idea that the columns were an encyclopedia enjoyed considerable fortune. Hrabanus's version was still being quoted by Werner Rolevinck and Hartmann Schedel in their fifteenth-century universal histories, which reached an extremely wide audience in early printings. But Peter Comestor (d. ca. 1178), whose sobriquet, "the Eater," celebrated his appetite for books, had recorded the most bizarre variant in his *Historia Scholastica*, a Bible paraphrase and commentary that was one of the "best sellers" of the later Middle Ages.[59] Deliberately or not, Comestor collated Pliny's idea that Zoroaster invented magic with Josephus's Sethian columns and Cassian's account of Ham's tablets, thereby creating a strangely ambivalent story: "Ninus conquered Ham, who was still living and reigned in Bractia [*sic*], and was called Zoroastres, the inventor of the magical arts, who wrote the seven liberal arts on fourteen columns, seven of bronze and seven of brick, against both kinds of flood [i.e., of water and of fire]. However, Ninus burned his books."[60] Even more weirdly, the knowledge Comestor's Ham/Zoroaster rescued from the flood was *not* his own invention of magic, as Cassian had claimed, for medieval theologians still considered magic evil. Acting completely out of character, Comestor's Ham preserved the seven liberal arts—not merely "good" knowledge but the sum total of it. In another passage of the *Historia Scholastica,* Comestor also copied Hrabanus's story of Jubal's two stelae, thus listing a total of sixteen antediluvian volumes.[61]

Comestor's treatment crowns the accumulation of detail that impelled the anecdote over five centuries.[62] Being so dramatic and explaining so much, it was a particularly fortunate variant: to name but two of its encores, Vincent of Beauvais (d. 1264) repeated it in his enormous *Mirror of History,* as did Athanasius Kircher in his *Noah's Ark* in 1675. Despite his vastly wider knowledge of history, Kircher continued the tradition of creative attribution and Christianized the fourteen columns by tracing them (wrongly) to Saint Augustine's *City of God.*[63]

GODFREY OF VITERBO APPROPRIATES SCRIPTURAL AUTHORITY

Ham/Zoroaster's seven liberal arts were the ultimate transformation of Josephus's anecdote until the arrival of Godfrey of Viterbo, a chronicler who died about 1196. Godfrey traveled widely in Europe, being a retainer of two popes, Urban III and Gregory VIII (both d. 1187), and two Holy Roman Emperors, Frederick Barbarossa (d. 1190) and his son Henry VI (d. 1197). Godfrey's importance as papal and imperial administrator was more than matched by his ambitions as a writer.

His chronicles are extravagantly optimistic about writing. Like numerous other medievals, he composed a history of the world from God's creation of the cosmos to his own time. Godfrey wrote his universal history not once but twice, initially titling it *Memoria seculorum,* and later rewriting it as *Pantheon.* Each title implied that Godfrey's history was the most complete ever written; their sense translates respectively as "Memorable Things throughout the Centuries," and "Temple of Godlike Men."

Godfrey expressed his optimism about writing through unblushing self-aggrandizement. Proud of his career, he claimed it made him a superior historian. He was, he said, no pale and sedentary bookworm but a seasoned cosmopolitan traveler.

> But if someone should ask where I found all this, he should know that, for the space of forty years, I sipped [*deflorasse*] the sweetest and most necessary nectars from the flowers of all the collections of books I ever found in all the realms and churches I visited. Often Greeks from Constantinople and Saracens from Babylon and Persians from Persia and Armenians from Armenia, who came to the Imperial and Papal courts on important diplomatic missions, instructed me, and sometimes they entrusted their writings to me. Know also that I worked on this book for nine years and successfully finished it in the tenth: Amen![64]

Godfrey reinforced his boasting through sacred biblical numerology because he wanted readers to treat his works as *scriptura*. His forty years of research echo the Israelites' wanderings through the desert. Bringing his work into the promised land of written composition required ten years, which is 3 × 3 + 1 years, reflecting the trinity and unity of God. The holy numbers imply that Godfrey's history reveals the apparent chaos of history as actually being the product of God's providential art. Although his posturing seems absurd, it evinces a reverence for writing shared with Isidore and a scope that matches Dante's.

Godfrey had another stroke of inspiration when he remembered Flavius Josephus's declaration that authoritative pagan historians corroborated the Bible's version of primordial history. Rather than merely cite Josephus, Godfrey plagiarized his claim—not so as to corroborate or celebrate Scripture but to establish his own historiographic supremacy. Quoting the *Jewish Antiquities* almost verbatim, Godfrey boasted that "according to the annual record books of the ancient kings, there were other Barbarian or Gentile historiographers, who wrote down in its entirety everything that happened, nor did they leave anything unrecorded. We will state some of their names here: *Mamenot* [sic], *who made the description of the Egyptians, Berosus* [sic], *who excerpted all the writings of the Chaldeans, Mochus and Estius* [sic], *and Jerome the Egyptian: they and many other Barbarian or Gentile historiographers are in agreement* with my histories and chronicles."[65] Godfrey thus claimed—fraudulently—to have consulted a virtual Alexandrian library containing all history, a panoply of extant sources unknown to anyone but himself which he could easily incorporate into his account if only he had fewer administrative responsibilities. His egregious self-conceit nonetheless conveys a supremely optimistic ideal of historiography as *Memoria saeculorum*, "the memory of all the centuries."

He may not have expected his readers to recognize the source where he found the names of his mysterious predecessors, but he was not afraid to mention Josephus by name. On one occasion he even claimed the authority of "Moses and Ezra and Josephus and other authors of the Old Testament," as if the *Jewish Antiquities* belonged to the scriptural canon.[66] In some passages Godfrey even insinuates that Berossos and the other "barbarians" are more ancient than Moses or Abraham. He thus adopted Josephus as his model for "how to write Scripture and influence people."

Godfrey pushed the antediluvian origin of writing even further back than Josephus had. In the proem of the *Pantheon*, he declared that "Adam, the firstformed original [*primus protoplastes*] man, who, as we read [elsewhere], possessed all knowledge of things and of all the arts, learned many things during the long ages after his creation; and Adam both wished and was able to teach his offspring everything that he knew."[67] Godfrey claimed Adam, rather than Moses or divine revelation, as his very own source for the earliest history, declaring in verse that "Adam, 'tis said, formed great columns of brick, / And decreed all events be recorded on them; / From them we copy all our ancient history." Adam's memoirs are the ultimate source of ancient history, and Godfrey quoted "Adam's brick columns" directly, contradicting Josephus's assertion that the Flood washed the bricks away. At any rate, by identifying Adam's columns as historical records, Godfrey ratified the consensus of his late antique and medieval predecessors, who valued the antediluvian columns more as historiography than as science.

Medieval Muslim writers, including the astrologer Abū Maʿshar (Albumasar, d. 886), told many tales of Hermes and other ancients as writers or discoverers of antediluvian wisdom safeguarded on stelae, in pyramids, or even on tree-bark and sealed in individual stones.[68] Abū Maʿshar, who advocated the interpretation of history via astrology, told a story of this sort to claim an antediluvian source for one of his works.

> The ancient kings of Persia were much concerned to preserve scientific books from natural disasters, whether atmospheric phenomena or earthquakes. Therefore, they had their books copied on the bark of *khadank*-trees, which is called *tūz*.
>
> King *Ṭahmūrath*, after 231 years and 300 days of his reign had passed, heard of violent rain-storms in the West, and recalled that the astrologers, when he first had become king, had predicted the spread of these storms to the East. He therefore had his engineers build [the citadel] *Sārawīya*, and deposited within it manuscripts of various scientific works written on *tūz*. Among these was one ascribed to an ancient sage, in which are contained the years and cycles for determining the mean longitudes of the planets and the inequalities of their cycles. . . . [T]he wise men of India and their kings, the ancient kings of Persia, and the ancient Chaldaeans . . . preferr[ed] them over others because of their accuracy and brevity.[69]

Abū Maʿshar also has anecdotes in which the first of three Hermeses (the antediluvian one) safeguards knowledge by depositing it in the Egyptian pyramids.[70]

Some of these tales passed into Christian writing mythologies, particularly among astrologer-astronomers and alchemists.[71] A magical treatise by Pseudo–Thomas of Aquinas asserts that its source is a book sealed in a stone by Adam's son Abel.[72] But the subject of cross-cultural influences by Islamic scientific fictions is too rich for more than cursory mention here.

Altogether, the seven centuries between Isidore and Dante demonstrate an unparalleled enthusiasm, indeed a reverence, for the wonders of writing, for its ability to protect the codex of civilization against erasure and guard human society against amnesia.

To the generations after Dante, that enthusiasm would appear naive and wrongheaded, at times even hypocritical, in its failure to recognize—and to mourn—the losses of genuine ancient *scriptura* during the "dark" ages. The desolation would become increasingly apparent to scholars who imagined themselves as resuscitating the corpse of culture, or even enabling its rebirth, its "re-naissance." Both ages, the "middle" as well as the early modern, were in fact riddled with ironies and contradictions about what they succeeded or failed at saving of ancient culture, and in their attitudes to those losses and recoveries. After an age of paradoxical optimism would follow a period of pessimism that now seems equally excessive.

CHAPTER NINE

Pessimism in the Age of Rediscovery, 1350–1500

> The tomb will fall, and the epitaph carved into its marble will fall: then, my son, you will die a second death. Oh, the bright and eternal glory that falls with the blow of a stone! Add to that the destruction of books in which your name is recorded by your hand or someone else's. And although this seems slower to arrive, since the memory of books is more lasting than that of the tomb, nonetheless it is the inevitable outcome, owing to the numberless injuries inflicted by nature and fortune, to which books, like every other thing, are subject. . . . With the death of your books you will also die, so a third death awaits you.
>
> —PETRARCH, *My Secret*

In the modern popular imagination, the Renaissance was due to unrivaled excellence in the visual arts—to the Giottos and Michelangelos. That it was also a literary and philosophical movement—or rather, several of them—may come as a surprise.[1] Likewise, the Reformation is often considered simply another term for a specific instance of ecclesiastical reorganization undertaken by Protestant Christians. In addition to being simplistic, *Renaissance* and *Reformation* are loaded terms. When applied to movements that flourished in the fourteenth through the seventeenth centuries, they implicitly devalue the "dark" or "middle" ages and the "Counter-Reformation," casting them as periods of ignorance or obscurantism. Such definitions are quite unhelpful for understanding the emotional history of writing: whether one period is "better" than another is not our concern. What matters deeply is the continuity of their fascination with writing as the essential human art in the face of periodic interruptions and revolutions.

Still, the era often classed as "Renaissance and Reformation" had a new conception of writing and its preservability, which determined its attitudes toward truth and falsehood in ways that were often contradictory. As a result, the deepest significance of the age lay in its stance toward the value of history as a model and guide. The relationship of fifteenth- and sixteenth-century people to the Greek and Roman classics, to the Scriptures, to historical change—none of these could have changed as thoroughly as they did without transforming, and being transformed by, the appreciation and practice of writing. In fact, literacy strongly influenced the activities of visual artists as well as authors. The transition from mere artisan to genius artist in the fourteenth and fifteenth centuries depended in part on the expansion from Latin to vernacular in every form of communication, from business accountancy to artists' creative involvement in church and civic pageantry. "Revival," whether a renaissance of artistry or a reformation of religiosity, was ever more dependent on writing and reading, and those who practiced any of these activities were more aware than ever of the fact. (The printing press, Gutenberg's new way of writing, was inseparable from all these trends but, for reasons that will become apparent, is more properly addressed in the next chapter.)

The late medieval revival did not happen in a sudden burst of optimism and exuberance. Indeed, the emotional history of writing sank into depression around 1350, when medieval authors' confidence in the stability of written culture was quickly overtaken by a pessimism worthy of Rodulfus Glaber, the French monk we discussed in chapter eight. Although most writers demonstrated faith in the integrity of Holy Scripture, several prominent authors broke with the previous tradition of general optimism about writing and began asserting that cultural life had undergone widespread devastation. They had traveled enough to be familiar with the uncertain fortunes of old manuscripts and their uneven availability to readers, so scholars' pessimism was partly a reaction to the overconfidence of earlier medieval authors. As the loss of confidence spread, scholars began describing veritable "genocides" of books.

Their attitude could seem anomalous. Over the previous three centuries, expanding trade had created flourishing towns with a significant bourgeoisie and a vibrant civic life, and raised urban living conditions to levels not seen since Roman times. Major universities were thriving at Paris, Oxford, Cambridge, and Bologna, while others were emerging; law, medicine, theology, and the liberal arts were thoroughly professionalized by 1300. And yet, harking back to Ralph Glaber, fourteenth-, and fifteenth-century pessimists dwelled on the cultural devastation of their era.

Both blithe medieval optimism and the exaggerated pessimism that succeeded it are symptomatic of the emotional history of writing. But there were objective reasons for the unhappiness of the fourteenth century. Although now considered the threshold of the Renaissance, the fourteenth century was afflicted by every form of social turmoil and cultural upheaval; indeed, the 1300s have been characterized as "calamitous." Weather and agriculture were in crisis, for Europe was caught in what is known as the "Little Ice Age." The century began inauspiciously with a series of major famines that left many Europeans weakened and vulnerable to disease. Ever-expanding trade networks brought prosperity to some but spread disease more efficiently among disparate locations. When, at mid-century, the bubonic plague, or Black Death, arrived in Italy aboard ships from the Eastern Mediterranean, five years of pandemic disease (1347–1352) sufficed to wipe out 30–50 percent or more of inhabitants in locales from Byzantium to Iceland.

There were spiritual epidemics as well. Beginning in the eleventh century, Europe had been increasingly haunted by the specter of heresy. Churchmen commonly referred to the spread of heresy as a *lues*, or plague, for dissident interpretations of Christianity had become a phenomenon of the masses, spread by unlicensed, frequently semiliterate preachers. Major heretical movements extended from the Waldensians and Cathars of Southern France and Northern Italy to the Lollards of England and the Hussites of Bohemia. So, beginning in the twelfth century, increasingly formal, centralized forms of Inquisition undertook to expose and repress "heretical depravity." In the early thirteenth century, the Franciscan and Dominican orders were organized and papally commissioned to combat heresy by preaching in the languages of ordinary people and to furnish the ranks of inquisitors. There were even military crusades against heretics: the Cathars, also called Albigensians, were all but exterminated by the early fourteenth century, several crusades targeted the Hussites in the early fifteenth, and there were other anti-heretical crusades.

Ironically, one of the principal stimuli to heresy—at times the main one—was the desire of laypeople to read the Bible in their own everyday languages. The process was related to the rise of the merchant class, which needed to keep accurate records of inventories and financial transactions. The spread of commerce had created the need for literacy and numeracy in the vernacular languages, and as merchants learned to keep written records of their transactions and correspond with foreign counterparts, they developed the desire to read and understand Scripture for themselves. Rising levels of literacy from the twelfth through the fifteenth centuries inspired the massive dissident religious movements. Waldensians, and later Hussites, demanded access to the Bible in their own everyday languages. Other groups, preeminently the Cathars, radically reinterpreted the biblical message. In Cathar theology, as in the earlier gnosticism, the stern god of Genesis, who created the material universe, was identified with Satan, while the New Testament foretold the eventual victory of a good, completely spiritual god.

A momentous transition was under way: intellectuals educated in Latin were losing their stranglehold on reading and writing. The Catholic insistence on Latin as the language of Scripture and liturgy had created a clerical monopoly on doctrine, creating a cycle of unrest and repression. In his *Decameron* (ca. 1353), Giovanni Boccaccio wrote—in Italian rather than Latin—"What books, what words, what letters are holier than those of the Holy Scriptures? And yet, there have been those who, by interpreting them in a perverse manner, have led themselves and others to perdition."[2] Beginning among twelfth-century Waldensians, the call for vernacular translations of Scripture would eventuate in Luther's German translation of the

Bible and a broader Protestant emphasis on developing a personal relationship with God through independent, even solitary reading of Scripture.

HUMANISM

It is undeniable that Europeans in the years 1350–1500 rediscovered the importance of writing by becoming acutely aware of its fragility. The period has been accurately described as an age of humanism, although the term should not be construed as indicating a "humane" philosophy of life, as it usually does now. And humanism certainly should not be interpreted as atheism or hostility to religion, an anachronism that fundamentalist Christians since the 1980s have proclaimed in their diatribes against "secular humanism," by interpreting various "humanist manifestos" since the 1930s as an existential threat to Christian hegemony in the United States. No humanist of the fourteenth through sixteenth centuries would have advocated, or tolerated, hostility to Christianity, much less atheism, which was almost literally unthinkable except as a baseless slander. Humanism was not founded on either humaneness or atheism; rather, as the historian of philosophy Paul Oskar Kristeller maintained, humanism was at its core a method of study.

In fact, as Kristeller and other historians reveal, late medieval and early modern humanism was fundamentally an attitude toward writing. In the decades around 1350, writers and intellectuals—most of them Italians, initially—became interested in texts that had been neglected, forgotten, or lost over the previous millennium. Initially, humanists' interest focused on the Age of Philosophy, or classical period, especially on pre-Christian Roman authors. Italians in particular felt especially strongly the urge to rediscover and recover pagan Latin literature: they considered Romans, even those who had not lived on the Italian peninsula, as their ancestors. At the same time, interest in early Christian authors such as Augustine and Jerome was expanding in new directions, with scholars considering them as historical authors and not only as timeless pillars of Christian orthodoxy.

Until the Age of Optimism faded around 1350, competence in Greek language and culture had become a rare and fragile accomplishment in the Latin West. Yet the Romans had recognized Greek as fundamental to a proper education. So increased interest in the Latin heritage led to a revival of curiosity about Greek. Between 1360 and 1397, thanks initially to Boccaccio, the city of Florence hired native teachers of Greek who inspired passionate enthusiasm for classical literature and philosophy among Florentine youth. By the mid-fifteenth century, Christians' interest in writing was leading some to learn Hebrew from Jews, sometimes without even attempting to convert their teachers to Christianity. Like their interest in Roman history and literature, Italians' cultivation of Greek and Hebrew for scholarly purposes would inspire northern Europeans to imitate and sometimes surpass them in the sixteenth century.

The preoccupations of this age differed substantially from those of previous times. The Age of Optimism had seen itself as the legitimate inheritor of antiquity; Dante pictured himself among the greatest poets of Greece and Rome, "sixth among such wisdom," and thus worthy to be guided through the Christian afterlife by the prince of pagan poets, Vergil. By contrast, scholars two generations later saw a catastrophic break, a great erasure, rather than a serene continuum. As historians have observed, the Age of Rediscovery stopped seeing Greeks and Romans as their cultural neighbors and ceased—both literally and figuratively—dressing the ancients in medieval costumes and customs. There was increasing interest in artistic and intellectual change rather than a timeless pantheon of greats assembled by divine providence. Scholars were motivated by the desire to understand the practices of language and writing, from calligraphy and orthography to grammar, syntax, rhetoric, and poetics, that had prevailed in classical antiquity. Not that the ideological and religious content of pagan literature received less attention than in previous centuries, but the importance of specific linguistic and literary forms to particular kinds of intellectual content became more noticeable to scholars, and the forms that attracted their attention often differed from those valued in previous centuries.

In the fourteenth century, the survival of writing was perceived rather suddenly as an urgent problem. One who worried deeply was the Englishman Richard of Bury (d. 1345), the Bishop of Durham,

High Chancellor of England, and Treasurer of King Edward III. Richard, a fanatical book collector, enshrined his passion in the *Philobiblon,* a treatise whose Greek title was intended to mean "On the Love of Books." It is one of the most eloquent expressions of wonder and devotion ever dedicated to writing, books, and libraries. Several chapters demonstrated Richard's bibliophilia through diatribes against neglecters and abusers of books, whether slovenly monks, possessive scholars, or tyrannical politicians. The chapter titled "Complaint of Books against Wars" passionately deplored the destruction of the Library of Alexandria.[3] Richard recalled Aulus Gellius's (d. ca. 165 CE) accusation that Caesar's troops had burned "seven hundred thousand bookrolls (*volumina*), which the Ptolemaic pharaohs had collected over long periods of time."[4] His mournful enthusiasm for the subject inspired him to a litany of extinct knowledge: "the religion of the Egyptians, which the Book of the Perfect Word so commends; the excellent polity of the older Athens, which preceded by nine thousand years the Athens of Greece; the [incantations] of the Chaldaeans; the [astronomical] observations of the Arabs and Indians; the ceremonies of the Jews." Missing also were "the architecture of the Babylonians," as well as "the antidotes of Aesculapius; the grammar of Cadmus; the poems of Parnassus; the oracles of Apollo; the *Argonautica* of Jason; the stratagems of Palamedes, and infinite other secrets of science."[5]

Some of Richard's sources, like the *Timaeus,* are already familiar, others must be conjectured. But he suspected that the following sources used by the very authors of the Bible had been incinerated:

> The secrets of the heavens, which Jonithus [an apocryphal son of Noah] learnt not from man or through man but received by divine inspiration; what his brother Zoroaster [also known as Ham],[6] the servant of unclean spirits, taught the Bactrians; what holy Enoch, the prefect of Paradise, prophesied before he was taken from the world, and finally, what . . . Adam taught his children of the things to come, which he had seen when caught up in an ecstasy in the book of eternity . . . the agriculture of Noah; the magic arts of Moses; the geometry of Joshua; the

enigmas of Samson; the problems of Solomon[, ranging] from the cedar of Lebanon to the hyssop.

All these, Richard said, "are believed to have perished at the time of this conflagration," destroyed by "those horrid flames."[7]

Richard imagined Ptolemy's library as a ghost archive of erasures, not a museum but a necropolis of ideas or, rather, irretrievable *secrets*. He was a poet of obliteration, a collector of deletion. Whereas previous writers expressed wonder at familiar knowledge, such as the seven liberal arts, attributing discoveries to sages from the world's infancy, Richard assumed that such heroic figures must have accumulated even more marvelous treasures, which later perished. The Library of Alexandria was a Paradise Lost, and scholars had inherited a fallen, unredeemed world bereft of wisdom.

Richard's library fire was as dramatic a symbol of disaster as the biblical Flood his predecessors imagined. However, he did not blame the pitiable state of learning in his day solely on the destruction of one library, for "in truth, infinite are the losses which have been inflicted on the race of books by wars and tumults." Richard's numbers were consistently hyperbolic: long before Alexandria, he thought, there must have been other great biblioclasms. Homer would have been shocked and puzzled to hear Richard exclaim, "How many thousands of thousands of [books] did the ten years' war of Troy dismiss from the light of day!"[8] In Richard's mythic view, the world itself was a vast cemetery of books, with nothing left but a few teetering headstones and mutilated corpses, amid a myriad of desecrated graves.

As his catalog reveals, Richard was interested mainly in "scientific" secrets—Books of Mastery rather than Books of Memory. Thus, he mentioned only two works of history, and only one was a real book: Apollonius of Rhodes's *Voyage of the Argo* (3rd c. BCE). Richard envisioned it as a lost chronicle from humanity's childhood, starring the original seafaring explorers who "made Neptune marvel at the shadow of the Argo," as Dante imagined.[9] Ironically for Richard, the *Argonautica* still existed: he would never know that in 1416 a surviving Latin adaptation (by Valerius Flaccus, d. ca. 92 CE) would be discovered in the Swiss monastery of Saint Gall, nor that, a century later, Apollonius's Greek original would reach Italy from Constantinople.[10]

At Avignon in 1333, Francesco Petrarca (d. 1374), the celebrated "founder of the Renaissance," spent a few hours talking with Richard, who, he said, claimed to have "an extraordinary number" of books.[11] The young Petrarch shared the Englishman's dolorous pessimism about the recoverability of ancient writings. Petrarch knew whereof he spoke, having discovered several ancient manuscripts, long neglected and believed lost. He was far from sanguine about the impact of his discoveries, however, and was genuinely distressed over the number of manuscripts destroyed along the centuries. He could touch the evidence with his hands, in quotations from unlocatable books enshrined in writings that had escaped destruction.

Reflecting his own career as a Scholastic and a dignitary of the Church, Richard imagined the vanished wisdom of the biblical patriarchs and the harm these losses inflicted on Aristotle. Petrarch had no use for Richard's legendary, eclectic bibliographies, and bitterly deplored his contemporaries' worship of Aristotle as *the* Philosopher, a dispenser of timeless, invariable truths, whom Dante called "the master of all who know." Instead, Petrarch mourned the destruction of Italy's Roman patrimony by barbarian hordes—not only the marauding tribes of late antiquity but also the modern cultural "barbarians" who preferred arid Scholastic commentaries and pedestrian compilations over the legacy of Roman intellect. The architectural ruins of Rome reminded him of the biblioclasms that had reduced its literature, politics, moral philosophy, and even the Latin language to rubble.[12] All his productions, in both Latin and the vernacular, reflect this attitude: although his reverence extended to Augustine and the Church fathers as well as Cicero and Vergil, his mourning for the classics was more poignant. Whereas Patristic culture appeared more or less intact, too many pagan masterpieces were mutilated or missing.[13]

Petrarch expressed his sorrow most affectingly on the heels of the Black Plague, in the early 1350s. Imitating Cicero's (d. 43 BCE) collections of letters to friends, he imagined the souls of the ancients surviving in the afterlife (not so much a Christian hell as a Homeric Hades or Dante's Limbo) and wrote to them as friends or relatives, praising their accomplishments, yet criticizing their personal failings. Writing to Asinius Pollio (d. 5 CE), renowned as founder of the first Roman public library, Petrarch ironically neglected this benefaction in favor of carping at Pollio's foibles. Historians have frequently been charmed by such avuncular, "man-to-man" criticisms, while neglecting the analogy that makes Petrarch's "encounters" possible. Because he treats books as distillations of great minds, for him mutilated books are like ghosts, and lost books suffer a fate worse than death. A lost book is an oxymoron, as shocking as if God revoked a person's entire existence, "as if he had never been born." Petrarch's concern for lost writings is more sincere than anything he ever writes to or about any particular person; it can be genuinely moving.

His grief over the destruction of ancient writings reflects the discouragement he experienced when rummaging in the libraries of several countries. Closing his second letter to Cicero, he changes tone abruptly: "You have heard my opinions about your life and your talent. Do you also wish to hear about your books, how fortune has treated them, and how the public and scholars view them?" Cicero should know that although some of his works survive, they are more praised than read; others are surely disintegrating even as Petrarch "updates" Cicero: "Some of your books, I suspect, are lost for us who still live, and I know not whether they will ever be recovered." Modern greed, careerism, and negligence have destroyed "many works of other illustrious authors," as well. "Even of the surviving books, large portions are missing; it is as though after winning a great battle against oblivion and sloth, we now had to mourn our leaders, and not only those who had been killed, but those who had been maimed or lost. This we deplore in many of your works . . . which have reached us in such fragmentary and mutilated condition that it would perhaps have been better for them if they had perished."[14] Petrarch invites us to imagine books as mangled veterans who regret not dying. It would take a number of centuries before scholars could celebrate the recovery of a fragment rather than deplore the loss of the whole that once surrounded it.

To the prolific Roman polymath and encyclopedist Marcus Varro (d. 27 BCE), Petrarch writes: "But so that nothing be kept from you regarding your circumstances, despite the fact that you are known to have read 'so many things that we marvel that you had time to write, and wrote [*sic*] so many things that hardly anyone could read them all,' still none of them survives or at best they are only in fragments." The only consolation in this catastrophe, says Petrarch, is the cold comfort of comradeship in defeat: "There is a throng of illustrious men whose works resulted from dedicated labor similar to yours and who enjoy no greater fortune." He names seventeen lost authors, mentioning "others too numerous to list," who "were once famous men but now are obscure ashes, and . . . their names are barely known." Varro was "very intimate" with Julius and Augustus Caesar, so Petrarch sends them his greetings since they "were devoted to studies and very learned." Deploring modern ignorance, Petrarch admits that "it would seem more fitting" to relay his greetings through "Holy Roman" emperors, but only "assuming that [they] are not ashamed of having destroyed the Empire."[15] The loss of Rome's cultural patrimony mirrored the destruction of its world-historical empire. As he writes to Cicero, "You will wish to know about the condition of Rome and of the Roman state. . . . But it is truly better to pass over such subjects in silence."[16]

Writing to Quintilian (d. ca. 100 CE), Petrarch recalls that his *Education of an Orator* was as esteemed as the works of Cicero. But "when it came into my hands," it was a colossal disappointment, "alas, mangled and mutilated." He remembers thinking angrily that his own "insolent and slothful age" neglected ancient works of genius to preserve books that would be "better lost." But in a rare burst of optimism, he concludes that the defective Quintilian manuscript at least allowed him to correct his earlier lack of proper esteem: "I am now pleased to have put an end to my error." The experience inspired Petrarch to a powerful analogy between the mangled corpse of a person and a defective literary corpus: like some Homeric or Vergilian survivor, he keens that "seeing the dismembered limbs of a beautiful body, my mind was overcome by admiration and grief." Despite the butchery, he discerned that "in these books—I know not the number, but doubtless they were many," Quintilian's achievement had once rivaled the better preserved treatises of Cicero's old age.[17]

To the arch-historian Livy (d. 17 CE), Petrarch confesses that "I should wish . . . either to have lived in your age or you in ours so that either the age itself or I as a person would become better through you." Livy's literary remains are pathetically small: "Now I am allowed to behold you in your books, not indeed in your entirety, but as much as has not yet perished through the sloth of our age. We know that you wrote 142 books on Roman affairs. Alas, with what enthusiasm and labor. Scarcely thirty of them survive." Still, handling the relics provides consolation, since "I busy myself with these few remains of yours whenever I desire to forget these places or times, as well as our present customs, being filled with bitter indignation against the activities of our contemporaries, who find no value in anything but gold and silver and pleasures." Petrarch thanks Livy for providing a time machine: "You often make me forget present evils by transferring me to happier centuries."[18] Yet the machine has a disappointingly limited range: "If I could only possess you in your entirety, with how many other names [of great Roman heroes] would I seek solace for my life and forgetfulness of this hateful age! Since I cannot find them in your works, I read of them here and there in other authors, and especially in that volume [a later epitome or summary of Livy's history] where you are to be found in your entirety but so abridged that most of the subject matter is lost though the number of books is correct."[19] Paradoxically, Petrarch's letter emphasizes *memoria literarum* as a means of forgetting, erasing the horrors of fourteenth-century life.

Petrarch's gloomy prognosis for culture, and human enterprise in general, his analogy between human death and biblioclasm, reminds us that he began writing to the ancients in 1350, only two years after the Black Plague reached Florence. Yet his letters are not without occasional glimpses of hope. In his letter to Varro he indulges in a moment of time-traveling alternative history: Varro's "books dealing with divine matters" make Petrarch wish the prolific Roman could have personally consulted Saint Augustine, "a very holy man with a divine intellect." The high esteem expressed by Cicero and other contemporaries makes Petrarch fantasize that Varro would have rivaled the Bishop of Hippo had he only heard the Christian message: "You would surely have become a great theologian since you dealt so accurately and analytically with the theology that you knew."[20] Petrarch traces his admiration for Varro's writings about the gods to quotations of Varro's *Human and Divine Antiquities* in Augustine's *City of God,* as well as in Cicero. The Christian bishop had criticized Varro's work severely, but his quotations preserved large expanses of it.

Augustine was a Roman citizen, but Petrarch wrote no letter to him for this collection. Instead he imagined a long dialogue with the stern old churchman in *Secretum,* or *My Secret.* There, he abashedly confessed that his love of pagan literature was as overwhelming as his amorous fixation on an unattainable woman, the iconic Laura. Laura's name symbolizes Petrarch's cult of writing, echoing the ancient belief that the poet's laurel crown represented "the immortality of the poet's own name," as well as "the names of those whom he celebrates."[21] Petrarch's indomitable love of pagan literature derives in large part from this idealization of writing as *fama,* or renown, an almost literal afterlife. In a well-known letter, he admitted treating Augustine's *Confessions* as others used the Bible or the *Aeneid,* opening the book randomly to see what its words would "prophesy." Ostensibly he hoped Augustine might spur him to abandon his sinful loves and become wholeheartedly Christian. But instead, this bit of literary augury allowed him to reenact an incident in the *Confessions* when Augustine, also anxious to be converted, thought he heard children chanting "pick up and read" and took it as a divine command to open the Bible randomly. But Petrarch was imitating an incident that was already an imitation: Augustine had tried *sortes biblicae* because he remembered that Saint Anthony had previously tried the same experiment successfully. Rather than sparking a spiritual conversion, Petrarch's experience remained a mere literary exercise, what critics call a *mise en abyme,* like the picture of a person holding a picture of himself holding an even smaller picture of himself . . . ad infinitum.[22] Despite this religious dead end, however, Petrarch's letter about his failed conversion is an important document of his meditation on the powers and weaknesses of scriptures, and of writing itself.

Still, the emotional impact of that letter pales before the threnodies in his comradely letters to the ancient Romans. There, Petrarch occasionally expresses an uncharacteristic, wary optimism that more complete texts of his beloved authors will be found. To Quintilian, he expresses the hope that "perhaps someone now possesses you in your entirety, who is doubtlessly unaware of his guest's renown." If this should happen, "may whoever had the good fortune to discover you know that in his possession is this object of great value, which, if he is at all wise, he will consider among his greatest treasures" and preserve for the future.[23] He reveals to Vergil (d. 19 BCE) that, thanks to Augustus Caesar, the *Aeneid* has survived, despite Vergil's despondent insistence that the unfinished epic be destroyed after his death. The emperor's decision, says Petrarch, snatched Aeneas from "the brink of destruction" in the "second flames" that would have destroyed knowledge of Rome's founder more thoroughly than the marauding Greeks destroyed his Trojan home.[24]

Petrarch intended to be the exception that proved the rule of universal "modern" decline: pride in his writings, and desire to have posterity idealize them, shine through all his works. But the episode that summarized his ambition was his coronation as poet laureate of Rome on Easter Sunday, 1341. Even more than Dante had, Petrarch craved acknowledgement as the literary heir of Vergil. By manipulating powerful patrons in Paris, Naples, and Rome, he contrived to have himself crowned with laurel by the Roman senate. The senators' motivation for bestowing this honor was supposed to be Petrarch's *Africa*, a Latin epic describing Scipio Africanus's defeat of Hannibal in 202 BCE, a poem that, ironically, he never finished.

His *Coronation Oration* demonstrates that Petrarch envisioned himself as the neo-Vergilian protagonist of a "renascence" of ancient Roman glory. He claimed to be the first *poeta laureatus* crowned since antiquity, after a hiatus of twelve hundred years. He reminded his audience that Romans reserved the laurel crown "for Caesars and for poets," and that "both were wont to be called sacred."[25] The "sacred" laurel crown acknowledged the creation of a scripture, "set apart" from other poems. Petrarch's coronation for his abortive secular scripture, discussed for nearly seven centuries, is one of the most successful publicity stunts ever performed.

But scholars—and at least one of Petrarch's contemporaries—have long recognized that he was *not* the first poet laureate in twelve hundred years. That honor had been bestowed on another writer when Petrarch was eleven years old. In 1315, the city of Padua crowned Albertino Mussato, a statesman and historian as well as a poet, for his *Ecerinis*, the first Latin tragedy since antiquity to observe the dramatic structure and verse forms of the classical genre. This conformity made Mussato's *Ecerinis* more a rebirth of ancient literary values than Petrarch's *Africa*, despite the tragedy's Christian worldview and modern subject (the sanguinary Veronese tyrant Ezzelino III da Romano, d. 1259). In fact, both coronations fit within a medieval tradition of academic exercises in European universities, which neither Vergil nor other ancients would have understood.[26] But if we look beyond or around Petrarch's massive ego, we see that he was also celebrating the mystique and the genuine power of writing.

Petrarch has been called a "pre-humanist" or "proto-humanist" by scholars who recognize that the fifteenth-century scholars and poets he inspired came closer than he to reviving ancient Latin culture. The historian Ronald Witt describes five successive generations of early Italian humanists, with Petrarch belonging to the third, Mussato to the second, and Mussato's mentor, Lovato de' Lovati (d. 1309), as arguably the "father of humanism." Lovato reintroduced several Roman authors to his contemporaries and championed works already familiar—in particular, the tragedies of Seneca, Mussato's model for *Ecerinis*. In this sense Padua, rather than Florence, probably merits the title "cradle of humanism." Witt stresses that these early Paduan intellectuals concentrated on reviving classical Latin language and verse forms at a time when Florentines, foremost among them Dante, were cultivating antiquity largely through vernacular translations and adaptations of Roman works. Petrarch was a bit late to this party, as well as the one hosted by Dante. He would doubtless be chagrined to learn of his other coronation, shared with Dante and Boccaccio, as one of the "Three Crowns of Florence."[27]

Coluccio Salutati (d. 1406) and Leonardo Bruni (d. 1444) were fourth and fifth generation humanists, respectively. Although neither was a Florentine, they became chancellors of the Florentine Republic, and this civic engagement brought them closer in spirit to the Paduans than to Petrarch, a "secular monk" who eschewed public service (though not politics) to cultivate his muses. Salutati was responsible for bringing Manuel Chrysoloras (d. 1415) from Constantinople in 1397 as the first long-term teacher of Greek, making Florence the Western pioneer in Greek studies. Witt remarks that, like Lovato and Mussato, Salutati and Bruni worried less than Petrarch and Boccaccio about incompatibility between pagan literature and Christian religion. But Petrarch's problematic quest to reconcile the two strains left a deep mark on Florentine intellectuals, whereas the intense cultivation of Aristotelian philosophy at Padua's university regularly produced alarmingly secular humanists, so faithful to philosophical inquiry that some later Paduan professors were denounced as "atheists."

BOOK HUNTING

In fairness to Petrarch, his self-promotion inspired others to emulate him, not least by recovering and copying manuscripts of Roman classics. His discoveries at Liège in 1333 and Verona in 1345 added two previously unknown orations of Cicero, as well as neglected personal letters from Cicero to his friends. The letters gave medieval Europe its first glimpse of the Roman statesman-philosopher as a private individual, making him a model for Petrarch and his admirers. A conscientious editor, Petrarch was the first scholar since antiquity to attempt reassembling Livy's vast but mutilated history of Rome. Petrarch's surviving personal copy, which he edited from manuscripts in France, contains thirty of Livy's first forty books; Petrarch's corrections and emendations became the basis for future editions, and very little of Livy was recovered after him.[28]

Boccaccio, a younger contemporary of Petrarch and Richard of Bury, shared their pessimism about the survival of ancient writings. A protégé tells how Boccaccio once visited a singularly renowned monastery near Rome. Founded by Saint Benedict (d. ca. 547), Monte Cassino was the mother house of the Benedictine order, and the birthplace of its centuries-long mission as transcriber and preserver of manuscripts. Boccaccio was eager to visit the convent, which he imagined as a sacred shrine to learning. But when he went there and asked to visit the library, a surly monk snapped that it was open and he should go on in by himself. After climbing a steep staircase, Boccaccio

> found the place of such great treasure without door or key, and as he entered he saw weeds growing through the windows and all the books and tables thick with dust. Marveling, he began to open and turn over one book after another, and he found there many different volumes of ancient and exotic works. From some of them several gatherings had been removed; from others the edges of the pages had been cut away; and thus they were mutilated in many ways. At last, he went away grieving and in tears, regretting that the toil and effort of so many famous intellects had come into the hands of such corrupt and wasteful men. Running into the cloister, he found a monk and asked him why those precious books had been so foully mutilated. He replied that some monks, hoping to make a few *soldi*, would scrape [the writing] off a gathering and make cheap psalters to sell to boys, and that they made gospels and [charms] [*brevia*] out of the margins to sell to women.[29]

The most recent teller of this tale confirms that "Monte Cassino in the fourteenth century was no longer the bustling intellectual and religious center it had been some two hundred years earlier when . . . manuscripts were produced in the scriptorium."[30] It is unclear how far the anecdote exaggerated the monastery's decline, but it aptly illustrates the emotional disposition that Petrarch and Boccaccio bequeathed to their early humanist heirs.

Like Petrarch, Boccaccio claimed to be prouder of his Latin scholarship than of his vernacular Italian works, a judgment rarely ever approved by admirers of the *Decameron*. Boccaccio's massive, erudite Latin treatise, *Genealogies of the Pagan Gods*, discussed pagan mythology as a euhemeristic allegory (a genre that treated pagan gods as primeval men and women, divinized for their superhuman accomplishments). Whereas Richard of Bury lamented lost primeval wisdom, and Petrarch the deprivation of personal contact with great minds, Boccaccio grieved more generally for the loss of human history. He eloquently described the obstacles a fourteenth-century scholar faced when reconstructing histories so old that their heroes were mistaken for gods. Protesting himself unequal to the task, he affected to warn the envoy of his patron, King Robert of Naples, that it was objectively impossible: "Let us, then, suppose that books in Greek and Latin such as you yourself mention, have really existed. Next, consider, please, how many enemies these books have had during the passing centuries. Not to cite particular instances, you will admit that a great many collections have perished by fire and flood." In words that resemble Richard of Bury's, he continued: "Even had the Alexandrian Library, which [Ptolemy] Philadelphus long since collected with utmost care, been the only one lost, the loss would have been appalling, since it contained, according to the Ancients, any book you might want."[31]

Boccaccio ambivalently described biblioclasms perpetrated by early Christians, writing that "as the pure and radiant light of Truth drove away the shades of the deadly Gentile error . . . and the messengers of Christ cried out against the doomed religion and drove it to extermination, there is no question that these zealots destroyed ere they died many books replete with material on this subject." In his own times, the consuming desire for wealth, which Petrarch also railed against, was as devastating as zealotry, so that "books dealing with our subject have fallen into disuse, and straightway quickly perished."[32] Historically, many books had been destroyed by "the hatred of princes who conspired against books as against their enemies. The number which perished thus, not merely on mythology, but on various arts, could not be easily computed." The forces of time, however, dwarf any damage humans inflict on books: "If all these enemies had relented, [books] would never have escaped the silent and adamantine tooth of fleeting time, which slowly eats away not books alone, but hardest rocks, and even steel. It has, alas, reduced much of Greek and Latin literature to dust."[33] With a resonance Boccaccio must have intended, his words recall Ovid's phrase *tempus edax rerum*: "O Time, thou great devourer, and thou, envious Age, together you destroy all things; and, slowly gnawing with your teeth, you finally consume all things in lingering death!"[34]

To some extent, Boccaccio's pessimism was rhetorical. But he was a far better poet of obliteration than Ralph Glaber or Richard of Bury.

To enhance readers' appreciation of his philological prowess, he chose the image of a shipwreck: "I will find and gather, like fragments of a mighty wreck, strewn on some vast shore, the relics of the Gentile gods. These relics, scattered through almost infinite volumes, shrunk with age, half-consumed, well-nigh a blank, I will bring into such single genealogical order as I can."[35] Recalling Petrarch's metaphor of the literary corpus as corpse, Boccaccio protested that his patron must not "expect, even after great outlay of time and midnight oil, that a work of this sort will have a body of perfect proportion. It will alas, be maimed—not, I hope in too many members—and . . . distorted, shrunken and warped." Only an Aesculapius, the mythical resuscitator of dismembered Hippolytus, could "collect these fragments, hither and yon, and fit them together." Given a subject "torn limb from limb and scattered among the rough and desert places of antiquity and the thorns of hate, sunk almost to ashes," only a godlike author could "bring to light and life again minds long since removed in death."[36]

The gloomy obituaries for writing and cultural memory expressed by Richard, Petrarch, and Boccaccio certainly reflect their calamitous century, wracked by the Great Famine, the Black Death, and the Hundred Years' War. Ironically, the increasing number of manuscript discoveries in their lifetimes probably contributed to their pessimism: the more works scholars rediscovered, the clearer it seemed that, to paraphrase Boccaccio, they were the flotsam and jetsam of a great shipwreck. Petrarch, in particular, may owe his fame as "father of the Renaissance" as much to the emotional impact of his pessimism on subsequent generations, as to their appreciation of his actual literary salvage. Much more than his irritating self-regard about leading the revival of ancient literary standards, the sorrow expressed in his letters to the Ancients merits an honored page in the emotional history of writing. His attitude had a notable impact on the best scholars of the following generations. At the end of the fifteenth century, Angelo Poliziano (Politian) showed how durable the imagery of Petrarch, Boccaccio, and their early followers remained. Lecturing on Quintilian to Florentine students, Poliziano asked, "But what's the use of rehearsing the calamities of past times? It's impossible to do so reverently without enormous pain, since it was then that those excellent

PLATE 1. The Flood. Tablet 11 of the epic of *Gilgamesh*. 7th c. BCE. Neo-Assyrian. Nineveh, Iraq. ID 00396940001.

PLATE 2. Erasable writing tablet (ivory and wax). Ca. 650 BCE.
Early Etruscan alphabet incised on frame.
Reproduced by permission of Museo Archeologico Nazionale di Firenze (Direzione regionale Musei della Toscana), inv. n. 93480. Any further reproduction expressly forbidden.

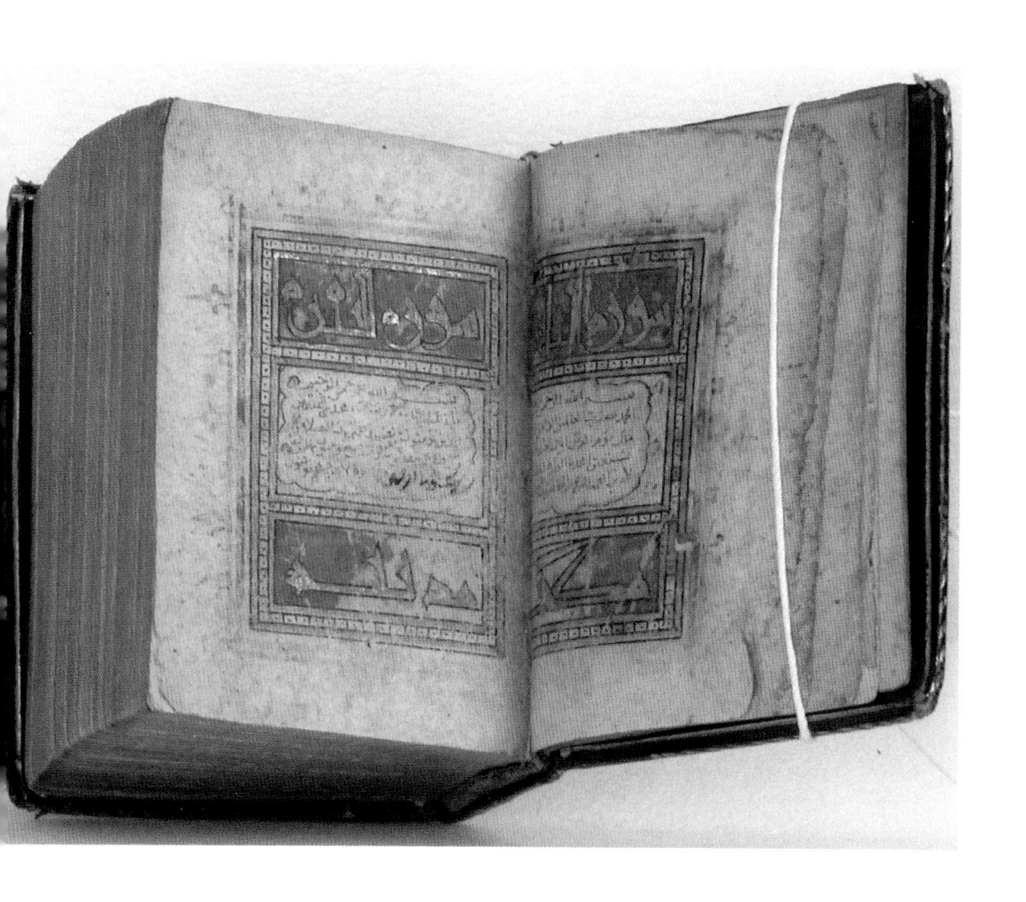

PLATE 3. The Qur'an. Manuscript. Probably 14th c. CE. Metropolitan Museum of Art, New York, Gift of Joseph W. Drexel, 1889.

PLATE 4. "Deuteronomy 5.1–33. Moses explains the Law of God to the Israelites" in front of the Tabernacle, although the Tabernacle is not mentioned in the text. Moses appears to read from a paper scroll, yet the only textual reference (v. 19; Vulgate v. 22) is to the "two tablets of

HEILIGE TAFEREELEN.

HET EERSTE TAFEREEL

Van Moſes Vijfde Boek, genoemt Deuteronomium.

Iet te onregt word het Vijfde Boek van Moſes het Verhaal der Woorden genoemt: om dat de voornaamſte wetten en geboden, met bijdoening van énige andere regten en inzettingen, voor die genen de welke, toen de Wet op Sinai wierd gegeven, of nog niet waren geboren, of om hunne jongkheid niet verſtonden, in het zelve zijn begrepen. Zulk een goede zaak mogt niet alleen andermaal worden verhaalt, maar kon ook voor de óren van een ſtug volk, dat telkens den onweg op wilde, niet te dikwils herzegt worden. Hierom riep Moſes gantſch Iſraël bij een, de hoofden der ſtammen, de oudſten, de amptlieden, en alle man. Aanſtonds openbaarde zig een wonderwerk, dat 's Mans ſtem van duizend duizenden wierd gehoort, zo als door de kragt Gods, het volk ten gevalle, meermaal was geſchied. De Heere onſe God, zeide Moſes, heeft op Horeb met ons een Verbond gemaakt. Met onſe vaderen (te weten die in Egypten zijn geſtorven) heeft de Heere dit Verbond niet gemaakt. Want de vaderen hadden voorheen het Verbond des geloofs, ende van den Meſſias; maar niet van de Wet, die na vierhonderd en dartig jaren, van Abrahams roepinge uit Ur der Chaldeen, is gekomen. Te regt dan zegt de Godstolk, met ons, die voor Sinai hun twintigſte jaar niet hadden bereikt, en tegenwoordig zijn en leven. En hij noemt de voorwaarden, in de Tien Woorden beſchreven, het Verbond; om dat God in de Wet voorſtelt, waar in het Verbond tuſſchen Hem en zijn volk beſtaat. Van aangezigt tot aangezigt heeft de Heere geſproken, dat is, duidelijk, klaar, gemeenzaam, en zonder tuſſchenbode, met een onderſcheidentlijk begrijp en gehoor van woorden. Bij welke gelegentheid Moſes zijn ampt als Middelaar ontvouwt. En daar op de Wet, op den berg, uit het midden van het vuir, der wolke, ende der donkerheid, aan de gantſche ménigte gegeven,

Nn 2 ver-

stone, which He gave to me" (in Exodus 31:18). *Taferelen der Voornaamste Geschiedenissen van het Oude en Nieuwe Testament.* The Hague: P. de Hondt, 1728. Illustration by Gerard Hoet.
Sheridan Libraries Special Collections, Johns Hopkins University.

PLATE 5. "Exodus 40.17–19. Erection of the Tabernacle, showing the sacred vessels." The Ark of the Covenant with its golden cherubim is in the foreground. *Taferelen der Voornaamste Geschiedenissen van het Oude en Nieuwe Testament*. The Hague: P. de Hondt, 1728. Illustration by Gerard Hoet. Sheridan Libraries Special Collections, Johns Hopkins University.

RIGITUR TABERNACULUM, CUM VASIBUS SACRIS.
n dresse le Tabernacle, avec les vaisseaux Sacrez.
et oprechten des Tabernakels met de heilige Vaten en't gereedschap. *J. de Later fecit.*

PLATE 6. "Genesis 5:24. God takes up Enoch." *Taferelen der Voornaamste Geschiedenissen van het Oude en Nieuwe Testament*. The Hague: P. de Hondt, 1728. Illustration by Gerard Hoet.

Sheridan Libraries Special Collections, Johns Hopkins University.

HET VI. TAFEREEL.

Dit tafereel vertoont ons eene godtsdienstige vergadering van heiligen der eerste werelt, die, uit Seth, Adams zoone en Abels plaetsvolger, herkomstigh, en ten tyde van Seths zoone Enos met den titel van Godts kinderen, in spyt en tot onderscheit der Kainyten, van hooger hant vereert, ook den naem huns hemelschen Vaders, van hun in hun gansche leeven en bedryf opentlyk en heimelyk beleeden, in openbaere en plechtige byeenkomsten begosten aen te roepen. Deeze, van allerlei sexe en ouderdom, bezigh met hunne slachtofferande, tot eene dagelyksche verversching en betuiging van hun gelove op het beloofde en verhoopte Vrouwenzaet ten altaere gebraght, vertoonen alle, hunne oogen en handen om stryt ten hemel heffende, in dit gelaet, in deeze gebaerden, de blyken eener zeldzaeme aendacht en wonderlyke verbaestheit. Iemant naemelyk van hun godtvruchtigh gezelschap, de aenvoerder van hunnen offerrei, een eerwaerdige gryzaert, wordt onder het oeffenen der gewyde offerdiensten niet alleen van de aerde, waer mede zyn lichaem zoo naeuw eene verwantschap heeft, in de lucht opgehangen, maer vaert ook met en door de opklimmende vlammen en zweevende wolken ten hemel; en (zoo veel men bespeuren kan) zonder eenige beroerniffe over dit nooit gehoort en nooit beleeft wonderwerk, dat, aen hem gewrocht, allen den aenschouweren en getuigen, niet hem zelven, eenen rechtvaerdigen en heiligen schrik aenjaegt. Dees is Enoch, de zevende van Adam, gelyk hem de godtspraek volgens de aertsvaderlyke geslachtrekening noemt; een man, die, uit kracht van zynen naeme Gode gewydt en toegeheiligt, van kintsbeen aen en in zyne jongkheit meer om de altaeren, dan om de wiegen plagh te hengelen, en in eenen volwassenen ouderdom, gelyk een zoon met zynen vader of een vrient met zynen vrient, hant aen hant wandelde met dien Godt, die op de altaeren zyne vier- en haertstede had; zonder rusten, zonder ophouden arbeidende, om den Godt des verbonts door alle behaegelyke godtsplichten te eeren, te dienen, te believen. Uit welken hoofde de opperste en onfaelbaere Waerheit deezen helt en zyne verkeeringe, waer door hy zynen Heere behaegde of poogde te behaegen, met recht de waerachtige getuigenis gegeeven heeft, die niemant onthouden wordt, dien het gevalt zynen geheelen leevens loop naer het bevel, na ingenomen raet, en onder de achtbaerheit des Aertsrechters te regelen. O zalige wandel in dien wegh, in wien zich de Hoogste aen den zondaer openbaert, in het gelove des beloofden Verlossers, hoe lang hebt gy geduurt? Schoon misschien uw omkring niet van volle drie hondert en vyf en zestigh jaeren geweest is, nochtans zyn de meeste dagen van dit groote jaeren jaer daer in doorgebraght. Immers geduurende de leste klonk alomme door velden en wouden Enochs predikaetsi van de eeuwige gerechtigheit, en zyne profesy van het toekomstige oordeel. En het lustte deezen onbesprookenen en onbevlekten hemeling, op een bedorven aerdryk naer de wyze der zoete visschen in zou-

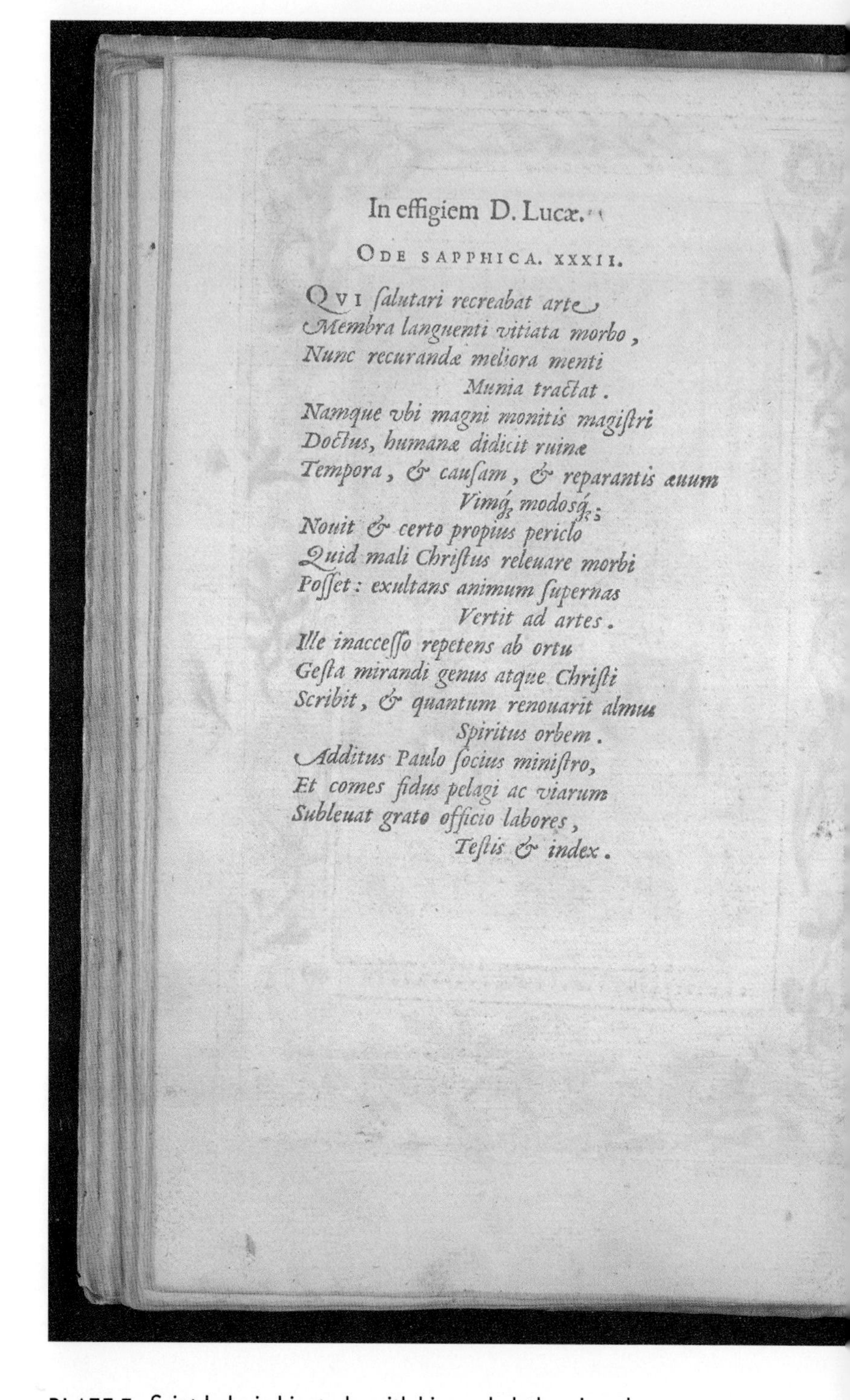
In effigiem D. Lucæ.

ODE SAPPHICA. XXXII.

QVI ſalutari recreabat arte
Membra languenti vitiata morbo,
Nunc recurandæ meliora menti
Munia tractat.
Namque vbi magni monitis magiſtri
Doctus, humanæ didicit ruinæ
Tempora, & cauſam, & reparantis æuum
Vimq; modoſq;:
Nouit & certo propius periclo
Quid mali Chriſtus releuare morbi
Poſſet: exultans animum ſupernas
Vertit ad artes.
Ille inacceſſo repetens ab ortu
Geſta mirandi genus atque Chriſti
Scribit, & quantum renouarit almus
Spiritus orbem.
Additus Paulo ſocius miniſtro,
Et comes fidus pelagi ac viarum
Subleuat grato officio labores,
Teſtis & index.

PLATE 7. Saint Luke in his study, with his symbol, the winged ox. Benito Arias Montano, *Humanae salutis monumenta* (Antwerp: Christopher Plantin, 1571).
Sheridan Libraries Special Collections, Johns Hopkins University, PA8457.A4 1571.

LVX EXPEDITA.
Vsus erit magnæ mentis linguæq; disertæ
Optimus, in sacros incubuisse libros.
P.B.
IH.W.
VERITATIS EXPOSITORI DISERTO. P.
HSD

PLATE 8. A famous palimpsest. Cicero, *De re publica*. Vat. Lat. 5757. 4th- or 5th-c. manuscript, overwritten in the 7th c. with commentary on the Psalms by Saint Augustine of Hippo. Discovered by Vatican Librarian Cardinal Angelo Mai in 1819. Discoloration caused by Mai's chemical reagents.

Public domain, Wikimedia Commons.

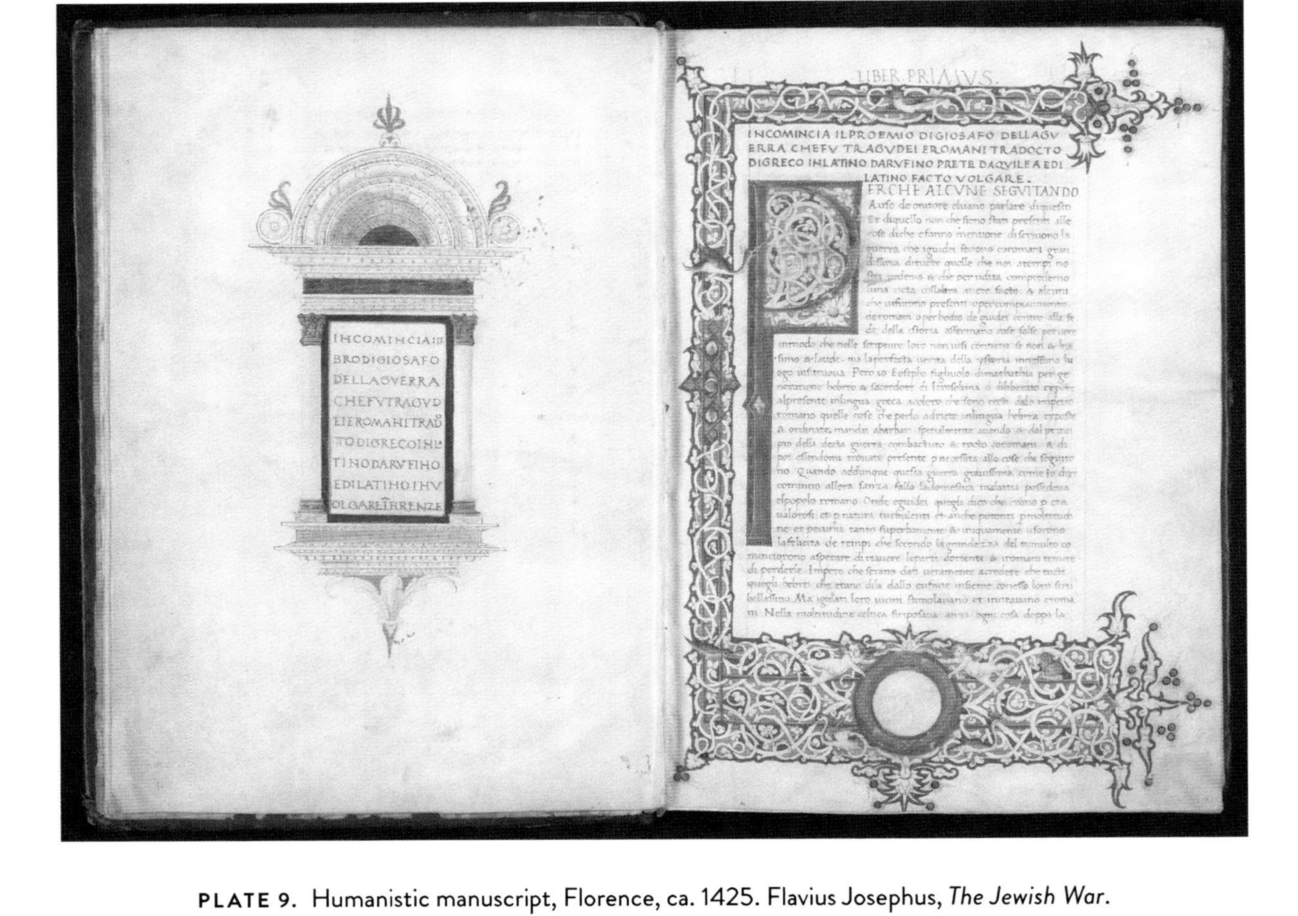

PLATE 9. Humanistic manuscript, Florence, ca. 1425. Flavius Josephus, *The Jewish War*.
"Translated from Greek into Latin by Rufinus, a Priest of Aquileia, and from Latin into the [Italian] vernacular."
Johns Hopkins University Garrett Library, Gar. 15 c. 1.

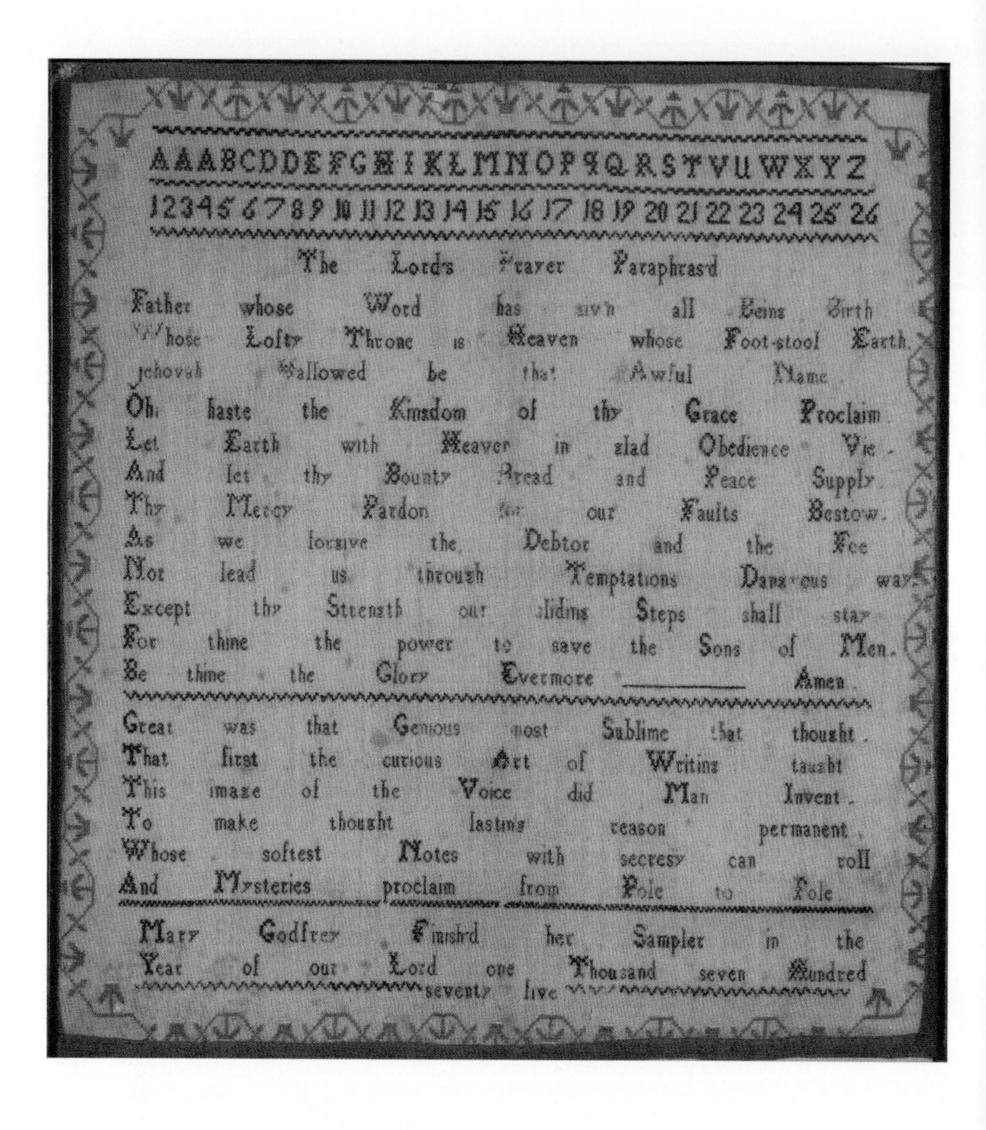

PLATE 10. Mary Godfrey's needlework sampler. England, 1775.
Sheridan Libraries Special Collections, Johns Hopkins University, Bib #8687493.

PLATE 11. Sequoyah poses with his Cherokee syllabary.
Oil portrait, ca. 1830. Henry Inman (1801–1846) after Charles Bird King (1785–1862). Smithsonian National Portrait Gallery.
Photograph by cliff1066, CC Attribution 2.0, Wikimedia Commons.

PLATE 12. Inscription of Darius I at Behistun, Iran (6th–5th c. BCE). Bisotun.
Photograph by Hara1603, public domain, Wikimedia Commons.

writers, the worthiest of immortality, were in part lost and in part evilly corrupted by the barbarians. A whole troop of writers came to our fathers as if thrown in prison and kept in chains, until eventually, with great effort, some of them returned, but lacerated and truncated, and oh so different from themselves, to this their homeland."[37]

On the other hand, Petrarch's heirs infused a fresh, though less explicit, strain of optimism into that emotional history. To take the most basic example, the typeface we call "Roman" was partially modeled on the handwriting of the early humanist Poggio Bracciolini, while "Italic" derives from the hand of his friend Niccolò Niccoli. Living in Florence before printing, at the turn of the fifteenth century, they perfected Petrarch's attempt to recreate the ancients' handwriting as well as their literature. Although the bookhands these humanists chose as models were actually produced in the age of Charlemagne rather than—as they thought—of Augustus, the misidentified *littera antiqua* was clearer and easier on the eye than the spiky, quirky "Gothic" hands of medieval Europe. The visionary humanist printer Aldo Manuzio of Venice (d. 1515) adapted Niccoli's compact "Italic" hand for his typefaces, reducing the amount of paper needed, and his smaller-format "pocket-book" editions made the classics accessible, in both quantity and cost, to students and impecunious scholars. The two hands, "Roman" and "Italic," eventually became the visual signature of the printing press for Western Europeans.

Poggio and Niccoli's interests did not stop at calligraphy. Together, they became the most efficient book-hunting team of their age, Poggio racing to locate neglected or forgotten classical Roman texts and transcribe them before the last few copies (occasionally only a single exemplar, such as Boccaccio's discovery of Tacitus's *Annals*) succumbed to obliteration. Travelling with the papal retinue to the Church Council of Constance, which was negotiating questions of heresy and Church governance, Poggio made forays into Swiss monasteries between 1415 and 1417, where he was allowed to rummage through library collections in search of rare or unique copies of classical texts. These he often transcribed personally, sending copies back to Florence, where Niccoli awaited them impatiently and then publicized them among intellectuals. Niccoli, who wrote very little of his own, worshiped at the altar of antiquity, even eating from ancient crockery. The Florentine Vespasiano da Bisticci (d. 1498), whose thriving secular scriptorium furnished luxury manuscripts to the noble intelligentsia of Europe, thought it a pleasure just to watch Niccoli eat, looking "just like one of the Ancients." Niccoli even had bookcases built to house the manuscripts that Boccaccio had bequeathed to the monastery of Santo Spirito in Florence.[38]

Poggio's efforts to preserve written culture extended to content as well as script, for he was an early translator of Greek classics into Latin. One of his most ambitious projects was translating Diodorus of Sicily's *Bibliotheca,* that encyclopedia of ancient writing lore. Poggio and Niccoli were important heirs of Petrarch, but they were far from alone. Leonardo Bruni, reputed to be the best scholar of Greek in his time, was perhaps the most thoughtful of the early Florentine humanists who considered the history of writing and culture. His *Dialogues for Pier Paolo Vergerio* paint a vivid portrait of Coluccio Salutati (a humanist in the generation after Petrarch), Salutati's disciples (including Niccoli), and their triumphs over obstacles to rediscovery.[39]

Fourteenth- and fifteenth-century humanists inaugurated a massive change in discussions of writing, foreshadowing philological discoveries that would be made by much later scholars. As their rediscoveries increased, humanists emulated the ancients' interest in the mundane history of writing materials and practices, which had faded in the centuries between Isidore and Dante. Humanists' interest in secular history and politics inspired a return to Pliny's preoccupation with the *human* history of writing rather than its mystical symbolism or metaphorics. In 1416, Cencio de' Rustici, who accompanied Poggio on his book hunts, ecstatically described their discoveries in the library at Saint Gall. Among a number of recognizable classics, they found "one book made of the bark of trees; some barks in the Latin language are called 'libri,' and from that, according to Jerome, books got their name." Like many scholars as late as the seventeenth century, Cencio was unfamiliar with papyrus, confusing it with *libri arborum.* Although the volume was "filled to overflowing with writings which were not exactly literature," Cencio's mistaken identification inspired him to quasi-religious piety: despite its negligible contents, "because of its pure and holy antiquity I greeted it with the utmost devotion."[40] Even when recording banalities, writing could display the aura of scripture.

Echoing Boccaccio's outrage at Monte Cassino, Cencio railed that at Saint Gall, "countless books were kept like captives and the library neglected and infested with dust, worms, soot, and all the things associated with the destruction of books," so that "we all burst into tears." He imagined that "if this library could speak for itself," it would protest, "Let me not be utterly destroyed by this woeful neglect. Snatch me from this prison in whose gloom even the bright light of the books within cannot be seen." It reminded him of all the "innumerable libraries of Latin and Greek books in ruins," a crime as great as "demolishing the Amphitheater or the Hippodrome or the Colosseum or statues," artifacts that embodied "that old and almost divine power and dignity" of Roman civilization. Less inhibited than Boccaccio, he blamed "these priests of our religion" for the neglect and destruction of books and monuments, declaiming "let us pursue such inhuman, such savage stupidity with curses."[41] Although the history of writing would remain largely mythical until the age of Champollion, these early humanists did not consider writing a symbol of anything but itself. Its mystique was in no sense a metaphor for mystical encounters with the godhead.

A SWASHBUCKLING SCHOLAR: GIOVANNI PICO DELLA MIRANDOLA

But the romantic, fantastic ideas about writing that left humanists like Poggio and Poliziano unmoved were far from dead. Richard of Bury would have been entranced had he lived long enough to meet Giovanni Pico della Mirandola (d. 1494). Famed (and occasionally infamous) from the moment he appeared on the Renaissance scene, this hereditary prince of a small state in northeastern Italy, a kinsman of other prominent nobles and intellectuals, was both very wealthy and on fire to discover all the most arcane ancient secrets lurking in the most exotic languages. Intellectually, he was the heir of Richard and the prince of those enthusiasts who asserted that "an ancient, pre-Platonic religious tradition" of pre-Christian and non-Jewish luminaries, including Orpheus, Pythagoras, and Plato himself, had foreshadowed the fundamental Christian dogmas, notably monotheism, the trinity, and the creation as recounted in Genesis. The list frequently included the Jewish patriarchs down to Noah, Zoroaster, Hermes Trismegistus, the Sibyls, and others still.[42] Between the ages of twenty-one and twenty-four (late 1484 to early 1487), Pico lived a legendarily rich life. First, he met the luminaries of Florentine intellectual life: the de facto ruler Lorenzo de' Medici, Marsilio Ficino—the Latin translator of Plato's complete works—and Poliziano, who was the most accomplished classicist of his day.[43] Strangely (or not), learning was not Pico's sole passion: he soon lusted after the wife of a Medici kinsman, abducted or eloped with her, and was practically killed when her husband's retainers captured them. While recuperating from his injuries, he planned what should have been his major intellectual coup.

He had decided that Plato and Aristotle were in fundamental agreement with each other, and moreover that they corroborated the truth of Christianity, in company with most of the legendary sages mentioned above. In this elite fraternity he even included the Jewish Talmudists and Kabbalists, whom Christians traditionally considered so inimical to Christ that they attacked and burned the Talmud, and sometimes the Talmudists themselves. Ablaze with bibliophilic enthusiasm and blessed with abundant wealth, Pico hired important philosophical and theological tutors to supplement his university education at Bologna and Padua.

In late 1486 he published one of the most remarkable documents in European history; following the tradition of university debates, he proposed to defend no fewer than nine hundred theses drawn from ancient philosophy, Christian theology, and Kabbalah. He would prove publicly, in Rome, that the entire illustrious panoply supported the truth of Christianity. He scheduled the event for January 6, 1487, the feast of the Epiphany, when the Magi, those Persian wisemen, bowed before the Christ child in recognition of his divinity. Pico offered to pay the travel expenses of anyone who cared to debate with him.[44]

Rarely has naive adolescent arrogance been so perfectly matched with fervent, productive study. Rarely, too, has it blossomed at such an inopportune time and place. Unfortunately, Pope Innocent VIII, on whose hospitality Pico presumed, was in no mood for such shenanigans. This was the pope who decided that witchcraft, that is, demon-worship and harmful sorcery, was running roughshod over Christendom. In his bull *Summis desiderantes*, Innocent endorsed the paranoid fantasies of the *Malleus maleficarum*, or *Hammer of Witches*, published in 1486. The promulgation of the bull and the initial sales of the *Malleus* took place in the same interval of time (December 1484–February 1487) as Pico's three great adventures. Among the nine hundred theses, Pico included several about magic, construing it as testimony to Christian truth. Thus, although he eschewed demon-worship and embraced "natural" or protoscientific magic, the pope shut down Pico's planned debate. A number of Pico's theses were condemned, and then eventually all of them were. Pico's *Apologia* defending the theses was banned, most copies were confiscated and destroyed, and Pico was forced to abjure both the theses and the *Apologia*. Subsequently he fled to France, was imprisoned there, freed by King Charles VIII, and allowed to return to Florence under the supervision of Lorenzo de' Medici. In his last years he turned to Savonarola, the anti-intellectual firebrand preacher who lit bonfires of cultural "vanities" and hypnotized Florence before being burned himself as a heretic in 1498. Pico was preparing to become a friar but was poisoned, along with his philologist friend Poliziano, and died on the very day in 1494 when Charles VIII, who inaugurated the wars that ravaged Italy for another sixty years, entered Florence.

Of Pico's many erudite works, the oration he composed as a prolusion to his aborted debate is still the most read. Known somewhat misleadingly as *Oration on the Dignity of Man*, it is actually a call to the kind of wide-ranging immersion in philosophy and theology that Pico himself undertook. It claims that not only Plato and Aristotle, but Orpheus, Zoroaster, and Hermes Trismegistus[45] should be cultivated, along with Jewish and Islamic sages; this would expose the deep concord between Christianity and other religions, and lead to universal peace. Pico accepted 4 Esdras's story that after the Babylonian captivity, Ezra and his scribes had reconstituted not only the twenty-four books of the Hebrew Bible, but seventy others containing the proper interpretation of the Bible; Pico identified the latter books with Kabbalah, based on God's command that Ezra reveal them only to a few worthies. Like writers since Enmerkar, Pico put little faith in oral transmission, saying that Ezra "clearly realized that, following the exiles, the massacres, the flights and captivity of the people of Israel, the mysteries of this divinely conveyed celestial doctrine would fade into oblivion in the absence of written documents, since they could no longer endure in memory alone."[46]

Pico's enthusiasm inspired a vogue of Christian Kabbalah that lasted into the seventeenth century.[47] If he had a single overriding fixation, it was recovering all such wisdom, whether pagan, Jewish, Christian, or Muslim. Nothing set his blood racing like such intellectual salvage, which, he thought, could return humankind to the Golden Age fabled by the ancients. For him, "ancient," "secret," and "wise" were synonymous. He dared to believe that his own agents had brought him both Esdras's secret wisdom and writings of comparable value from other ancient sages, including "Zamolxis, who was imitated by Abaris the Hyperborean," and Zoroaster, who, Pico cautioned his audience, was "not the one of whom perhaps you are thinking," that is, Ham the son of Noah, notorious among churchmen for centuries, but instead "the son of Oromasius."[48]

REDISCOVERED SCRIPTURES? HOAXES AND FORGERIES

The polar opposite and evil twin of *scriptura* is literary forgery. One is sacred because "It is written," the other despised for unconvincingly claiming the same privilege. Despite being written, a forgery is not considered *sacer*, not worthy to be "set apart" from other writings, except as an abomination. Both categories reflect a value judgment ultimately founded on emotion. No argument that God or Shakespeare did or did not author a particular text will ever be accepted by everyone. Modern scientific methods can only determine whether the material support of a text—ink, paper, papyrus, stone, or other materials—is probably authentic. (This happened recently when the scrap of parchment inscribed "Jesus said 'my wife'" was proved to be ancient, while the ink—and thus the text—was modern.) The same applies to handwriting and signatures. A written document must be judged by the same fallible sensory criteria that determine whether currency and works of art are counterfeit.

But this is only the *documentary* level of forgery. At the level of *text*, a forgery, like any other written work, is a discrete body of words that can occur on any material support (including, since the nineteenth century, photography, Morse telegraphy, sound recordings, and electronic media). Ultimately, only philology can distinguish between authentic and forged *texts*, by reference to the history of a language—its vocabulary, stylistics, rhetorical gambits, and relation to other texts. Yet philology is no infallible guide either: even if we discount scholars' emotional investments, a genuine text could show numerous exceptions to philological norms that a forger might cannily respect.

The fifteenth century was the first great age of literary forgery, the time when the problem of authenticity first received systematic attention. As Cencio de' Rustici illustrates, fifteenth-century humanists valued texts according to their antiquity, for they knew how easily materials degraded, how often texts needed recopying, and how insidiously words could be miscopied or falsified. So their controversies overwhelmingly regarded textual rather than documentary forgeries. The most famous early modern exposé of a forgery is Lorenzo Valla's devastating attack on the "Donation of Constantine" (1440). The "Donation" purported to record the Emperor Constantine's gift of the entire western half of the Roman Empire to Pope Sylvester I when the emperor moved his capital city from Rome to Constantinople in 330 CE.

Although Valla was not the first to question the document's authenticity, his exposé was a model of deconstructive thoroughness.[49] His proofs of the document's fraudulence ran the gamut from psychology to linguistic history. He argued the following points: (1) The Emperor Constantine was not the kind of man "to want to make a donation" of such magnitude and responsibility to Pope Sylvester, nor did the pope have good reason to accept the gift. (2) Neither man was "in a legal position" to transact the donation: their hands were tied by Roman law. (3) "Nothing was given by Constantine to Sylvester" after the emperor's conversion to Christianity; instead, history showed that Constantine made "modest gifts of places" to Sylvester's predecessor. (4) Claims that Constantine's decrees and Sylvester's official biography originally contained the text of the "Donation" were demonstrably false. (5) "Abundant evidence," including "donations of certain other emperors," demonstrated that "if Sylvester had ever taken possession" of the Empire, "once he or some other pope had been deprived of it" by later emperors, "after so great an interval of time it could not be recovered by any legal claim" in secular or canon law. (6) The pope's current possessions, the "Papal States" in Italy, were created far too late to fall under Sylvester's authority.

Valla built his seventh proof on the properties of writing materials. He recalled that Judas Maccabaeus proposed to Roman authorities a treaty "incised on bronze"; Valla added slyly, "I say nothing about the stone tablets of the Decalogue, which God gave to Moses."[50] The "Donation," by contrast, claims proof "by no document at all" on gold, silver, bronze, marble, "or, finally, in books, but only," according to its own text, a single manuscript. Exploiting Christian fondness for Josephus's Sethian columns, Valla paraphrased the tale. "Since the opinion was handed down from generation to generation that human achievements were to be destroyed at one time by water and at another by fire, Jobal, the inventor of music (according to Josephus), inscribed his teaching on two columns—one in brick against fire, the other in stone against water, both of which lasted, by his own account, to Josephus' own time—in order that his benefaction to mankind should last forever."[51] Interestingly for a humanist concerned with sources, Valla cited the contaminated ninth-century version rather than Josephus's original. Was this an uncharacteristic flub or a convoluted jab at the Church for honoring the "Donation," which Valla revealed as an early-medieval clerical forgery?

Valla promised to consider only "ecclesiastical examples, rather than secular ones" because "my argument is with priests and not with laymen." But he did mention ancient Roman law. "Even among the Romans," he said, "when they were rustic and uncultivated, when there was slight and scarce literacy, the laws of the Twelve Tables were nonetheless incised on bronze," so that "after the city was captured and burnt by the Gauls, they were subsequently discovered intact."

(Modern scholars disagree.) All Valla's examples proved how "prudent foresight overcome[s] the two greatest forces in human affairs, the length of time and the violence of fortune." Given such precedents, "did Constantine really sign a donation of the world only on papyrus and with ink?"[52] Wouldn't that have been foolhardy?

All these critiques relied on external criteria, whether legal, historical, or psychological. But the proofs that assured Valla's lasting reputation were internal. The text overflowed, he said, with "contradictions, impossibilities, stupidities, barbarisms, and absurdities."[53] Deriding, on page after page, the "Donation's" anachronistic and misapplied terminology, Valla asks, "Does not that barbarous way of talking attest that this nonsense was not concocted in the age of Constantine but later?"[54] Without mentioning Numa or Dictys, he demolishes the "Donation's" claim that its original was entombed with "the venerable body of the blessed Peter." Valla scoffs that "because the Donation of Constantine cannot be exhibited, [the forger] has therefore said that the grant is not on bronze tablets but on paper sheets." The original was supposedly "hidden with the body of the most holy [Peter], to keep us from boldly looking for it in the venerable tomb, or, if we were to do so, we would suppose that it had been ruined by decay."[55]

The "Donation" committed a number of what Susan Stewart calls "crimes of writing."[56] But for Valla, it seriously aggravated its historical and religious trespasses by mutilating the Latin language: "May God destroy you, wickedest of mortals, for ascribing barbarous speech to an age of learning."[57] Like other early Italian humanists, Valla expressed a patriotic, even religious veneration for classical Latin; he went so far as calling it "a great sacrament" in another landmark treatise: *magnum latini sermonis sacramentum*. Ignominiously, he said, the peoples of Italy "lost Rome, we lost rule, we lost power."[58] Since then, Latin had sunk to "the same condition that Rome was in when occupied by the Gauls." Owing to medieval neglect, "everything is overturned, burned up" such that "no one has spoken Latin in centuries, nor do they understand when they read it."[59] Like Italic blood, Latin eloquence had been diluted by invading Goths and Vandals, who left "a great quantity of codices in Gothic characters." Handwriting and language reflect each other. "So if these peoples were able to corrupt Latin *scriptura*, what must we think of their language? . . . After their invasions, there were no more eloquent writers" because "whereas the ancients mixed their language with Greek, these mixed theirs with Gothic."[60]

Yet Valla remained optimistic and patriotic because "the Roman Empire is wherever the language of Rome rules." Roman hegemony was no longer defined by land, bricks, and marble, so "we should celebrate our age" because "if we strive a bit more, I am confident that we can soon restore, even more than the city of Rome, its language, and with it, all the disciplines."[61] Genuine knowledge depended on a language capable of precision and nuance: Valla exemplified his age's abhorrence of corrupted language and its ambition to repair damaged human institutions by reforming writing. He professed himself a Christian, but for him the meaning of *sacra scriptura* was not limited to the Bible. Because classical Latin was a *sacramentum*, its *scriptura* was also sacred.

THE ARCH-FORGER

Ironically, by sensitizing their contemporaries to the mechanisms of forgery, Valla and other humanist debunkers taught scholarly malcontents how to claim that such biblioclasms as Richard of Bury had deplored, along with malicious campaigns of suppression and erasure, had bequeathed to early modern Europe an extensively or totally false picture of previous times. The ambition to "set straight" the historical record drove scholars with a historical or cultural axe to grind to create far more sophisticated falsifications than the eighth-century "Donation of Constantine." In the first century of printing, a generalized appreciation, even a reverence for writing and for the technology that had rescued it "from all threat of oblivion," in one humanist's phrase, replaced the previous century's pessimism. But by vastly increasing the availability of ancient texts, both familiar and rediscovered, printing made possible a veritable flood of literary forgeries that demonstrated an ambition to rewrite history.

The sixteenth century would inaugurate a long tug-of-war between propagandists and more idealistic, humanist-trained scholars, aptly described by Anthony Grafton in *Forgers and Critics*.[62] Annius of Viterbo (Giovanni Nanni, d. 1502), the superstar among forgers, was, like Valla, exceptionally thorough. But he did not share Valla's veneration for the "sacrament" of classical Latin nor for the culture that produced it; his early opponents mocked his Latin as rustic, inelegant, and obtuse. Moreover, they quickly perceived that he had no respect for the mission of history as reputable humanists defined it: for him, historiography was not an instrument of truth, but of partisan propaganda.[63] (This is not to claim, of course, that reputations for scholarly probity never hid religious or cultural biases; Annius was able to exploit and amplify the prejudices of the sources he utilized.)

Annius's complex, programmatic, and deeply cynical forgeries were uniquely appropriate to this age of literary revival. He claimed that a pair of monks had brought him two precious relics of antiquity: the long-lost chronicles of "Berosus Chaldaeus" and "Manetho Aegyptius," the two historians whom Flavius Josephus had praised most for their antiquity and their neutrality. These two books, Annius claimed, were the ultimate historical sources, transported all the way from Armenia, the cradle of reborn civilization, where Noah's Ark landed after the Flood.[64] Annius understood that neither Josephus nor the historical Berossos had proposed a dispassionate, neutral version of cultural history; but he was also impressed by the respect that classical authors accorded Berossos's astronomy and by Christians' enthusiasm for the narrative supplement to Genesis that Josephus provided.

Annius was as ambitious as his contemporary Pico della Mirandola, but he lacked even a shred of Pico's intellectual and religious idealism. With unheard-of audacity, he forged the works of eleven ancient authors, headlined by "Berosus" and "Manetho," and then wrote massive interpretations of them, publishing the collection at Rome in 1498 under the title *Commentaries on Various Authors Who Spoke of Antiquity*. Annius's big book imitated the scholarly format of traditional commentaries on the Bible, the law, and other documents crucial to civilization, but his interpretations grossly distorted key passages in foundational texts (including the Bible) by pairing them with the falsehoods he attributed to "Berosus" and his other pseudo-authors.

Annius was clearly inspired by Godfrey of Viterbo's claim to have read "Berosus" and other long-disappeared authors; he even mentioned Godfrey but never in connection with world history or Josephus's claims about it, thus paying his fellow townsman only a faint and backhanded homage. Going beyond what Josephus and a few other ancients had written, Annius created an intriguing biographical backstory; he claimed that "Berosus" had been the prefect of the Library of Babylon, a fictitious institution that Annius modeled on the Mouseion of Alexandria, the library of Sais in Plato's *Timaeus*, and, perhaps, the Vatican Library, which had expanded impressively since 1450. Pseudo-Berosus himself "revealed" that in the time of Alexander the Great, the Babylonian mega-library had contained myriads of historical documents dating back to the beginning of time. Another of Annius's pseudo-authors disclosed that the library was destroyed soon after "Berosus's" time, tragically denying posterity the chance to consult the invaluable archive firsthand. But luckily, as one Annian pseudo-author divulged, "Berosus" had been a consummate, farsighted bookworm. Before the resources in his charge were obliterated, he had managed to compile a highly concentrated *defloratio*, or abridgment, of all the most important historical data they recorded (a shameless appropriation of Josephus's exaltation, "*Berosus, qui Chaldaica* defloravit"). The *Defloratio Berosi* itself was lost for many centuries, Annius claimed, but he had had the great good fortune to receive it from the two visiting Armenians, who were friars of his own Dominican order.

To imagine the antediluvian sources that "Berosus" allegedly condensed, Annius drew inspiration from both Godfrey's and Peter Comestor's biblically grounded, universal chronicles. Combining their references to the Sethians' antediluvian stelae, crossing them with Josephus's original, and superimposing the Epistle of Jude's quotation from the Book of Enoch, Annius "proved" the traditional assumption that Enoch's generation—when music, tents, and metal-working were also invented—was the most logical era for the creation of the antediluvian stelae, the ultimate, original source of history. "As is clear from chapter five of Genesis, the holy prophet Enoch was born 1034 years before Noah's Flood. The apostle Jude . . . testifies in his canonical Epistle that Enoch prophesied a future Judgment, by means of both . . . flooding and a final conflagration. And Flavius Josephus testifies in the first book of the *Jewish Antiquities* that Enoch wrote these things on two columns, one of bronze and the other of brick."[65] The inaccuracies in this passage exemplify Annius's habit of using Josephus's name to accredit tales that distorted the *Jewish Antiquities*. Unlike whoever wrote the Epistle of Jude, Annius had little interest in Enoch as a prophet, which was the only orthodox way a Christian could still mention him; instead, he needed Enoch to be a historian. He knew from Augustine and other ancient authors that the early Church had destroyed the Books of Enoch, owing to the heretical theology behind their "revelations." Although Annius could not have read "Enoch," he invented a primeval world that rivaled its antediluvian fantasies but with the added advantage of connecting Old Testament history directly to present-day Europe.

Annius's forgeries were a thundering response to Valla's attack on papal power a half-century earlier. He invented a *translatio imperii* (transference of political authority across the ages) that should revindicate the papacy's historic entitlement to absolute power, political as well as spiritual. "Forget the latecomer Constantine," his forgeries argued: "High Priests have ruled over the Roman heartland since the Flood, making popes the rightful heirs of Noah and thus legitimate emperors of the entire world."

As proof of his claims that pre-Roman Italy was the center of world history, he adopted and exaggerated Josephus's attack on Greek civilization as late, derivative, decadent, and based on myths that lacked the provability and authority of written history. Even more vehemently than Josephus, he argued that Greek historians were nothing but a dynasty of chauvinistic forgers. Using the recent Latin translations of Herodotus and Diodorus, he exaggerated their praise of pre-Hellenic civilizations to demonstrate that the glory of the Greeks was based on lies: Greek historians had knowingly appropriated the achievements of far more ancient barbarians such as the Egyptians, turning their heroes, notably Hercules, the primeval civilizer, into *graeculi*, or Greeklings.[66]

To reinforce his fraudulent textual reconstructions of "Berosus" and "Manetho," Annius forged nine other texts by renowned ancient

Jo: Williams

Antiquitatũ variarũ volumina. XVII.

A venerãdo & sacræ theologiæ: & prædicatorii ordĩs ꝓfessore Io. Annio hac serie declarata.

Contentorum in aliis voluminibus. Liber primus.	Fo. I.
Institutionum Annianarum de æquiuocis. Liber. II.	Fo. IX.
Vertumniana Propertii. Liber. III.	Fo. XXIX.
Xenophontis Aequiuoca. Liber. IIII.	Fo. XXXIIII.
Fabii Pictoris de aureo sæculo. Liber. V.	Fo. XLI.
Myrsili. Liber. VI.	Fo. LII.
Catonis fragmentum. Liber. VII.	Fo. LVII.
Itinerarii Antonini fragmentum. Liber. VIII.	Fo. LXXII.
Sempronii de Italia. Liber. IX.	Fo. LXXVI.
Archilochi de temporibus. Liber. X.	Fo. LXXXII.
Metasthenis. Liber. XI. *ie Megasthenis.*	Fo. LXXXIIII.
De Hispanis. Liber. XII.	Fo. LXXXVI.
De Chronographia Etrusca. Liber. XIII.	Fo. XCI.
Philonis. Liber. XIIII.	Fo. XCIIII.
Berosi. Liber. XV.	Fo. CIIII.
Manethonis. Liber. XVI.	Fo. CXLVI.
Anniarum. XL. quæstionum. Liber. XVII.	Fo. CLIII.

N.IV.6.

Venundãtur ab Joanne Paruo & Jodoco Badio.

FIGURE 9.1. Title page with contents of Annius of Viterbo's forged ancient chronicles. 1498, reprint Paris: Jean Petit and Josse Bade, 1515. A reader has changed the name of Annius's Persian pseudo-author "Metasthenes" to "Megasthenes," an attested ancient Greek author (4th–3rd c. BCE), although Annius explicitly distinguished the two.
Johns Hopkins University Libraries, Bibliotheca Fictiva Collection, no. 158.

authors, claiming to have discovered them himself in a manuscript collection compiled by a contemporary of Dante. Despite spanning a millennium and several disparate societies, the eleven texts of Annius's forgeries agreed uncannily well with each other, as he explained in three hundred pages of meticulous commentaries (figure 9.1).

In addition to the books he claimed to rescue from oblivion, Annius also "discovered" a wealth of monumental inscriptions, including "hieroglyphics" that he clumsily assembled from recycled medieval friezes. Annius's picture-writing bore no resemblance to anything genuinely Egyptian; but he and his contemporaries had never seen ancient specimens (compare to figure 9.4 below). Even the obelisks that emperors transported to Rome in antiquity remained buried and neglected for another century after Annius.[67]

Annius's astonishing facsimile of philology used forgeries of both histories and inscriptions for the same mendacious purpose: to demonstrate that little Viterbo, fifty miles north of Rome, had been the Etruscan capital of Noah's theocratic world empire. But the ancient Romans gradually deserted Noah's pious proto-Christian religion, lured away by the idolatry and impious, rationalistic quibbling of Greek philosophy. Optimistically, Annius prophesied that Noachian civilization would soon be restored to glory by none other than his own patrons, Pope Alexander VI, his son Cesare Borgia, and the monarchs Ferdinand and Isabella, who were purging Spain of Muslim "infidelity."

Some contemporaries, motivated by sincere love of the classics and an allegiance to philological neutrality, saw that Annius's "discoveries" were too good to be true and began attacking them shortly after publication.[68] But his forgeries became an enduring propaganda boon to rulers all over Western Europe. The twenty-second and last edition of them was not printed until 1659, and mendacious "histories" based on them were still being published in the eighteenth century. Simultaneously the emperor of the entire world and its first Pontifex Maximus, Noah was the most prestigious ruler imaginable, and the ironclad proofs of Etruscan, Roman, Italian, and papal primacy contained in Annius's "rediscovered" texts and meticulous commentaries made contestation impossible (figure 9.2).

FIGURE 9.2. Early Rome's "Etruscan Quarter." Annius, forgeries. 1498, reprint Paris: Jean Petit and Josse Bade, 1515.
Johns Hopkins University Libraries, Bibliotheca Fictiva Collection, no. 158.

Or so it seemed to Annius, except that dozens of rival pseudo-historians easily saw how to twist the authority of "Berosus Chaldaeus and his commentator Johannes Annius of Viterbo" to exalt patrons ranging from Ferdinand and Isabella to Elizabeth I of England and Louis XIII of France.[69] Annius opened a Pandora's box of fraudulent claims to political hegemony and cultural primacy; no one's genealogist convinced anyone else's, so the sixteenth century and much of the seventeenth were flooded with claims, counterclaims, refutations, and rebuttals, some of them spanning hundreds of pages. None of these books convinced anyone who was not already a committed partisan. Few writers actually believed Annius's solemn folderol, but every political faction steadfastly pretended to and defended its own re-forgeries of "Berosus" and company.

To confuse matters even more, Annius interspersed nuggets of genuine biblical and classical learning in his outlandish travesties of scholarship, though he misinterpreted them shamelessly. He even made positive contributions: despite fraudulently claiming that the mysterious Etruscan language was an older, Noachian dialect of Hebrew, he made a few genuine discoveries. He has been called, without sarcasm, the first Etruscologist for discoveries he made about the alphabet from genuine inscriptions.[70] Anthony Grafton has shown that some of Annius's boasted knowledge of Hebrew is not imaginary, although he derived it indirectly from biblical commentaries in Latin and not, as he claimed, from rabbinical training.[71]

FUN WITH PHILOLOGY: A SERIOUS LARK

Another fin de siècle book is now more renowned than Annius's forgeries, though the two have much in common. In 1499 the *Hypnerotomachia Poliphili* was published at Venice.[72] The rather monstrous title of this bizarre work translates as *Poliphilo's Dream about the Strife of Love*. It had apparently been composed thirty years earlier by a humanist who signed himself "Francesco Colonna" in an acrostic. Bizarre it definitely was, but it was gorgeously printed by Aldus Manutius, the Venetian humanist and great friend of Erasmus. Stuffed with elegant, erudite woodcuts of temples, fountains, coins, and stelae, it has been called the most beautiful book ever printed and has been for centuries a supremely coveted prize of bibliophiles.

For our purposes, though, it is an invaluable witness to what has been called the "romance of antiquity."[73] It documents the fifteenth-century fascination with writing, epigraphic inscriptions, and the monuments of classical antiquity, both architectural and linguistic.[74] Poliphilo's name reflects his love of Polia, a personification of wis-

FIGURE 9.3. Three doors, "crudely excavated in the living rock, a work of the ancients, and old beyond belief," showing "inscriptions in Ionic, Roman, Hebrew, and Arabic characters." *Hypnerotomachia Poliphili*. Venice: Aldus Manutius, 1499. Sheridan Libraries Special Collections, Johns Hopkins University, Incun. 1499 C699 FOLIO c. 1.

FIGURE 9.4. "Hieroglyphic" frieze, *Hypnerotomachia Poliphili*. Venice: Aldus Manutius, 1499. Sheridan Libraries Special Collections, Johns Hopkins University, Incun. 1499 C699 FOLIO c. 1. Translation herein, p. 253.

dom; the multifarious meanings of her name all point to the book's "polyamorous" infatuation with nature, pre-Christian antiquity, and writing itself. Falling asleep, Poliphilo dreams of searching for Polia in a landscape depicted by woodcuts of colossal ancient monuments and isolated or fragmented inscriptions in Latin, Greek, Hebrew, and Arabic (figure 9.3). Among these are fanciful hieroglyphics that resemble Annius's. Like the forger's, Poliphilo's "sacred letters" were not inspired by real monuments but derived from ancient verbal descriptions, popularized among humanists by the *Hieroglyphica* of Horapollo, a text discovered in 1419 and known only in manuscript until 1505. The "hieroglyph" Poliphilo saw was, he says, "ancient and sacred." He translates it into a scripture of nebulous affirmations meaning "from your labour to the god of nature sacrifice freely. Gradually you will make your soul subject to God. He will hold the firm guidance of your life" (figure 9.4).[75] The insipidity of the message was clearly determined by the need to use familiar descriptions or make logical extrapolations from Horapollo.

Like practically everyone before the late seventeenth century, Poliphilo's author assumes that hieroglyphics were not a writing system but a set of arcane pictorial symbols whose significance had to be learned through person-to-person initiation.[76] Thus, wandering through landscapes by turns pastoral and menacing, Poliphilo encounters a variety of personages celebrating rituals of imaginary ancient mystery religions, including worship of the ancient fertility god Priapus, which a woodcut portrays in all his ithyphallic glory. By their abundance, these illustrations allow the reader to experience Poliphilo's adventure as something like a graphic novel. The clear message of the *Hypnerotomachia* is that one can, indeed must, love learning about antiquity with the same erotic ardor that bedazzled men feel for a beautiful woman.

Every aspect of this love is founded on enthusiasm for the writing and languages of the ancients. Technically, the *Hypnerotomachia* is written in Italian, but its idiolect contains so many invented words based on Latin and Greek that it hardly registers as the same language used by Dante or Boccaccio. The book's extreme linguistic hybridization performs the same blurring of boundaries that characterizes every other aspect of it: the *Hypnerotomachia* is suspended between antiquity and modernity, waking reality and nostalgic dream, eros and erudition. A few years earlier, Pico della Mirandola assembled his syncretic vision of "ancient theology" from ideas he found in Latin, Greek, Hebrew, and "Chaldean" (Aramaic) texts; likewise, Poliphilo's dream-world magnifies antiquarian fantasies about rediscovering the aesthetic splendor of Roman, Greek, and even Egyptian artifacts. The *Hypnerotomachia* is graphomaniacal: whatever it happens to narrate, its implicit subject is always writing, not only when Poliphilo encounters enigmatic inscriptions and ruins from antiquity but also because the sheer hedonism of substituting grand multilingual assemblages for everyday Italian words marks his profoundly erudite language as inherently *written*. For centuries, scholars had imagined linguistic and graphic systems, such as allegory and hieroglyphics, that would exclude everyone but an elite few: among these schemes, the Egyptian "symbols" were thought to be the most rigorous instrument for separating privileged inductees from the vulgar, uncomprehending herd. Like allegorical writings, inscription by means of pictures would safeguard ultimate truths from desecration, while initiates could enjoy belonging to a philosophical meritocracy such as Pico della Mirandola imagined for the Kabbalists.

The *Hypnerotomachia* plays with such fantasies of intellectual superiority by basing its language on the rhetorical principle of *contaminatio*, or inappropriate mixing. Words like *perpolito et faberrimo* ("precious and ingenious" or "fine and skilled")[77] collapse historical differences and stretch semantics to entertain scholars enamored of antiquity, while hiding their meaning from anyone who lacked Latin and at least a smattering of Greek. For such scholars, even *scriptura* (with a small *s*) had to be sacred, set apart: like the Bible itself, writing worthy of the name should not be an "open book" to just anyone. This kind of scholarly elitism in language, with its avoidance of plain speech, would appeal to some sixteenth-century humanists, but humorists such as Rabelais would brilliantly mock its excesses. Linguistic exaggeration made the *Hypnerotomachia* an appropriate prelude to the sixteenth century's manifold interests in languages and linguistics.

PART IV

Toward Modernity

CHAPTER TEN

Alternating Currents, 1450–1550

Every true humanism delights simultaneously in the world and the book.

—E. R. CURTIUS

PESSIMISM EVOLVES

After 1350, an increasing sensitivity to the fragility of the literary heritage, coupled with the often spectacular recovery of Latin and Greek books that were previously mourned as lost or were completely unknown, inaugurated an age of speculation. It was exciting to imagine the myriad works that might yet be recovered, or better yet, *dis*covered for the first time. In the progression from Dante's symbolic books to the mythical, essentially magical tomes imagined by Richard of Bury, to the humanistic classics hunted down and rescued by Petrarch, Poggio, and their heirs, books became the objects of a desire that was often both intense and unfocused.

LOST BOOKS

But excitement about possible discoveries depended on knowledge and estimates of what had been lost, and for two centuries after Petrarch's death, European scholars lamented not only the losses he had mourned but numbers of others about which he could have known little or nothing. Beginning with Richard of Bury, such threnodies assumed that the damage from each biblioclasm was incalculable, that known losses were a tiny fraction of the actual total, and that every loss was tragic, even if it had no relation to the Bible or the classics. Frequent repetitions of the rhetorical commonplace *ubi sunt* ("where are they now?") reminded readers that transience and death were inevitable for books and libraries just as for great heroes and their larger-than-life exploits. In 1540, Agostino Steuco, prefect of the Vatican Library, declared, "There is no language that does not deplore the disappearance of great authors," and wondered, "Where are all those books of Hermes Trismegistus, that most ancient of men? Iamblichus says he wrote 110,000 books, that is, papyrus bookrolls, on theology and natural philosophy."[1]

Sixtus of Siena (d. 1569) was a converted Jew who joined the Dominican Order and specialized in the study of the Bible. But like Steuco he eloquently deplored the disappearance of many nonbiblical "monuments of the most ancient antiquity." "Innumerable writings of various Gentile peoples were sometimes engulfed by the destructiveness of Time, in other cases were destroyed by the invasions of barbarians or the madness of tyrants when empires fell, most notably in the destruction of the Holy City, when so many thousands of volumes in the Library of Jerusalem were burned up, first by the Assyrians, later by the Greeks, and finally by the Romans."[2] Although Sixtus cared most about books in the Jewish and Christian traditions, here he speculated what else might have perished in the successive catastrophes of Jerusalem. In so doing, he romanticized the library of the Jerusalem Temple into a competitor of the Library of Alexandria as described in the *Letter of Aristeas*. Just as Pseudo-Aristeas had exploited the reputation of the Alexandrian library to exalt the Hebrew Bible, and as Josephus manipulated the reputations of Chaldea and Egypt to defend its historicity, Sixtus imagined holocausts of untold riches, both sacred and not.

Despite their stronger investment in the uniqueness of the Bible, Protestants were susceptible to the same pessimistic bibliographic urges. Michael Neumann was a late sixteenth-century Lutheran humanist who translated his surname into Greek and became a "new man"—Neander—in that language as well. In 1565, Neander published a Greek grammar to prepare Lutheran youth for reading the New Testament in the original and for becoming "new men" with Saint Paul (Ephesians 4:23–24). But Neander was not uniquely fixated on the Bible. He prefaced his 420 pages of language instruction with a 340-page treatise that expounded his personal canon of the best books ever written, in every field, from antiquity to his own time. However, before optimistically listing the hundreds of books that his readers should seek out, Neander led off with a remarkable threnody for books lost throughout history:

> My dear young people, innumerable thousands of excellent books, in every kind of science, studies and discipline, have died a sad and miserable death through the injury and iniquity of time, the long passage of years, age which consumes all, the turmoil and malignity of wars, the devastation of lands by floods, the unexpected accidents of fire, and sometimes through shameful, detestable hatred and envy (which has moved the dull and stupid burners of books even in our own time), or by being eaten by roaches, worms, and other plagues of literature. . . . Nowhere today can be found the infinite treasures of books and history written before the Deluge, even though there is no lack of writers who claim they are still extant among the Indian philosophers called Brachmans or Gymnosophists.[3]

Nothing reveals the attraction of lost books and libraries more powerfully than Neander's martyrology of disappeared books and his wan hope that some antediluviana may yet survive far away from European eyes. Even his mention of Ozymandias inscribes the old pharaoh in a history of lost libraries, very appropriately omitting his name: "Diodorus Siculus mentions the library of a certain man in Egypt, which was called sacred, in whose vestibule was inscribed the phrase *iatreion psuches* 'medicine for the soul.'" Neander seems to protest that, in effect, all the most ancient writings were *scriptura*: "Diodorus also writes that in [Deucalion's] flood of ancient Greece, many illustrious monuments of letters perished along with the people."[4]

Writing at Rome in 1565, the Portuguese humanist Gaspar Barreiros took the holocaust metaphor further still and lamented the destruction of good books in an extended analogy between vanished heroes and obliterated books.

> There is nothing in human affairs more wonderful or salutary than the monuments of wise and learned men that are entrusted to letters, through the apt counsel of which we learn how to live well and happily, acquire the knowledge of human and divine matters, and learn the course and revolutions of history. Likewise I consider that there can be no more awful or calamitous loss than the loss of books. And when I think how many priceless books of great philosophers and illustrious poets, eminent orators and profound theologians have perished, I often very nearly experience what is related of a certain ancient and powerful king of the Persians. When he had gathered a numerous and proud army . . . [Xerxes] is said to have wept because that infinite multitude of men was fated to succumb to death before a hundred years should pass.[5] But he wept over the future death of men who were trained and ready to kill, whereas we more rightfully mourn that time has utterly killed wise thoughts of illustrious souls and divine researches that were wonderfully beneficial. Who does not bitterly contemplate the vast, immense number of Marcus Varro's books, cut off by the injuries of time? That man about whom the emperor Augustus said that it was wonderful he had time to write, and who wrote so much that no one else had time to read it?[6]

This dirge helps us understand why the fascination of lost libraries remained so vigorous even amid the remarkable bibliographic recoveries of the Renaissance. For Catholics and Protestants, both the very religious and the more secular, the very idea of even small *s scriptura* was precious, and the more ancient a writing, the more it was cherished, even when unconnected to Holy Writ. Recall the reverence with which Cencio de' Rustici described the "venerable antiquity" of a book that had little value from a literary standpoint but was, he thought, written on *libri arborum*, the fabled early writing-support. Here must be a miraculous survival from the earliest times!

Like Petrarch two centuries previously, Barreiros blamed biblioclasms on human negligence above all, denouncing sloth as a major symptom of literary decline. And as if quoting Lorenzo Valla's *Elegantiae* from a century earlier, he complained, "Nor did iniquitous time only obliterate the innumerable writings of so many exquisitely learned men, for it also destroyed the elegance of the ancient Latin language, and completely extinguished every light of eloquence."[7] But he diverged from Petrarch and Boccaccio by seeing vernacular writings as another indication of carelessness and decline rather than mere trifles of scholarly leisure. To his mind, vernacular books by definition could not be eloquent or good: they were worthless and were dragging culture to hell in a handbasket. Despite his admiration for the erudition of "Italy, which all alone after Greece, had fostered good literature," Barreiros denounced the Italians of his day for perversely wanting to make their vernacular language and literature rival those of ancient Rome. No writings in Italian—not even Dante's, Petrarch's, and Boccaccio's, nor any recent trash—were worthy of comparison to the classics. If Italians did not come to their senses, their hereditary Roman culture would soon revert to "that extreme barbarity that begrimed it in the time of the Goths." Worse yet, other nations would follow the Italians' example and cultivate their own barbarous languages. Then "all Greek and Latin books will either be corrupted by being neglected and covered with dirt in private homes, or will be . . . dirtied by the hands of the riffraff in the way Latin eloquence is now being sullied by Italian youths."[8] Barreiros's message was clear: the vulgarization of Latin and Greek culture was going to have the same disastrous effect on the stability and permanence of *litteratura* that fires, floods, insects, barbarians, and time had wreaked in the past. Barreiros considered forgery the third great symptom of decline. He authored the first truly systematic refutation of the forged library created by Annius of Viterbo (whom we met in chapter nine), marrying eloquent outrage to a meticulous deconstruction of the *Antiquities*'s language as well as its bogus pseudo-biblical history.[9]

RISING OPTIMISM: THE PRINTING PRESS

But there were countercurrents of optimism. The "Gutenberg revolution" that began in the 1440s inspired confidence that further biblioclasmic damage could be staved off. As we shall see, printing became crucial to the concept of *scriptura* in the sixteenth century, especially in the restricted theological sense. But more secular-minded writers embraced the press at least as strongly. In 1499, an Italian humanist published an encyclopedia titled *On the Inventors of Things* (*De inventoribus rerum*). This tremendously popular book was printed over a hundred times in three hundred years, both in its original Latin and in modern-language translations. From his wide readings in classical literature, Polydore Vergil had gleaned stories of grand inventions—not only writing, but ships, houses, even marriage and religion—discovered by ancient personages who were often as mythical as Enmerkar or Thoth. Whether one believed the stories were literally true or not, knowing and appreciating them was the mark of educated Renaissance writers and readers. As encyclopedias have often done, Polydore's could serve as a crib, a shortcut to literary and cultural sophistication for everyone from schoolboys to scholars and poets.

One story Polydore told was mostly factual and absolutely up to date: he extravagantly praised Johannes Gutenberg's "newly devised way of writing." Thanks to the printing press, he exulted, one man on one day could produce "the same number of letters that many people could hardly write in a whole year." Better yet, Polydore declared, "this invention has freed most authors, Greek as well as Latin, from any threat" that their writings would be destroyed and forgotten. (Being a conscientious humanist, he, like Valla and Barreiros, had little respect for writings in everyday modern languages, which required no elite education.) According to Polydore, even libraries were "nothing in comparison" with Gutenberg's achievement. Although libraries were "a great boon to mortals, to be sure,"[10] in 1499 they were still overwhelmingly collections of unique or rare copies, painstakingly handwritten over weeks or months, that could be suddenly obliterated by fires or floods or gradually consumed by vermin and mold. Everyone knew that the best way to preserve priceless works of history, science, and literature was to make as many copies of them as possible, and printing presses now turned out hundreds of identical exemplars of the same book.[11] Not only were the treasured works safer, they could be consulted by many more readers (figure 10.1).

But Polydore was not completely starry-eyed: he knew that familiarity breeds contempt. He recalled that when Gutenberg's invention was brand-new, it inspired wonder (*admiratio*); but already, after only half a century, it was clear to Polydore that it would be undervalued (*vilior*) in future times.[12] His prophecy soon came true. We saw that, far from being a discovery of modern times, "information overload" was already deplored by some writers in the Middle Ages. But owing to the easy multiplication of copies and the huge reductions in their cost, overload became a serious menace in the sixteenth century.[13] Now, some scholars complained, there really were too many books, too much to read, too much of it not worthwhile.

The problem of "too much to read" had another, more disturbing—indeed polarizing—aspect. Coming after the religious turmoil that racked Europe between the twelfth and fifteenth centuries, Gutenberg's invention heightened the alarm among heresiophobic Catholic authorities, particularly after 1517. Martin Luther's rebellion against the Church, especially his translation of the Bible into German, coupled with the reproductive power of the printing press, institutionalized heresy to create a new, "Protestant" religion. Luther and his heirs transformed reading the Bible for oneself, in one's mother tongue, from a centuries-old taboo into the prerequisite for eternal salvation. With more texts circulating than ever, religious censorship became increasingly centralized and formalized, culminating after 1559 in the Index of Prohibited Books. Henceforth, certain books and authors would be off limits to all but a few privileged, licensed Catholic readers. Copies would be expurgated, either by burning them in their entirety or by obliterating the offending words and passages, somewhat as classified governmental documents of today are "redacted" by blacking-out politically sensitive passages. Protestant sects were equally zealous in quarantining or erasing doctrinally objectionable writings, continuing the long Christian tradition of biblioclasm that Boccaccio deplored but that the frescoes of Pope Sixtus V's Vatican Library would commemorate positively alongside their extravagant celebrations of heroic moments in the history of writing.[14] Since the Renaissance, other religious and political sects have continued to ban certain books, hoping to minimize or eliminate their influence by keeping them from everyday citizens and schoolchildren, or even to burn them.[15]

But in general printing added a new dimension to Europeans' enthusiasm for the art of writing by vastly multiplying their ability to

Typographus. Der Buchdrucker.

Arte mea reliquas illustro Typographus artes,
Imprimo dum varios ære micante libros.
Quæ prius aucta situ, quæ pulvere plena iacebant,
Vidimus obscura nocte sepulta premi.

Hac veterũ renouo neglecta volumina Patrum,
Atq; scholis curo publica facta legi.
Artem prima nouam reperisse Moguntia fertur,
Vrbs grauis, & multis ingeniosa modis.
Qua nihil vtilius videt, aut preciosius orbis,
Vix melius quicquam secla futura dabunt.

reproduce texts. When Polydore Vergil eulogized Gutenberg's "newly devised way of writing," he was repeating a boast that printers themselves had spread.[16] There is no direct evidence of Gutenberg bragging about his invention, even about his iconic forty-two-line Bible of 1455, but two years later, Johann Fust and Peter Schoeffer, who had appropriated Gutenberg's operation after he defaulted on loans, did anticipate the Italian humanist, when they printed a Psalter.

Title pages would not become normal for decades, so Fust and Schoeffer ensconced their boast in the volume's colophon, or sign-off paragraph; colophons, originally an invention of the manuscript age, now contained the printers' names, the completion date, and the place of a book's publication. Without mentioning Gutenberg, his successors stated that their Psalms had been produced by "the artful invention of impressing and lettering, without any action of the pen." Three years later (1460), someone (Gutenberg, according to discredited conjectures) printed a famous thirteenth-century dictionary. The colophon declared that "this noble book, the *Catholicon* ... has been printed and put together, not with the aid of reeds, styluses, or feather-pens, but by the wonderful harmony, proportion, and measure of punches and matrices."[17] The colophon piously acknowledged "the aid of the Almighty, by whose command the tongues of the speechless are loosened, and who frequently reveals to children what He conceals from the wisest men." Even nonsacred *scriptura* excited printers' awe and pride in their new tool. A manual for confessor-priests proudly declared that it was newly written "not by means of

---

FIGURE 10.1. "Typographer. The Book-Printer." "As a Typographer, I illustrate the other arts / when I print a variety of books with the shining bronze. / Things that formerly grew dirty, which lay covered in dust, / we saw buried in darkest night. / I renew those neglected volumes of the ancient Fathers, / and assure that when published they are read in schools. / It is said the art was first invented in Mainz, / a serious town, ingenious in many ways. / Nothing seems more useful or more precious to the world; future centuries will scarcely give us anything better." Hartmann Schopper, *De omnibus illiberalibus sive mechanicis artibus.* Frankfurt: Apud Georgium Corvinum, 1574.
Sheridan Libraries Special Collections, Johns Hopkins University, 831.4 Sch659 1574 c. 1.

pens, as the ancients did, but with letters artificially sculpted from bronze through a strange and wonderful research."[18] Eight years earlier, Nicolaus Jenson's 1471 edition of Quintilian's *Education of an Orator* had given purely humanistic praise to the "wonderful art of creating books, not by writing them with a reed, but as if impressing them by means of a signet-ring or seal"; the latter phrase likened products of the press to the uniformity of waxen seals that asserted regal or pontifical authority over a document.

References to penless writing were repeated for decades, often in world or "universal" histories as an homage to God the Creator; after all, these works usually began with an acknowledgment or paraphrase of "sacred" or biblical history, and many, like Godfrey of Viterbo's *Pantheon* or Peter Comestor's *Historia Scholastica*, explicitly followed the chronology laid out in Genesis all the way through to their own times. The colophons of Werner Rolevinck's *Fasciculus temporum* (*Little Bundle of the Ages*, 1474) and Giacomo Filippo Foresti's *Supplement to the Chronicles* (1483), each reprinted many times, are representative. One edition of Foresti declared that "nothing in the world was ever more noble, more praiseworthy, or more useful, indeed more godlike or more holy"—despite its fundamentally secular nature. A printing of Rolevinck in 1481 erroneously referred the invention of printing to the year 1457 but called it "that most subtle science of printing books, unheard-of through all previous centuries," adding, "this is the art of arts, the science of sciences; by the exercise of its rapidity the desirable treasure of wisdom and science, which all men by natural instinct desire, leaps out from its deepest shadowy hiding-places, both commanding and illuminating this world sunk in wickedness." Nor was celebration of printing absent from the famous, massive *Nuremberg Chronicle* (1493), published simultaneously in Latin and German editions, with woodcuts lavishly illustrating every age since God's "Let there be light." It even left several blank pages so that readers could supplement its chronology, down to the undatable but obviously imminent end of the world.

But awe at the mystique of writing did not overshadow printers' patriotism. In the brilliant but jealous fifteenth century, printers seized every opportunity to broadcast the intellectual, political, and genealogical excellences of their homelands. Jenson's Quintilian attributed mechanical printing to himself, "Magister Jenson the Frenchman," who was "another Daedalus, to speak the truth." Comparing printing to the mythical Greek's inventions (notably the Cretan labyrinth and wings for human flight) was hardly inappropriate, but Jenson's crediting himself with discovering the art was barefaced hype. In 1483, a printer who published Saint Jerome's authoritative universal history exulted that printing was "being spread through practically every part of the world" and that thanks to it, "all of antiquity has been gathered together in a short time and is read by infinite numbers of moderns." But the same printer crowed that "the great debt scholars of literature owe the Germans can never be adequately expressed!" claiming that Gutenberg had invented printing around 1440.

In 1460 the *Catholicon*'s printer, whoever he was, boastfully eulogized "the noble city of Mainz in the famous German nation, which God in his mercy has deigned to favor and ennoble above other nations of the earth, by the gracious gift of such high and shining genius." This jab repaid the Italians, who, since Petrarch, often expressed scorn for "Goths" and other "barbarians." Italian praise for other nations' achievements could be faint and backhanded at best; in a 1471 Venetian printing of the younger Pliny's letters, Ludovico Carbone acknowledged "the very opportune assistance lent by *the most noble minds of the Germans*, who thought up those most useful forms for printing books, so as to stockpile and sell many copies of the wisest authors at the same time. . . . All the most useful books can now be had in great numbers and with less expense." But Carbone concluded naughtily that "Roman and Greek eloquence," the pride of Italian scholarship, "has spread so far that *even the Gauls and Britons* now seem to have good orators and poets."

Such enthusiasm for scriptive progress derived in part from pessimism about writing over the preceding century. Jenson's 1472 edition of the elder Pliny's encyclopedia put printing into the overall cultural context when it appended a concluding letter exalting several modern inventions as superior to those the old Roman had praised: since Pliny's time, navigation had expanded to the Atlantic, to Africa, and to India thanks to the magnetic compass; sugar, "which Galen called 'honey from a cane' (*mel cannae succarae*)" had spread sweetness everywhere. But "what invention could ever be properly compared with the book-arts of our printers?" Thanks to printing, "nothing of the genius and accomplishments of the wise will ever perish, even in the old age of the world, and every future age will cultivate letters to an incredible extent. If this activity had existed among the ancients, the excellent accomplishments of our Pliny and other divine men would not be lacking." This may be the most nostalgic daydream ever recorded about lost learning: "If only the Ancients had had printing!" But in the same breath the letter mentioned cannon and gunpowder (*bombardis pyridibusque*), "which mountains—not to mention siege-machines and walls—can hardly withstand," with apparent approval.

Despite enthusiasm for printing, manuscripts remained a primary means of publication even in the seventeenth century, particularly among scientists and religious dissidents. But although some purists blamed a drop in books' aesthetic appeal on mechanical reproduction, printers strove from the start to maintain the look and feel of "real books," the most elegant examples of manuscript production. From Gutenberg onward, typefaces carefully imitated contemporary bookhands, often beautifully. After all, a book had to look like a book, so the construction and appearance of printed books evolved conservatively. Even after paper became common, scribes had used parchment for deluxe manuscripts; thus, early in the history of printing, a few prized copies of certain works were committed to parchment. Like parchment, luxurious illuminations—illustrations, chapter headings, decorative capital letters, and other customized hand-painted adornments—were a creative anachronism that mimicked the material presence of manuscripts. Even utilitarian books of the period often added rudimentary decorations of the sort after printing (see plate 9).

However, the bloom did not long outlive printing's "cradle" age of incunabula, the books printed before 1501. We saw that in 1499 Polydore Vergil observed that Gutenberg's invention had inspired wonder (*admiratio*) when it was new, but that after less than half a century, it was becoming so familiar that it would be less esteemed in future times.[19] Anticipating the science fiction writer Arthur C. Clarke (who was quoted in the preface), Polydore understood that any advanced technology initially seems magical but gradually becomes mundane.[20]

Polydore's prophecy came true on many levels. In 1492 an illustrious defender of manuscript publication had written that "the printed book is made of paper, and, like paper, will quickly disappear. But the scribe working with parchment ensures lasting remembrance for himself and for his text."[21]

ERASMUS OF ROTTERDAM

Fourteenth- and fifteenth-century religious trends among readers of vernacular languages inspired rebellion against Church authorities' continued insistence on Latin as the language of worship. Equally crucial to controversies over the Bible was Italian and Northern humanists' cultivation of Latin and Greek as historically situated, constantly changing languages, subject to decline and restoration. In 1504–1505, before Luther became prominent, Desiderius Erasmus (d. 1536) found and published a manuscript in which Lorenzo Valla (d. 1457), the debunker of the "Donation of Constantine," had criticized the accuracy of Jerome's Latin Vulgate. Like Valla, Erasmus approached Scripture as a text necessarily transcribed by humans, disputing the possibility that Jerome's translation was either infallible or transmitted immutably across the centuries. Erasmus espoused *philologia sacra*: although the Bible was somehow inspired by God, its words could be distorted by human error—translations could be inaccurate, or correct translations could be miscopied and corrupted by sleepy or puzzled scribes. So Erasmus retranslated the Greek New Testament, publishing five gradually improved editions between 1516 and 1535.

Like Valla, Poliziano, and the other pioneers, Luther and Erasmus relied on improvements in philology to recover the most accurate possible text and relied on printing to stabilize it through multiplication. Both reformers massively influenced Bible study, but they differed over the question of language. By preferring Latin and Greek to the modern vernaculars, Erasmus remained closer to traditional orthodoxy, acknowledging the Church's authority over Scripture as *written* language. He remained Catholic, never revolting openly against the Latin Church. Yet he earned the enmity of conservative Catholics: presuming to correct Jerome's Latin as if it were just another text seemed like sacrilege.

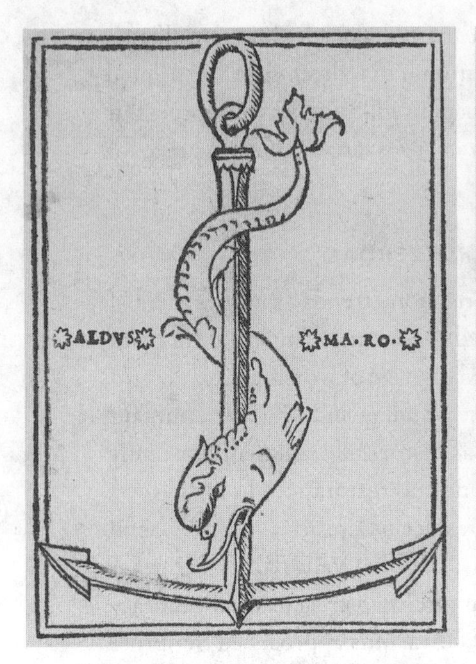

FIGURE 10.2. Printers marks. *Left*: "Hieroglyph" of *festina lente*, "make haste slowly." Aldus Manutius. *Right*: Josse Bade (Jodocus Badius Ascensius).
Sheridan Libraries Special Collections, Johns Hopkins University (left panel). Sheridan Libraries Special Collections, Johns Hopkins University. Photo by Earle Havens (right panel).

Erasmus's approach to the Bible reflected his broader humanist reverence for Latin and Greek writings as guardians of ultimate human concerns. His major scholarly work, the *Adages*, saw eight editions between 1500 and 1533.[22] Collecting several thousand proverbial sayings from Latin and Greek, he annotated them with erudite essays. An entry might be a mere paragraph or cover many pages. Some of the most memorable adages concern writing, but not all are triumphalist celebrations. One of the shortest examines the saying, "You write in water," meaning, "You're wasting your time." Although the proverb seems to contrast water unfavorably with ink, Erasmus traces it to the most negative moment in Plato's *Phaedrus*, where Socrates disparages writing, saying it is preserved "in water, or in that black fluid we call ink." Writers use their pens "to sow words that can't either speak in their own defense or present the truth adequately."[23]

In his adage *Festina lente*, Erasmus notes that Aldus Manutius borrowed his famous dolphin-and-anchor printer's device from two esteemed Roman emperors (figure 10.2 *left*). Erasmus interprets the image as a hieroglyph expressing "make haste slowly." Like his contemporaries, he described hieroglyphics as arcane symbols rather than a system of linguistic signs and agreed that hieroglyphs restricted interpretation rather than facilitating it: "The priest-prophets and theologians in [ancient] Egypt . . . thought it quite wrong to express the mysteries of wisdom in ordinary writing and thus expose them, as we do, to the uninitiated public." Without knowing "the properties of individual things and the special force and nature of each separate creature," no one could "interpret the [hieroglyphic] symbols and put them together, and thus solve the riddle of their meaning."[24] Thus hieroglyphics would protect wisdom from the clumsy hands and grubby minds of the rabble. This elitist, humanistic myth chimed with the Catholic priestly monopoly on interpretation, based on the conviction that when layfolk interpreted Scripture independently, heresy and doctrinal chaos inevitably resulted. For Erasmus as well as conservative Catholics, Latin served a hieroglyphic function—common folk should perceive it as obviously meaningful but mysterious and tantalizing, rather than an "open book" for their perusal.

By analogy, Erasmus assumed that anyone who properly understands and appreciates writing finds the classics invaluable. Speaking of Aldus's printshop, where he had worked as an editor, Erasmus praised its editions of ancient Greek and Roman texts as benefactions to "the cult of liberal studies," that is, the studies proper to a free man (and a few exceptional women): "I think especially of all those who . . . aspire to that true knowledge stemming from antiquity, for the restoration of which this man [Aldus] was surely born—was made and modelled, so to say, by the fates themselves." Erasmus juxtaposed the labors of "my friend Aldus" to "the barbarous and uncouth learning of our own day," meaning the Scholastic theology and philosophy taught in universities. Enthusiasm for Aldus's enterprise led him to predict that, "in the next few years," scholars would find "everything in the way of good authors in four languages—Latin, Greek, Hebrew, and Chaldaean [i.e., Aramaic]—and in every field of study . . . complete and correct, through this one man's efforts." Praise military heroes if you must, said Erasmus, but they are inconsequential compared to one "who restores a literature in ruins (almost a harder task than to create one)," for "he is engaged on a thing sacred and immortal, and works for the benefit not of one province only but of all nations everywhere and of all succeeding ages." Aldus's activity "was in old days the privilege of princes, among whom Ptolemy won special glory" for the Library of Alexandria. But whereas Ptolemy's manuscripts were "contained within the narrow walls of his own palace," Aldus was building "a library which knows no walls save those of the world itself."[25]

As the essay proceeds, Erasmus's tone grows steadily hotter and more polemical about the state of the humanities. Aldus's efforts are a blessing to the learned but a rebuke to slothful or selfish scholars: "A wealth of good manuscripts is still lying hid, either concealed by neglect or kept secret by some men's ambition, whose only motive is to have an apparent monopoly of knowledge." Aldus's new editions would show "what prodigious blunders disfigure the text, even of those authors who are now thought to be in a pretty good state." The Venetian printer was undertaking "a labor indeed worthy of Hercules, fit for the spirit of a king, to give back to the world something so heavenly," rescuing it from "a state of almost complete collapse." Erasmus's description of Aldus's arduous and sacred mission echoed the thoughts of Petrarch, Boccaccio, and Poggio, for like those heroic manuscript hunters, Aldus was striving "to trace out what lies hid, to dig up what is buried, to call back the dead, to repair what is mutilated, to correct what is corrupted in so many ways." Posterity confirmed Erasmus's prophecy about Aldus's unique contribution to Latin and Greek studies. The small format of Aldus's editions, printed in compact, elegant humanist book hand, made the classics available to students and scholars more cheaply than ever before.

Still, all was far from well in the world of print publishing; in fact, the situation was dire. Slovenly and cynical shortcuts, false economies, and greed were more norm than exception, causing a flood of grossly inaccurate texts. At fault were "those common printers who reckon one pitiful gold coin in the way of profit worth more than the whole realm of letters." Why aren't there laws to protect printing, Erasmus asked, as guild regulations govern shoemakers and carpenters? "Certain sordid printers" are abusing the fame that Aldus has won for Venice, flooding foreign markets with inferior books "merely because they bear that city's imprint." Venetian printers are so ignorant, lazy, and mercenary "that scarcely any city sends us more shamelessly corrupt editions of the standard authors." Granted, printing is a commercial venture; but "is he to enjoy his profits, or rather the proceeds of his thieving, who at one stroke imposes on so many thousands of his fellow creatures?" Printers have exacerbated the inverse ratio between quality and quantity of books: "In the old days the copying of books by hand was subject to standards no less high than are now demanded of notaries public and sworn attorneys, and they ought certainly to have been even higher." Yet "how little is the damage done by a careless or ignorant scribe, if you compare him with a printer!" *Caveat emptor*, of course, but laws ought to punish printers "if the title page promises diligent accuracy and the book is stuffed with blunders." Printshops are in a race to the bottom: there has never been "any more portentous source of textual confusion than the entrusting of so sacred a responsibility to obscure nobodies and ignorant monks and lately even women." Beneath Erasmus's genuine indignation, we nonetheless detect an advertising campaign, elevating Aldus by disparaging his competitors, as if to protect his market share (figure 10.2 *right*).

The quality problem was bad enough in itself, railed Erasmus, but printing was also turning a blessing—the multiplication of copies—into a curse. "Is there anywhere on earth exempt from these swarms of new books? Even if, taken out one at a time, they offered something worth knowing, the very mass of them would be a serious impediment to learning, from satiety if nothing else, which can do far more damage where good things are concerned." The siren song of novelty makes overindulgence in reading as pernicious as overeating and equally hard to overcome: "men's minds are easily glutted and hungry for something new, and so these distractions call them away from the reading of ancient authors, than whom we have nothing better to show." Not that moderns could discover nothing new: possibly some fact escaped Aristotle; perhaps an insight was "hidden from Chrysostom or Jerome; but I do not think there will be anyone to do for us . . . what they did. Now these great men are virtually outdated, and we waste our time on indiscriminate rubbish, while honorable fields of study are neglected"[26]

There is so much more complaining in the essay that at moments, rather than describing his own time, Erasmus seems to be predicting our own—when the question, "What good are the humanities?" seems to be constantly asked, in expectation of a negative response. The frivolity of reading-matter and people's demand for it is such, says Erasmus, that "the authority of legislatures, councils, universities, lawyers, and theologians falls in ruins." "If things go on as they have begun, all power will in the end be concentrated in a few hands; we shall see in our midst the same sort of barbarous tyranny that exists among the Turks. Everything will be subject to the whim of one man or of a few; no traces will remain of civilian government, and all will be subject to the violence of a military junta. All honourable fields of knowledge will become a desert, and one law alone will be in force, the will of the dictator."[27] Rarely have the perils of mass communication been expressed in such apocalyptic terms, contrasting so vividly with pages that have just praised it to the skies. Clearly, Erasmus could not consider the benefits of printing without mourning the deleterious effects of commercial expansion on it and on society as a whole.

Although the reading public was still limited, Erasmus's words anticipate our times, when commercialized, unregulated, deceptively "individualized" media (designed more for watching than reading) provoke the most portentous forecasts of societal decay, including the threat—and occasional reality—of neo-fascist dictatorship. Who can say what Erasmus would think of "social media," with their capacity to spread disinformation and "fake news," or even of "self-publishing" through vanity presses, e-books, and blogs?

A BIG, NEW HISTORY OF WRITING

One of the most influential books on the history of writing was written by Theodor Buchmann, a Swiss follower of the radical reformer Huldrych Zwingli (d. 1531). Like Philipp Melanchthon, Michael Neumann, and other reformers, Buchmann declared his Protestant identity by translating his surname into Greek, the language of the

New Testament. His surname already declared him a "Book-Man," but his pen name, Bibliander, made him a "Bible Man" as well, allusively encoding his commitment to reading Scripture in its original languages. Like other Evangelical scholars, he cultivated Old Testament Hebrew; unlike many of them, he learned enough Arabic to revise and publish the twelfth-century Latin translation of the Qur'an. The book that made his reputation, however, was his *Commentary on the Common Principle of All Languages and Letters* (*De ratione communi omnium linguarum et litterarum commentarius*). Bibliander's title now seems overblown, and his information was certainly more valuable as emotional history than as a history of technology. But his book was a frequent reference for early modern historians of writing, despite being printed only once, in 1548, then practically forgotten after 1700.[28]

For Bibliander's contemporaries, the Bible remained the last word (as well as the first) in historical importance. Thus, most agreed that "all languages" numbered exactly seventy-two, because "it is written": Genesis listed six dozen descendants of Noah's three sons. Other, less common calculations recorded seventy-three or seventy-five, but seventy-two had the advantage of being a holy number: there were six days in a week, six days of creation, twelve tribes of Israel and twelve disciples of Christ, as well as seventy-two translators of the Septuagint.[29] Ergo, there must be seventy-two languages in the world.

Saying something substantial about all these languages was a different matter. Bibliander frankly admitted that his grandiose title seemed "a rash and incautious promise, or rather, a gamble, a vain attempt, and even a serious agitation of an unhealthy mind."[30] Nonetheless, Bibliander felt his quest for linguistic history was justified by his Protestant reverence for biblical authority. Acts 2:4 declared that when the Holy Spirit descended on Jesus's disciples, it inspired them to speak and preach in many languages and spread the Word throughout the world. Hence, Bibliander's age must, he thought, revive that mission and spread Reformed Christianity among Catholics, Jews, Muslims, and heathens. But the age of the apostles and their miracles was long gone, admitted Bibliander, so instead of Pentecostal inspiration, he would have to rely on his studies and "my talent, however small." His false modesty notwithstanding, Bibliander's talent and studies actually did make him a more thoroughgoing collector of writing lore than anyone previous.

On the subject of language, human reason concurred with "the most reliable history of blessed Moses" that speech was not a human invention because God must have initiated communication with Adam, not vice versa. Bibliander demonstrates that notions of evolution did not have to wait until Darwin's time to disturb Bible-loving Christians: rediscovered pagan historians troubled Bibliander, as when Diodorus of Sicily claimed that the human race was originally speechless, people needing to "express their thoughts like mute animals, nodding and uttering confused sounds."[31] Bibliander confirms Scripture's implication that Hebrew was the original God-given language, and he reaffirms the pious consensus that after the Babelic confusion, only the descendants of Eber still spoke "Heber-ew." Why did such ideas matter? Bibliander quoted a recent Christian writer, who claimed that "after the general resurrection, all people shall [again] speak one language, that is, Hebrew."[32] Again, why? Because the Babelic confusion demonstrated that other languages were a punishment for sin. Humanity's return to perfection demanded recovering the original universal language.

As an up-to-date scholar, Bibliander was not content to glean the history of communication solely from the Bible and its commentators. He also consulted modern travelers, such as his contemporary Guillaume Postel (d. 1581). This brilliant, eccentric Frenchman was "the man who put down for us the basics" of Arabic grammar and who wrote the pioneering *Introduction to the Alphabets of Twelve Different Languages* (both books printed in 1538).[33] The extinction of languages interested Bibliander as much as their beginnings: he quoted the humanist Volaterranus (Raffaele Maffei, d. 1522) that the Etruscan language and alphabet were "once, according to Livy and Pliny, held precious by the Romans" but had become "entirely undecipherable."[34] Ironically, just when Bibliander was publishing this affirmation, a clique of Florentine academicians was begging to differ, using forgeries and other questionable sources à la Annius to trace their beloved vernacular back to the Etruscans and demonstrate Florentine superiority to Rome through historical priority. Here, thanks to the increasing cultural chauvinism of courtiers and the spread of local academies, was a different kind of rebuttal to the notion that Latin was uniquely scriptural. As proud practitioners of both belles lettres and politics, sixteenth-century Italian, French, English, and Spanish authors did not simply drift away from writing in Latin. Many of them, Catholics as well as Protestants, actively apostacized, demonstrating that literary *scriptura* had no need of Latin to be taken seriously and that it should take advantage of increasing vernacular literacy.

Like his predecessors, Bibliander considered writing truly wondrous, but he went beyond all of them, collecting whatever he could find on its history, from the ancient Greeks to humanists of his own times. He recalls Cicero's praise of "the first person who [confined] the vocal sounds, which seem so interminable, in a few characters." "That gift," he says, "granted to mortals by God, was strived after with wonder and gratitude by all men gifted with judgement and virtue." Quoting the *Readings in the Ancients* (*Antiquae lectiones*) by the humanist Caelius Rhodiginus (Ludovico Ricchieri, d. 1520), Bibliander asks, "What is the nature of that greatness in writing, which furnishes us with the knowledge of so many great things, preserves our histories, does not allow anything to perish, and forces time, which wears away all things, to surrender to our hands? What is more magnificent? What could be more admirable than the fact that writing, over which even the eager rapacity of death has no power, has been granted to mankind?" Caelius outpaced other encomiasts of writing by considering the hand's indispensable role in enabling and preserving discourse. "I dare say that with our hands severed we would be not merely defenseless, but more unfortunate than beasts. What use would that divine light of reason be if deprived of the hands' help? I think that those who declared their hands the servants of reason and wisdom did the right thing."[35] Bibliander's reverence, based in reason as well as revelation, extends to all writing, *scriptura* with a small *s*, as well as Scripture.

Much of Bibliander's lore had been repeated endlessly, as previous chapters have shown. Hermes, Theuth, and, of course, Pliny's history of writing materials make their appearance. Like generations of writers before him, Bibliander put his personal stamp on familiar "facts" by recombining and embellishing them. Regarding Cadmus and the

Phoenicians, Bibliander notes that Plutarch identified the Phoenicians as Hebrews. He recalls Augustine's assurance that the Hebrew alphabet antedated Moses but also the Saint's doubt that Adam invented writing. Yet he does not forget Byzantine sources, which were increasingly available since the fifteenth-century exodus of scholars and books from Constantinople: he recalls that the Byzantine *Suda* (personified as "Suidas") attributed letters and all other inventions to Adam. He does not overlook the "columns" of Seth's sons nor the infamous Book of Enoch. Much of his lore came from erudite contemporaries: Polydore Vergil, of course, but also "Caelius Rhodiginus," and others not yet introduced here, with names like "Crinitus," "Ravisius Textor," and "Alexander of Alexander."

Bibliander's interest in church history fed his enthusiasm for Eusebius of Caesarea, who bequeathed him a story about writing that was strangely reminiscent of Plato's tale about the god Theuth:

> According to Eusebius, Philo Biblius, the Greek translator of Sanchuniathon of Beirut, who, in the era of Semiramis, wrote a history of the Phoenicians, writes the following: "Sanchuniathon was a very skilled and curious man. He seems not to have missed any fact worthy of [memorializing], but he inquired with more diligence about Thaatus's inventions than about anything else. For it did not escape Sanchuniathon's notice that Thaatus was the first amongst mortals to have invented the letters and to have dared to commit to eternity events worthy of memory. Thaatus was called 'Thoyth' by the Egyptians, 'Thoth' by the Alexandrians, and 'Mercury' [*sic*] by the Greeks."[36]

Because the Phoenicians were related to the Jews, Sanchuniathon possessed authority by association. Moreover, for monotheists interested in the history of writing, demotion from god to mere man made the pagan "Thaatus" more interesting still by exempting him from pretensions to divine revelation. After Bibliander, historians of writing frequently mentioned Sanchuniathon, some accepting Eusebius's depiction of him as a genuine archaic writer, others suspecting fraud. Indeed, whether Sanchuniathon's works were forgeries remains undecided.[37]

Bibliander did not accept Sanchuniathon's history of writing: "The opinions of those who claim that letters were introduced either by Gentile authors or Moses or Abraham, are sufficiently disproved by a single text [*scriptum*] of the prophet Enoch, who flourished some 1300 years before the Flood."[38] Bibliander would have known that throughout history the majority of Jewish and Christian exegetes rejected the Book of Enoch. But like numerous others, Bibliander assumed that, since the canonical Epistle of Jude quoted Enoch's prophecy of the Judgment Day, at least those verses must have been authored by the archaic patriarch.

Despite lacking biblical evidence, Bibliander thought the invention of writing should be attributed to Adam, not only in deference to Josephus's tale of the Sethians but also because

> it is consonant with reason that Adam, who had fallen into sin through the seduction of Satan, but was subsequently reinstated to his former position by divine grace, must have portrayed his thoughts and deeds in letters. This fact conforms with the excellence and dignity of the first man. For who else do we think could have been that god or divine man who, as Plato says, first invented writing? And to whom else should we attribute that divine mind, as Cicero and Pythagoras termed it, that brought so many different sounds under firm rules, and assigned the components of human speech to appropriate marks—to whom but Adam himself, who, as the sacred history testifies, imposed appropriate names on all things, with God's approval?[39]

This is a truly "Platonic" defense of Adam; he was the most appropriate inventor of writing, since God created him directly; conversely, such a wonderful invention deserved the archetypal inventor—QED (*quod erat demonstrandum*).

> No achievement of human reason seems to me greater than this: that by means of marks and signs that it creates, reason portrays itself, depositing itself in books as if preserving itself in strongboxes, and transposes itself into other minds whose bodies are separated from it by great tracts of time and space.

> Here, the mind finds an instrument suitable for its nature, with which to encompass matters both human and divine, corporeal and incorporeal. In brief, let me say that whatever a man can grasp with his mind, he can, through the medium of letters, take and present to the perception and judgment of others.[40]

Even without the allusion to Quintilian's strongbox, it would be hard to find a more lyrical description of writing than this passage.

Despite his enthusiasm for Adam, Bibliander knew that not everyone found the question of who *first* invented writing to be paramount. So, in apparent contradiction to his other arguments, Bibliander copied a catalog of inventors from the humanist Petrus Crinitus (Pietro Ricci or Del Riccio, d. 1507). In a philological compendium titled *On Honorable Learning*, Crinitus claimed he had found an extremely old (*pervetustus*) manuscript in a library, containing these verses.

> Moses first wrote Hebrew letters.
> The wise-minded Phoenicians founded the Attic ones.
> The Latin ones we write were invented by Nicostrata.
> Abraham discovered both the Syrian and Chaldean ones.
> Isis with no less art brought forth the Egyptian ones.
> Gulfila produced the ones of the Goths, which we see are
> the latest.[41]

Crinitus's catalog is noteworthy for including the historical Ulfilas among the mythical personages; even more interestingly, two of the mythical inventors, Isis and Nicostrata, are female. Like Sanchuniathon's Thaatus, Crinitus's Isis is human rather than divine. As for Nicostrata, some claimed that her son Evander (a prominent figure in Vergil's *Aeneid*) introduced her alphabet into Latium when he immigrated there. Thus, her Greek name was appropriately changed to a Latin appellation, Carmenta, to honor her gift of prophecy (*carmen*).

Despite his allegiance to biblical exceptionalism, Bibliander could not resist Crinitus's verses. Whoever composed them probably adapted them from Isidore's *Etymologies*—still an obligatory reference in the Renaissance—though Bibliander gave no sign of recognizing the source.[42] He had company: eighteenth-century historians of writing still quoted the jingle without irony. Perhaps there were other reasons for the verses' longevity besides the inertia of authority: whatever else it does, Critinus's catalog treats the invention of writing as a series of human discoveries rather than a theocratic Big Bang. Although Bibliander could accommodate both views, he had no respect for anyone who denied the benefactions of writing: "that great beast of an emperor Licinius, the persecutor of all things Christian, called letters 'a poison and a public plague'"[43] according to Sextus Aurelius Victor's *On the Caesars* (361 CE). Bibliander's Licinius had a long afterlife as the exception to universal gratitude toward the art of writing. Unfairly, Licinius's infamy was apparently owed to a campaign of slander by his co-emperor Constantine I. Though celebrated as the champion of Christianity, Constantine was also Licinius's rival, brother-in-law, and executioner.

PRINTING AND TECHNOLOGY

Predictably, Bibliander devoted several pages to a laudatory history of mechanical printing and began with Polydore Vergil's testimony that one man with a press could print more characters in a single day than many scribes together in a year. To Polydore he added an even fuller account of printing taken from Jakob Wimpfeling's *Epitome of German History* (1505), which, like earlier colophons, emphasized the contributions of Gutenberg and other Germans from Strasbourg. Bibliander also vindicated the Germanness of one particular printer, Ulrich Han. Han had been wrongly considered French because his surname was Latinized to *Gallus*, which could mean "Gaul" as well as "Rooster" (*Hahn*). "This [Latinizing] should never be done, according to [the Italian humanist] Ermolao Barbaro [d. 1493]. For the proper names and nicknames of towns, as well as those of people, should be retained in their own languages. Thus, when referring to Ulrich Han, who guided things literary from paucity to massive abundance by means of this divine undertaking, Antonio Campano jokingly composed four verses."[44] Alluding to Han's establishment of a press in Rome about 1468, Campano (d. 1477) had punned on the printer's ambivalent surname by contrasting Mr. Rooster with the Capitoline geese, whose screeching alerted the Romans to an attack by invading Gauls in 387 BCE. "Goose of Jupiter Capitolinus, by squawking and flapping your wings, you defeated the *Gallus*; in revenge Ulrich Gallus taught us to do without your feathers, which are no longer necessary" for writing.[45]

Wimpfeling had omitted Campano's two final verses, which paraphrased Polydore's "more printing in a day than scribes in a year," concluding more grandiloquently that *omnia vincit homo*, humans conquer all. Wimpfeling replaced the suppressed comparison with verses by another Italian humanist Filippo Beroaldo (d. 1505), who declared, "O Germany, Antiquity invented no gift more useful than yours: you teach us to write books with the Press." The ancient inventors of vehicles, ships, plows, weaving, architecture, and astronomy were mentioned among the gods, but Germany "gave the world a fast and ingenious means by which people can write books, that is, create instruments of wisdom."[46] More payback, perhaps, for all the times Petrarch and his heirs had derided the scholarship of the "Gauls" and "Goths."

But printing is a technology, and technology can bring evil as well as good. Like Jenson's 1472 Pliny, Bibliander recalls that Germans also invented "that brazen tormentor they call the bombard or *sclopetum*." Nothing "more terrifying or dire could be thought up for the detriment of mankind, since one cannon ball, skilfully fired against an army, could annihilate a hundred or more people in the same instant." The alchemist monk Berthold Schwarz, says Bibliander, may have invented gunpowder while experimenting with mercury, sixty years before the printing press. Not even the worst tyrants of antiquity—not the barbarous Scythians nor Homer's Lestrygones and other cannibals—could have created anything more awful than the "noisily horrific vessels" known as cannon.

Circling back to *scriptura*, Bibliander admits that, on balance, "Germany has one invention in common with Tubal-cain the son of Cain, and another with the sons of God, whether we believe that Moses, Abraham, Seth, or even Adam invented writing."[47] The nonchalance of this admission belatedly plants the suspicion that perhaps the present and future of writing are more important than its ancient history, that perhaps the old legends are ultimately a daydream, an emotional homage, rather than a reconstruction of "how it really was."

Ever since Lucian of Samosata (2nd c. CE) and Heliodorus, themes of writing had often been on display in romances and novels. Geoffrey of Monmouth, Benoît de Saint-Maur, and other twelfth-century writers who appropriated and continued Vergil's *Aeneid* to create sequels to the Trojan War, told tales about the documentary provenance of the fictional stories they recounted. Like *The History of the Kings of Britain* or the *Roman de Troie* that they prefaced or commented on, these tales of rediscovered manuscripts were clearly intended to induce wonder and provoke a certain kind of belief or, at least, suspension of disbelief, in the veracity and historicity of the main story. The inspiration for such paratexts came from ancient responses to Homeric myth, such as the *Diary of the Trojan War*, ascribed to "Dictys of Crete." But in the fifteenth and sixteenth centuries, methodical historical and theoretical consideration of writing, scripture, and books was widespread enough that authors of fiction could playfully disrespect their era's most cherished attitudes to writing.

As the fifteenth century faded into the sixteenth, an essentially new genre emerged, the pseudo-forgery, a parodic or lighthearted text which could simultaneously advance and subvert strong claims to historicity and scripturality. A pseudo-forgery had two potential audiences: one that was naive enough to accept its claims and another able to see through the pretense.[48] Ideally, the same reader could perform both roles, "monocular" naivety and philological, "binocular" sophistication, thus maintaining a skeptical suspension of belief toward the assertions of the text. When the satirical or playful tonality was absent or muted, the pseudo-forgery would eventually develop into the early modern realistic novel, such as *Robinson Crusoe*. Unlike both solemn or "genuine" forgeries and fully novelistic fiction, fake forgeries keep their readers at emotional arm's length. Rather than programmatically seek the reader's complicity, sympathy, or suspended disbelief, at crucial points these narratives are calculated to elicit suspicion, to strike the judicious reader as untrustworthy.

Unlike novels of later centuries, fake forgeries rarely explore the pathos of real emotional conflict and often narrate events that are inherently unrealistic or implausible. They are parodic, and often satirical, yet they differ from most satire in two ways. First, the targets of their caricature include themselves; second, in order to lampoon a political, religious, or philosophical subject, they display an exaggerated or inappropriate esteem for it. It is no accident that authors of this period, from Leon Battista Alberti (d. 1472) to Erasmus, Thomas More, and Rabelais, displayed the unmistakable influence of Lucian, especially his *True History*, a tissue of lies more outlandish than Odysseus's tales of bedding amorous goddesses and slaying famished monsters.

Like the most transparent of earnest forgeries, the exaggerations or naivety of fake forgeries allow the reader to hear the well-defined voice of an unreliable narrator who "speaks" the text. Like their earnest progenitors, fake forgeries imply, indeed depend on, the persona or "voice" of a narrator obsessed by the specter of a skeptical reader. Such fictional authors constantly overcompensate and "protest too much," repeatedly stressing their own veracity; thus they arouse the very suspicion of unreliability that they seemingly aim to sedate. They are the playful literary descendants of solemn forebears such as Pseudo-Dictys and the forger of "Numa's Books." Fake forgeries are self-reflexive to an extreme degree: they emphasize their own writtenness and, like the Old Testament pseudepigrapha, are often deeply preoccupied with their own processes of composition. Indeed, both fake and "real" forgeries claim, either implicitly or explicitly, the status of scripture, of "gospel truth," even when their objectives are secular.

Among the most successful fake forgeries of all time is Thomas More's *Utopia* (1516), a whimsical takeoff on Plato's Atlantis fantasy (figure 10.3 *left*). Speaking in his own voice, More relates an account of New World exploration which he supposedly heard from the fictional Raphael Hythlodaeus. Like other proper names in the book, the explorer's is paradoxical; while his Christian name recalls the archangel, his surname means "expert in trifles" or "well-learned in nonsense."[49] Likewise, the message behind his portrait of utopian society is ironic; it forces the reader to wonder: How far is it a critique of Christian Europe, of rationalist utopia, of both, or of neither?

Hythlodaeus painstakingly describes every aspect of utopian society, including philosophy and religion. He notes that utopians are "trained in all learning" and "exceedingly apt in the invention of the

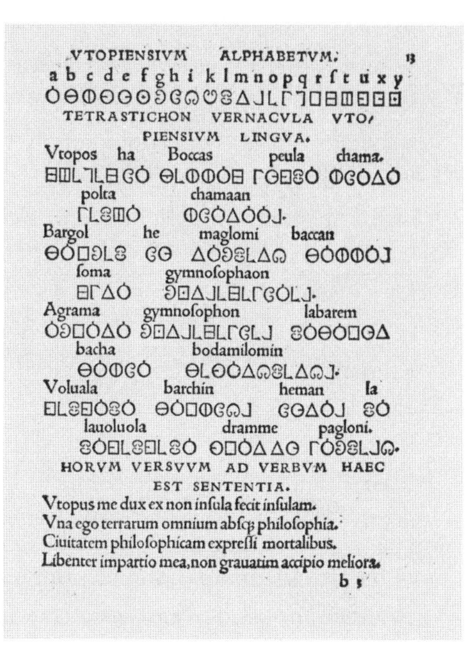

VTOPIENSIVM ALPHABETVM. 13

a b c d e f g h i k l m n o p q r ſ t u x y

TETRASTICHON VERNACVLA VTO-PIENSIVM LINGVA.

Vtopos ha Boccas peula chama.
polta chamaan
Bargol he maglomi baccan
ſoma gymnoſophaon
Agrama gymnoſophon labarem
bacha bodamilomin
Voluala barchin heman la
lauoluola dramme pagloni.

HORVM VERSVVM AD VERBVM HAEC EST SENTENTIA.

Vtopus me dux ex non inſula fecit inſulam.
Vna ego terrarum omnium abſq; philoſophia.
Ciuitatem philoſophicam expreſſi mortalibus.
Libenter impartio mea, non grauatim accipio meliora.

b 3

FIGURE 10.3. *Left*: "A true description of Utopia." Map from Jakob Bidermann, *Utopia didaci bemardini*. Dillingen: Joannem Casparum Bencard, 1691. *Right*: The Utopian Alphabet. Thomas More, *Utopia*. Basel: Johannes Frobenius, 1518.
Sheridan Libraries Special Collections, Johns Hopkins University, Bib# 9328160 (left panel). Biblioteca Nacional de Portugal. Public domain, Wikimedia Commons (right panel).

arts which promote the advantage and convenience of life." However, not until the arrival of the Europeans did they know anything about printing and the manufacture of paper. Although Hythlodaeus and his companions "showed them . . . paper books" and "talked about the material of which paper is made and the art of printing," they could not give "a detailed explanation, for none of us was expert in either art." Rather than revere the Europeans as gods or the art as magical (since they lived in an otherwise advanced society), the utopians rediscovered or reverse-engineered papermaking and printing for themselves. "With the greatest acuteness they promptly guessed how it was done. Though previously they wrote only on parchment, bark, and papyrus, from this time they tried to manufacture paper and print letters (figure 10.3 *right*). Their first attempts were not very successful, but by frequent experiment they soon mastered both. So great was their success that if they had copies of Greek authors, they would have no lack of books."[50] Of course, Hythlodaeus knows of a Venetian printer who could supply those copies. . . . Like Erasmus, Hythlodaeus praises Aldo Manuzio, revealing that the classics he and his companions brought to utopia were "Aldines."

The master of fake forgers was François Rabelais, who, despite his substantial expertise in Latin and Greek philology, had ironically been deceived by a couple of deliberate or accidental forgeries of Roman documents (though he never knew it). His comic novels are encyclopedias of lore about writing and scripturality.[51] In his tall tale *Pantagruel* (1532), he masqueraded anagrammatically as "Alcofrybas Nasier" and outrageously compared the throwaway chapbook that inspired him to the Bible itself, declaring that

> if only each one of us would abandon his own tasks, stop worrying about his vocation and cast his own affairs into oblivion so as to devote himself entirely to [*The Great and Inestimable Chronicles of the Enormous Giant Gargantua*], without letting his mind be otherwise distracted or impeded until he had learnt them off by heart!—so that, if ever the printer's art should chance to fail or all books to be lost, every father could teach them clearly to his children, and pass them on from hand to hand to his descendants and successors as a religious cabbala![52]

The resonances with parascriptural mythologies ranging from Josephus and 4 Ezra to Pico della Mirandola and Annius of Viterbo are too obvious to need commentary. But in addition, "Alcofrybas" demands that preserving his book become a religious vocation, that his readers become secular Benedictines resisting the threat of a biblioclasm as devastating as Noah's Flood.

Like Annius, Alcofrybas orients his revision of the Bible and its lore toward the service of secular patriotism, meticulously tracing the genealogy of his patron "from Noah's Ark down to this age."[53] He acknowledges that his readers "inwardly raised a most reasonable doubt during the reading of that passage, and so ask how it is possible that such a thing could be, seeing that at the time of the Flood all mankind perished, except Noah and the seven persons who were in the Ark with him." In a preemptive rebuttal to these skeptical readers, Alcofrybas describes how a ninth person, Pantagruel's giant ancestor Hurtaly, being too large to fit inside the Ark, rode out the Flood sitting astride it, "like little children on their hobby-horses." Like Annius invoking the Kabbalists, Alcofrybas cites the Massoretes, medieval experts on the grammar and pronunciation of biblical Hebrew, as his authorities for Hurtaly.[54]

Rabelais ensures that, like overly ambitious forgers in the real world, Alcofrybas reveals himself an unreliable narrator, unwittingly inviting the reader's amusement at his clumsy pretensions to arcane knowledge. As if predicting the scientific reality behind postmodern recoveries of quasi-obliterated texts such as Archimedes's treatises and the blackened scrolls of the Vesuvian Villa dei Papiri,[55] Rabelais has another character, the trickster Panurge, provide several imaginary sources for his vaunted expertise at "the art of reading invisible letters." This supreme exercise of cryptography has been prompted by an enigmatic ring sent to Pantagruel as a message, but it has so far resisted Panurge's best efforts. "I have employed a part of what is set out by Messer Francesco de Nianto of Tuscany, who wrote on ways of reading invisible writing, and what Zoroaster wrote in his work [*Peri grammaton acriton*] on hidden writing and what Calphurnius Bassus wrote in On Writings Unreadable [*De litteris illegibilibus*]."[56] The Tuscan "Mister Francis of Nothing" is likely a jab at Annius's pretended expertise at reading Etruscan, but the parody of scriptive—and scriptural—pretensions is the paramount issue here. Once Panurge's texts on reading invisible writing proved fruitless for deciphering the ring, "they found written inside in Hebrew: LAMAH HAZABTHANI . . . 'why hast thou forsaken me?'" This parody of insolent, fraudulent pseudo-scriptural scholarship degrades Christ's penultimate words on the cross into a reproach from an abandoned lover.

Like the narrators of other Renaissance fake forgeries, Alcofrybas sometimes presents the reader with a text allegedly written by someone else, as evidence of his claims. In such cases, the narrator poses as an innocent sponsor, a discoverer or translator who merely brings the text to the reader's attention, as the forgers of books attributed to Numa, Dares, and Dictys had done. This sponsorship is the telltale procedure of both real and fake forgery because the forger's first and fundamental task is to plausibly deny authoring the text. In the first chapter of *Gargantua* (1534 or 1535), the prequel to *Pantagruel,* Alcofrybas sponsors an anonymous document supposedly found in a gigantic bronze tomb, along with enigmatic objects reminiscent of the ones Pliny describes in "Numa's grave": at a spot marked by "the sign of a goblet, around which was inscribed in the Etruscan script, HERE ONE DRINKS [HIC BIBITUR]," a book was found, written "not on paper, not on parchment, not on wax [tablets], but on elm-tree bark." Unlike the Books of Numa, the manuscript Alcofrybas sponsors is badly preserved: it is so damaged by its great age, by vermin and rot, that "you could hardly make out three letters in a row," and it smells "more pungent . . . (albeit far less pleasant)" than roses. Despite once again employing "that art of reading indistinct writing," this time as "taught by Aristotle," Alcofrybas remains unsuccessful: even the undamaged portions he deciphers are complete nonsense.[57] Conversely, in the final chapter of *Gargantua* Alcofrybas sponsors another "rediscovered" text, inscribed on a plaque of bronze, that makes too much sense: it reads equally well as a prophecy of the apocalypse or the blow-by-blow description of a tennis match.[58]

Rabelais's sponsorial gambits had been prepared by a long tradition of spurious quotations of scholarship in chivalric romances, particularly those associated with the figure of Charlemagne. Already in the twelfth century, the patriotic saga of Charlemagne's battles with Muslim opponents in the vernacular *Song of Roland* had been retold in a Latin prose paraphrase that was even more hyperbolic than the original. Attributed to Turpin, the battling archbishop of Reims who plays a central role in the *Song*, the forged *History of Charlemagne and Roland* is preserved in over two hundred Latin and vernacular manuscripts and had a lasting influence on chivalric romances. By the fifteenth century, however, the naively improbable features of the story had led Pseudo-Turpin to be solemnly claimed as the "historical" source of entertaining romances, particularly in the Italian Roland-tradition leading from Luigi Pulci's *Morgante* (1478–1483) through Matteo Maria Boiardo's *Orlando innamorato* (*Roland in Love*, 1494) to Ludovico Ariosto's *Orlando furioso* (*Roland Gone Mad*, 1516–1532). All three authors were careful to milk the absurdity of their claims to absolute veracity by posing as hyperscrupulous (or hypercynical) editors of Turpin's "eyewitness account."[59] Strangely, although he wrote the earliest of these romances, Pulci struck the most explicit pose as pseudo-forger, for which Rabelais expressed his appreciation at several moments. When recounting the death of Margutte, the "half-giant" and boon companion of the titanic Morgante, Pulci's narrator confessed being unable to find sufficient information in "Turpino." Luckily, however, he found it in a certain Alfamennone, "he who wrote the Statutes of Women." The book had been discovered in Egypt, but "it was found in the Persian Language, / later translated into Arabic and Chaldean, / then into the Syrian language, / and afterward into Greek and then Hebrew, / then into the ancient and famous Roman tongue. / Finally, it was brought into the vernacular; / thus it circled the tower of Nimrod, / until it was translated into Florentine."[60]

It is well known that apparent forgeries are sometimes nothing more than incorrect attributions made by unbiased but ill-informed scholars, as was the case with the two pseudo-Roman forgeries Rabelais edited.[61] So despite extreme cases like *Pantagruel* and *Gargantua*, the difference between a fake forgery and an earnest one could theoretically be almost undetectable. In fake forgeries, we have to discern irony in the traces left by an author who shares our readerly amusement at the naive fraudulence of the text, the sponsor, or both. Telltale traces can be rhetorical—the sponsor's inept disclaimers of authorship or the suspicious, "too good to be true" circumstances of a text's discovery—or clues can be philological: anachronisms in the text's language or its evocation of known historical circumstances. Ultimately, the difference between fake and real forgeries boils down to whether the implied reader or target audience is discernibly expected to have some philological sophistication.

It is just possible to imagine a serious textual forger who would be so inept at disguising his proprietary enthusiasm for the text that he could be mistaken for a satirist (we shall see that this is very nearly the case with the Book of Mormon). The talent for feigning such incompetence, made possible by humanistic training in rhetoric and historical linguistics, is the hallmark of early modern fake forgers. The sponsored texts discussed here demonstrate that if the "Donation of Constantine" had not existed, a writer such as Rabelais could have invented it. Like the "Donation," the fragile document containing Gargantua's genealogy was supposedly buried in a tomb but is now unavailable for inspection, except in the sponsor's alleged transcription. Each sponsor is inept both rhetorically and philologically, to a degree that matches the improbabilities of the text he presents.

Like Rabelais, another of his favorite authors, Teofilo Folengo (d. 1544), invented several personas as authors or editors of his tales. Folengo went even a step further than Rabelais, setting his authorial personas in conflict with each other. In the preface to one of several successive editions of Folengo's mock epic *Baldus*, an editor-persona named Aquarius Lodola accuses a villainous competitor named Scardaffus of having stolen, "castrated," and ruined Lodola's scrupulous edition of *Baldus*. The supposed author of this priceless masterpiece of historiography is the bumbling sage Merlinus Cocaius, who, like Alcofrybas Nasier, is also a character in the tale. In practical terms, Lodola's diatribe against Scardaffus was Folengo's way of announcing a new edition of Merlinus Cocaius's epic; it allowed him to present his revisions entertainingly as Lodola's restoration of Baldus's "true history," which the rascally plagiarist-forger-editor Scardaffus had perverted in the previous edition. With equal slyness, Rabelais added the subtitle *restitué à son naturel* ("restored to its original state") to later printings of *Pantagruel*, insinuating that it, too, had previously been adulterated by parties unknown.[62]

Prefatory materials and other paratexts are essential to forgery, while language determines the success of Renaissance fake forgeries: Rabelais's outlandish coinages and Folengo's "macaronic" sabotage of Latin hexameters with words in rustic northern Italian dialect make laughter inevitable by transgressing the rules of proper literary decorum, thereby invalidating the scriptural, gospel-truth pretensions of the narrative. The message of a fake forgery can even have more than two layers. Folengo's fiction has only two strata; the straight-faced solemnity of the narrator or editor provokes a knowledgeable reader to derision. But Rabelais adds a third layer, which he calls "Pantagruelism." The term describes tacit or overt messages that some of the text's lies and impostures merit the reader's indulgence toward human imperfection because they are amusing, festive, and cause no deliberate harm to others. As a philosophy, Pantagruelism has two other essential elements, not only religious toleration but also the pleasures of wine and conviviality, which often occasion the narrator's tipsy befuddlement, as he loses his train of thought or his attention flits from pillar to post, piling up spoonerisms, puns, and other signs of distraction.

Another text, published on the cusp of the fifteenth century, showed similar possibilities. As we saw previously, the *Hypnerotomachia Poliphili* openly presents itself as a dream rather than a document with pretensions to gospel truth. Without this oneiric pretense, the *Hypnerotomachia* could seem to be either a real or a fake forgery. Rabelais implied as much when he plagiarized the *Hypnerotomachia* in his unfinished *Fifth Book*; there he copied Poliphilo's architectural and artistic descriptions to describe the temple of the Dive Bouteille Bacbuc (the "Divine Bottle Glug-Glug"), the holy oracle of Pantagruelism.[63]

Rabelais was also a deliberate creator of imaginary libraries, descriptions of book collections that, unlike Richard of Bury's Library of Alexandria or Annius's Library of Babylon, were not inherited from mythology but invented from whole cloth. In an early chapter of *Pantagruel*, the giant and his retinue visit the library in the Abbey of Saint Victor in Paris, historically a stronghold of religious conservatism; on its shelves Alcofrybas enthusiastically catalogs a wealth of learned books, which, unlike the library itself, never existed. The books' burlesque titles indicate that they are anything but priceless, except as stimuli to laughter. Rabelais fills page after page with titles that satirize Scholastic philosophy, poke fun at books by his contemporaries, or simply mock useless erudition. Instead of a *Mirror* of history or a *Summa* of theology, we find such gems as *The Codpiece of the Law*, *The Handcuffs of Devotion*, or *Of Peas and Bacon, cum commento* (with a commentary), ending with Merlinus Coccaius, *De patria diabolorum* (On the Homeland of the Devils), an homage to the final cantos of Folengo's *Baldus*.[64] Imaginary libraries and bibli-

FIGURE 10.4. "His imagination filled up with everything he had read." Don Quixote reading in his study. Workshop of Gustave Doré for edition of London and New York: Cassell, Petter, and Galpin 1866. Sheridan Libraries Special Collections, Johns Hopkins University, PQ6329.A2 1866 c. 1.

ographies would become a favorite vehicle of satire and fantasy in subsequent centuries.

The Renaissance pseudo-forgery is the most important ancestor of the novel, and perhaps its direct progenitor. As modern criticism acknowledges, every novel implies a theory of the relation between its own narrative and the world outside and, furthermore, between its own means of production—writing—and the external world. The novelist's need to establish this relation was inherited from pseudo-forgery. The best example of the transition came in 1605, when a narrator calling himself "Cervantes" claimed to edit the adventures of a certain Don Quixote de La Mancha from a hack's translation of an Arabic biography of the hero. "Cervantes" supposedly discovered the biography in a manuscript that a fishmonger was ripping up to wrap his wares.[65] The empirical Cervantes could have such fun with the fiction of the rediscovered documentary source because, for at least a century, the editorial pretense had been revealed as completely worn out, having been too long exploited to bestow spurious authenticity on documents intended to influence serious religious and political realities.

But Cervantes transformed the genre of the pseudo-forgery into the prototypical novel by adding a cast of largely realistic characters whose interactions with Don Quixote and his squire Sancho sharply delineated the premises of everyday existence from those of literary myth (figure 10.4). The Don's credulity about the heroes of his favorite romances contrasted jarringly with the various kinds of skepticism his interlocutors displayed: the censorship of the real books in Quixote's fictional library by the priest and the barber, Sancho's enduring peasant realism, not to mention the detachment and even cruelty of figures like the Duke and Duchess, who degraded Sancho and the Don into burlesques of themselves, stage-managing their adventures by adroit manipulations of literary personas and situations. The scenarios the ducal couple laid down, reminiscent of *commedia dell'arte*, functioned somewhat like the forged "second part of the *Quixote*" penned by a real-life impostor of Cervantes, for they seduced Sancho and the Don into self-parody or, we might even say, self-forgery.

CHAPTER ELEVEN

A Second Age of Scripture, 1500–1600

Before Luther's death in 1546, his German Bible went through over 400 total or partial reprints, amounting to about 200,000 copies.

—MARTYN LYONS, *A History of Reading and Writing*

PRINTING, *SCRIPTURA*, AND HOLY WRIT

Humanistic training in philology, literature, and history increasingly sensitized the fifteenth and sixteenth centuries to the complexity of truth-claims in ways that could implicate attitudes to the Bible. But despite humorous interludes in scholarship and entertainment, a passionate, sincere interest in the relation between the biblical text and the ideals of Christianity had never flagged and was indeed growing stronger than ever by the sixteenth century, thanks to the same expansion of humanist scholarship. According to legend, the Protestant Reformation began when Martin Luther nailed ninety-five theses critical of Church doctrines and practices to the door of All Saints' Church in Wittenberg, on October 31, 1517, the eve of that church's name-day. The Christian sixteenth century, sometimes called the Age of Reformations, is epitomized by Luther's controversial act of *scriptura*. In the emotional history of writing, his century was in fact a Second Age of Scripture.

Luther's colleague Philipp Melanchthon may have invented the nailing-up story, but the ninety-five theses were real and bore that date when printed.[1] Moreover, whether historical or not, Luther's provocative gesture was actually not the first. In late April, 1517, Andreas Bodenstein von Karlstadt, Luther's colleague at Wittenberg University and an even more radical reformer, had posted 151 theses on that same door. Like the 900 theses of Giovanni Pico, the manifestos of Luther and Karlstadt were invitations to an academic debate, a common event in medieval universities. All three proposed debates were in great measure about *Scriptura*, exploring the relevance of the Bible as written document to contemporary religious, intellectual, and political issues.

Luther's famous watchword, shared by Karlstadt and other reformers, was *sola scriptura*, a claim that the Bible was the ultimate—even the only—source of religious authority. *Sola scriptura* devalued writing in general (*scriptura* with a small *s*) in favor of Holy Writ (*Scriptura* with a capital *S*): no human-authored document could compete with the text of the Bible, for "God's word" was absolutely and uniquely authoritative. This claim contradicted the medieval Church's assumption that its "apostolic" or postbiblical doctrines and practices—enshrined in the writings of theologians, church councils, and popes—were divinely inspired and thus that Scripture was *not* the sole authority. Protestant reformers rebutted that extrabiblical traditions, no matter how old, were merely human unless the divinely inspired Bible obviously supported them. Doctrines about the sacraments, especially Eucharist and marriage, and about historical figures, such as the saints, must be rejected because they were not explicitly articulated in Scripture. Such refusals had long animated medieval dissidents—heretics—who demanded Bibles in everyday languages instead of Latin, so they could read it for themselves.

All writing is subject to multiple interpretations; thus, far from being as simple as it sounds, *sola scriptura* had a long history before and since Luther.[2] Early modern Catholics warned of the dangers of bibliolatry, a misplaced reverence for writing in itself, rather than for divinely inspired doctrine. The legacy of *sola scriptura* still flourishes among modern Evangelical Protestant sects that espouse radically simplified notions of the doctrine to defend Creationism, "Intelligent Design," and other "literal truths" of the Bible which clash with modern scientific and social principles. For such "fundamentalists," the fact that "It is written" makes Scripture inarguably, straightforwardly true, perfectly describing the world as it is and society as it ought to be, and needing no further explanation. An extreme example of modern scriptural literalism is the Ark Encounter attraction recently unveiled at the Creation Museum in Williamstown, Kentucky. The fundamentalist proprietors of this Disney-esque theme park have constructed a "life-size" replica of Noah's Ark, 300 "cubits" (137.16 meters) long and stocked it with animatronic facsimiles of "all the animals"—including appropriately sized dinosaurs, since fossil evidence of them is by now inarguable.[3] The argument seems to be that if the original Ark can be duplicated, Noah must have built it exactly as "it is written." Yet even its dimensions as described in Genesis 6:15 are ambiguous—exactly how long was a cubit? Increasing historical knowledge about the ancient world had rendered the question problematic by the seventeenth century.

Luther's idea was simpler: the scientific niceties that preoccupy modern fundamentalists arose after his time and are a matter of belief and believability rather than faith. He presumed that each Christian should read Scripture independently, guided only by the Holy Spirit, which would ensure a necessary degree of interpretive uniformity. His project required presenting the Bible to ordinary people in their own languages, since few could understand Latin. Moreover, *sola scriptura* did not presume that the Latin text of the Vulgate was authoritative, as the Catholic church had maintained for centuries. The Latin of the Vulgate as well as the Hebrew of the Old Testament and the Greek of the Septuagint were originally spoken languages, like German, yet the Latin Church had refused to authorize translations for a thousand years. The ability to read Scripture, as the Church knew from long experience, opened the door to multiple interpretations, among which the most dangerous was frequently the obvious and literal. Despite the prohibition, vernacular Bibles were by no means a novelty in 1517, for, as we have seen, the spread of commerce after about 1000 depended on vernacular literacy.

The most basic level of biblical interpretation before and after Luther was dedicated to resolving or explaining away contradictions between various books of the Bible or, indeed, within the same book.

Problems began in the very first chapter of Genesis: verse 27 declared, "God created man in His image, in the image of God He created him; male and female He created them." This implied that Adam and Eve were created at the same time; but Genesis 2:22 declared that God "fashioned the rib that he had taken from the man into a woman." So which was it? This contradiction could be and was explained away, but it *had to be explained*. Uninitiated and unsophisticated readers were a danger to their own and others' salvation. And there were other, far thornier difficulties: What did Jesus's "This is my body" mean at the Last Supper? Did it indicate a literal *transubstantiation*, completely replacing the physical reality of bread with the invisible reality of Christ's flesh and blood, leaving only a deceptive appearance of bread (the Catholic view)? Or did it imply a "consubstantiation" that added the invisible reality of Christ's body to the visible reality of bread, as heretics including John Wycliffe (d. 1384), the more radical Bohemian Hussites, and Luther himself asserted? Or did the words "do this in memory of me" imply that "this is my body" was a mere metaphor, a commemoration of the Last Supper rather than a transformation of the bread and wine? The problem had divided Christians repeatedly (and violently) for over five hundred years: in the fifteenth century, factions of the Hussites went to war among themselves over disagreements that included the nature of the Eucharist.[4]

As we saw in chapter nine, literate laypeople's desire for translation of the Bible into modern vernaculars was already a long-standing problem. Peter Waldo (d. 1205) a rich merchant of Southern France, had been something of an arch-heretic in his own time, in part because of his followers' demands for a Bible translation. The Waldensians' translation dates to 1175–1185, but even it was not the first. Translations into Frankish already existed in Charlemagne's time; closer to Luther's lifetime there were translations into Middle English ("Wycliff Bibles," ca. 1382–ca. 1395), Czech (Leskovec Bible, 1360; later redactions under the "arch-heretic" Jan Hus, 1402–1415), and Hungarian ("Hussite Bible," 1420s–1430s). The English and Czech Bibles served two related movements—Lollardy and Hussitism—that qualify as proto-Protestant, for characteristics that include their desire for vernacular Scripture.[5]

Such precedents inspired Martin Luther to publish his German New Testament in 1522, although eighteen German translations preceded it,[6] while a Gothic translation by Ulfilas or Wulfila (d. 383) actually predated Jerome's Latin Vulgate by two decades. However, Luther refused to translate from the Vulgate as earlier interpreters had done; instead, he adopted the humanistic mantra of *ad fontes* ("go back to the original sources"). Expertise in Greek had spread from Italy to northern European scholars, so Luther translated the New Testament from Erasmus of Rotterdam's new edition of the Greek text. In 1534, Luther and several associates went on to finish translating the Old Testament and apocrypha from Hebrew, but they also consulted the Septuagint, the Vulgate, and the *Biblia Germanica*, which had been printed in 1466.[7] To ensure his translation's comprehensibility to ordinary Germans, Luther had performed field work on their spoken language, listening and taking notes; he coined new words to bridge differing dialects of German and to render Latin and Greek concepts that vernacular dialects lacked. Presses rapidly produced thousands of copies of his Bible in several formats, including large-print for the visually impaired. Luther's Bible was so successful that its hybrid language made it the founding text of modern German literature, as fundamental to that culture as Dante's *Divine Comedy* is to the Italian.

Despite this enormous success, Luther claimed no originality. Even before publishing his New Testament, during 1519–1521, he printed Jan Hus's *On the Church*, declaring, "We are all Hussites without knowing it," and referred to Hus, whom the Council of Constance had burned for heresy in 1415, as a martyr.[8] The press was so crucial to Luther's mission that if the Czech reformer had had equivalent access to it, the Reformation might have begun a century earlier.

SCRIPTURE AND FORGERIES

A forgery was by definition a would-be scripture, whether its pretenses were sacred or secular—whether pseudo-biblical, like Annius of Viterbo's "ancient authors," or pseudo-historical in a secular sense, like counterfeits attributed to Cicero and other classical authors.

Thanks largely to the enduring example of Lorenzo Valla, the "Donation of Constantine" also kept forgery on everyone's mind. Both forgeries and the scholars who discuss them are susceptible to partisan bias, so, unsurprisingly, Protestants celebrated Valla and made the pseudo-Constantinian fraud a frequent target of their anti-Catholic, anti-papal scorn. Given his responsibilities as prefect of the Vatican Library, it was perhaps foreseeable that Agostino Steuco should staunchly defend the "Donation" against Valla's critiques (see below). Steuco was fighting a rearguard action; by 1547, when he published *Contra Laurentium Vallam,* Valla's attack was over a century old. Yet Steuco's book could be seen as an important moment in the Council of Trent's response to Protestant dogmatists' fixation on Scripture. The year before Steuco's attack, in its session of April 8, 1546, the Council refuted Luther's *sola scriptura* allusively but firmly, by declaring that the "truth and discipline" of Christianity "are contained in written books *and* in unwritten traditions, which the apostles accepted from the mouth of Christ Himself, or were dictated by the Holy Spirit and written down by those same apostles, and have come down to us as if passed from hand to hand."[9] By defending the "Donation," Steuco was vindicating not only papal supremacy but the authority of those extrabiblical, "unbroken" traditions.

A SYSTEMATIC COMMENTARY ON WRITING IN THE BIBLE

A landmark document on questions of *scriptura* was the *Bibliotheca sancta* by Sixtus of Siena, who made a brief appearance in chapter ten. Sixtus's "holy library," an encyclopedia of Bible study, was published several times between 1566 and 1742. Its books examine: (1) "The number, organization, and authority of the Holy Books"; (2) "The writings and writers mentioned in the Holy Books"; (3) "The art of expounding the Holy Books"; (4) "The Catholic expositors of the Holy Books"; (5) "Annotations and Censures of Commentators on the Old Testament"; and (6) "Annotations and Censures of Commentators on the New Testament." Sixtus's two final books (7 and 8) refute "those who attacked the New Testament" and "those who attacked the Old Testament."

Book Two, the most ambitious, catalogs "all volumes, books, writers and writings, extant and nonextant, both the authentic ones [*in-*

dubitatas] and those of doubtful or uncertain authenticity, that are mentioned in various places by the authors of the Holy Bible."[10] Abandoning all postbiblical fables about lost antediluvian masterpieces, Sixtus only includes items the Bible *refers to* as books (*libri*)—but he includes all of them. He even surveys books attributed to personages who were simply mentioned in the Bible. Some were considered forgeries, like the Books of Enoch; many were chimeras or "ghosts," books known only as titles, excerpts, or summaries. Undaunted by the lack of physical copies, when the Bible mentioned ghostly writings, Sixtus analyzed their "modes of Inclusion" in the Bible, "their Meaning, and their Authority." He subdivided "Inclusion" into "Transcriptions," "Abbreviations," and "Citations," defining each subcategory at length.

Sixtus approached *Scriptura* as a complex, multifarious, and layered conglomerate, a mound of gravel rather than the single smooth stele imagined by seekers of the Bible's "original text." "Transcriptions" were "faithfully copied from other, older writings, and inserted word for word into the Bible, just as they were first redacted by their authors." Rather than proto-biblical doctrines, other "Transcriptions" included "letters from kings and princes of the Assyrians, Medes, Persians, Egyptians, Greeks, Romans, Jews and other peoples, which are mixed into our holy books as an essential part of sacred history." These putative "Transcriptions" were, Sixtus presumed, scrupulously authentic copies of original documents that once existed.

"Abbreviations," abbreviated writings, were putative doctrinal *sources* of the Bible. "Although they were at first written down by their authors at great length . . . [they] were later condensed and shortened by saintly writers before inclusion among the books of the Bible." Here, if Sixtus were not Sixtus, we might expect to find items like the works of Adam. He does list enormous numbers of such putative books but only as phantoms or frauds, since the text of the Bible gives no hint that they existed. Other "Abbreviations" look patently imaginary to a modern eye: five thousand books (*libros*) of canticles by Solomon (like the Song that bears his name); his three thousand books of parables, and his immense work (*opus*) on questions of natural science (which Richard of Bury mourned as lost); along with three thousand Psalms of his father David. Mention in the Bible guarantees a work's authenticity: Sixtus never questions whether the thousands of Solomon's and David's books ever existed, presuming they were abbreviated by writers who were themselves "saintly." He examines large numbers of abbreviated books attributed to authors including Esdras (Ezra), Nehemiah, and Samuel. Sixtus even includes incidental, elusive mentions of works by obscure figures like the prophets Gad and Iddo "and very many other books (*volumina*) of this kind, from which sources it is believed that a large number of things now contained in the Holy Bible are derived."

Some "Abbreviations" differ only minimally from Sixtus's "Citations" (*Allegatae*), "writings about which the Holy Bible records only the names of their authors, or their mere titles, or condenses their arguments as briefly as possible, or merely cites or alleges testimonies from them." Apparently, a "Citation" contains no actual quotation and thus has no substance outside being mentioned. Yet "Citations" include the Epistle of Jude's actual quotation from 1 Enoch. Less disreputable "Citations" include works by historically attested pagans, including the Persian King Cyrus, the Greek poets Aratus and Callimachus, and the comic playright Menander, along with mysterious tomes like the "Annals of the Kings of the Medes and Persians." Generations of commentators had puzzled over mysterious mentions in the Old Testament, including the "Book of the Wars of the Lord" (*Liber bellorum domini*), the "Book of the Righteous," and the "Book of the Words of the Ancients."

Unlike modern fundamentalists, Sixtus never asserts that the Bible as a whole is "literally true." He divides the books it mentions according to modes of meaning, contrasting historical works with those that are mystical and symbolic. Historical works should be interpreted literally: they "set out only the sequence of events [*rerum gestarum*] or of writings: there is no need to search for a secret or symbolic sense beyond the historical information in order to understand them." They include "many letters of princes and kings of the Chaldeans, Egyptians, Jews, and Romans . . . in which only the surface or literal sense need be examined."

Historical works once existed in the physical world. Mystical and symbolic books, by contrast, only ever existed figuratively, since they "represent occult and sublime mysteries under all sorts of symbolic simulacra or semblances (whether visible or imaginary) of books, bookrolls, codices, and writings." Scripture offers numerous examples:

> the Signed Book, the Rolled-up Book, the Flying Book, the Book of Life, the Book of the Life of the Lamb, the Book of the Seven Seals, the Bookroll Written Both Inside and Outside, the Eaten Book, the Book of Remembrance, the Book of God's Repudiation, the Book of Death, the Book of the Two Wooden Tablets, the Letter that Kills, the Manuscript of the Decree, the Adamantine Scripture, the Epistle of Christ, the Scripture of the House of Israel, and many others of this sort of writings which, if you don't see into their mystical and hidden senses, you will not understand at all.

Sixtus's description of these symbolic, imaginary books anticipates the twentieth-century scholars E. R. Curtius and Eugenio Garin, who discussed "the book as symbol" as it appears in medieval and early modern works of literature and philosophy.[11]

ADAMOLOGY

The ultimate accolade to writing was the idea that it had existed since the birth of humanity, implying that God so valued it that he did not wait for humans to develop it. A millennium before Sixtus of Siena, Augustine had shown that one could revere Moses as transcriber of the Decalogue without needing to claim that he invented writing. But the urge to discover that writing and humanity emerged simultaneously or nearly so recurred many times across the centuries. Both Old Testament apocrypha and Flavius Josephus had portrayed the world's first family as authors and, at least since Freculf of Lisieux's ninth-century "proof," some Christians argued seriously that Adam was the first writer.

A majority of medieval writers limited themselves to repeating Josephus's anecdote of the Sethians. But in the early sixteenth century, thanks to Bibliander and two other much-read authors, Adam made a comeback as in some sense the ultimate source or author of Scripture and, eventually, as a writer. One author who was seduced by the dream of Adam's authorship was Agostino Steuco. As we saw above, Steuco was a Catholic theologian and philosopher but also a librarian. As prefect of the rapidly developing Vatican Library from 1538 to 1548, he was professionally curious about the origin and conservation of history. He convinced himself that the history recounted in the Old Testament, from the creation of the cosmos onward, was not uniquely sourced from God's inspiration of Moses. It was also a record of human experience that stretched back to the memoirs of Adam.

> The Founder (*Princeps*) of the human race saw himself being created by God, saw God with his own eyes, saw the beauty of the world, both while it was a-borning and once it was finished. He gave names to the beasts, and had absolute knowledge of everything else that he observed; *all this is proved by many-faceted and irrefutable reasoning*. Adam's grandchildren, who lived for long ages with him, were known by Noah. And *it is reasonable* to assume that Noah heard them describing the beauty of their grandsire's body, and the even greater beauties of his soul. And what is more pertinent to divinity, Noah heard them rehearsing the theology of their godlike ancestor: how he was created in paradise, how long he stayed there, how he was expelled; in what way heaven, earth, and living beings were created; what was fashioned first, what last of all; what powers were inherent in those trees, one of Life and the other of Knowledge of good and evil. *It is obvious* that everything the Father and Mother of humankind discussed was essentially theology, since they constantly spoke of things before their Fall, about the origin, habits, and shapes of the angels and demons. How could we suppose otherwise?[12]

"How could we suppose otherwise" than that Adam was eyewitness even to his own creation and left testimony to it? Interestingly, Steuco credits Eve with confirming, supplementing, or coauthoring Adam's proto-biblical theology, likely on the basis of such apocrypha as *The Life of Adam and Eve*.

In addition to proposing a radical version of "the Bible as history," Steuco's *On Perennial Philosophy* (*De perenni philosophia*, 1540) is an ambitious attempt to harmonize biblical and pagan cultures. Just as he deemphasized Moses's role as primal, inspired author of the Bible, so he downplayed the uniqueness of biblical theology. As his title intimates, Steuco wished to demonstrate the radical idea that pagan philosophy and Christian theology were not merely compatible, as Marsilio Ficino and Giovanni Pico had claimed, but practically identical. In greater detail than previous syncretists, Steuco argued that every important Catholic doctrine had been adumbrated by pre-Christian philosophers. The only notable exception was Aristotle, who presumed the world was eternal and did not explicitly support the immortality of the individual human soul or the reality of angels. For such faults, Steuco wasted little love on Aristotle; unlike other syncretists, he could not fit Plato's recalcitrant disciple into a proto-Christian scheme. Still, it has been observed that, by making "perennial theology" overlap so thoroughly with Christian doctrine, Steuco risked making Christianity superfluous.[13]

Steuco's vision of the perennial concord between theology and philosophy and of Adam as the original source of history strongly resembles the work of his contemporary, the philosopher Judah Leo Abravanel, better known as Leo Hebraeus or Leone Ebreo. Both philosophers were strongly influenced by the same Neoplatonic currents that inspired Ficino and Pico. The Jewish scholar was twice exiled, first from his native Portugal and later when Ferdinand and Isabella evicted all unconverted Jews from from Spain in 1492. Leone apparently wrote *Dialogues of Love* during the first years of the sixteenth century, but they were not printed until 1535, about a decade after his death. As the product of a necessarily cosmopolitan Jewish immigrant to Italy, *Dialoghi d'amore* was first published in Italian vernacular rather than Latin, but it was widely translated, including into Hebrew. There have been attempts to prove it was originally composed in that language, or in Spanish or Portuguese, but the Italian redaction is demonstrably the oldest, and its affinity to Italian Neoplatonists is undeniable.

Like Steuco, Leone Ebreo traced theology and philosophy to Adam. Both were pleased to imagine the father of humanity as the Bible's point of origin, but in slightly different ways. Whereas the

Christian relied on logic to demonstrate that an Adamic oral tradition lay behind the "history" of Genesis, Leone simply asserted that the original, oral Kabbalah acknowledged Adam as a source for its interpretation of Scripture. Somewhat paradoxically, both philosophers tended to assume that oral tradition had been completely adequate for recording centuries or millennia of complex history and theology until Moses put them in writing.

Dialogues of Love explores Neoplatonic conceptions of love by staging a conversation between the man Philo and the woman Sophia, characters whose interlocking names are reminiscent of Poliphilo and Polia. One of their main topics is the Renaissance Neoplatonic ideal of a fundamental agreement between pagan wisdom and biblical religion, the *prisca theologia,* or ancient theology, that Ficino and Pico pursued. Because love is the force that binds and animates the universe, God, the source of all love, must have revealed fundamental truths, including monotheism, to primeval philosophers. Philo explains the continuity of wisdom by reference to Kabbalah as a precursor of writing, asserting that

> divine authority [came] not only from Moses, the divine lawgiver, but [ever] since Adam, the first [man], from whom the unwritten oral tradition, called Kabbalah (signifying "reception") in the Hebrew language. [Kabbalah] came to the sage Enoch, and from Enoch to Noah the famous, who, after the Flood, on account of his discovery of wine, was called Janus, because Janus in Hebrew means wine. And they depict him with two opposite-looking faces, because he was alive before and after the Flood. He left this, with many other human and divine matters, to the wisest of his sons, Shem, and to his descendant, Heber, who were the teachers of Abraham. . . . Abraham also saw Noah, who died when Abraham was fifty-nine years old. According to what is said, from Abraham and his successors, Isaac, Jacob, and Levi, came the legacy of the Hebrew sages called Kabbalists, who say it was confirmed by Moses through divine revelation, not only by word of mouth, but also in many passages of Holy Scripture, with proper and authentic proofs.[14]

Leone's etymology of *Kabbalah* as "reception" was regularly interpreted to mean strictly oral transmission. Leone's pre-Mosaic Kabbalists are among those figures whom D. P. Walker called "textless ancient theologians," who preserved religious wisdom largely intact until Moses transcribed the Pentateuch.

We saw in chapter nine that the arch-syncretist Giovanni Pico was particularly fascinated by the possibility that Kabbalah, understood as a method of interpreting Scripture, prefigured or foretold Christianity somewhat as pagan doctrines had done since the time of Orpheus; so he hired Jewish scholars as tutors. But Leone Ebreo *was* a Jewish scholar, and in his hands, the genealogy of wisdom maintained its biblical Jewishness despite a tincture of Neoplatonism. Leone's Kabbalah remains oral at least until Moses, whereas Pico had dated the crucial transition much later, to Ezra. Ironically, however, while his notion of Noah as the original Janus and the etymology of the name from the Hebrew word for wine may have several sources, one of them was certainly Annius of Viterbo, who popularized both ideas. Annius based his forged ancient theology on the spurious evidence of ancient and modern Kabbalists whom he invented from whole cloth; he popularized the same genealogy of wisdom as Leone, but declared it to have been transmitted entirely through writing.[15]

RADICAL DISTORTIONS OF ADAMOLOGY

Sixteenth-century Adamology espoused the goal of demonstrating a providential, largely written continuity of scriptural truth down to the current day. Annius exploited Josephus's assertion that Berossos and other pagan authors had corroborated Scripture; but he understood that the cultural implications of Josephus's project extended far beyond reaffirming the Bible's status as God's own revelation and involved its claims to historiographic authority. With consummate duplicity, Annius adopted the self-effacing pose of humble editor and commentator, and presented his commentaries on pseudo-authors as an extension of his theological vocation. While pretending, like Josephus, to corroborate the Bible's account of history by quoting Pseudo-Berosus and his other invented authors, Annius consistently did the opposite, parasitically invoking the Bible to establish the credibility of his forgeries and, ultimately, to turn the Scriptures against themselves.

He adapted Godfrey of Viterbo's pseudo-Josephan genealogy of antediluvian chronicles to siphon authority from Genesis and transfuse it into the history of the world that Pseudo-Berosus supposedly abridged from the Library of Babylon.

> The Chaldeans depended on the history written by Adam, for Adam was the first to write, on the basis of divine revelation, about the creation of the world and his own creation. And Adam wrote a history of events down to the time of Enoch, to whom he bequeathed the task of continuing the narrative. Enoch bequeathed it to be continued by the prophet Lamech, the father of Noah, and Lamech bequeathed it to . . . Noah. After the Flood, Noah bequeathed it to the Chaldeans, from whom Abraham and the others derived the truth of the events that they wrote about.[16]

Annius concluded impudently that the Bible and Pseudo-Berosus were in near-total agreement about the earliest history of the world and invoked Josephus's already specious assertions as corroboration. "The Hebrew history of antiquity is as similar as can be to the Chaldean version, and for that reason Moses is cited as a witness by Maseas [*sic*] the Phoenician and Hieronymus of Egypt, as Josephus asserts in the first book of the *Jewish Antiquities* and in *Against Apion the Grammarian*. Thus it is no wonder if Moses and Berosus are in agreement, for they both drank from the same Fountain of History."[17] Annius slyly stole Diodorus Siculus's idea of a "fountain" of universal history for a precise objective: to redirect ideas about the Book of Genesis by alleging its consonance with histories of primeval times written by imaginary Gentiles like Pseudo-Berosus. At the same time, he distorted the historical claims of authoritative Roman writers such as Vergil and Livy to make them agree with the fantasies of "Berosus" as well. In this way, Annius's commentaries allowed him to rewrite all of history, beginning with Adam.

With disconcerting irony, Annius's cynical fantasy appeared to corroborate the idealizations of pagan ancient theology daydreamed by Pico and other Christian philosophers. The forger even based some of his crucial assertions on a certain "Rabbi Samuel" and other Talmudists, claiming to have consulted them in person or through their writings. One wonders what Pico would have thought of "Samuel the Talmudist"—not to mention Annius's Chaldean and Egyptian pseudo-sources—had he lived six more years and read the *Antiquities*. Would he have been astute enough to see through them or would his enthusiasm have overwhelmed his philology? Certainly, many scholars over the next three centuries professed, whether sincerely or (much more often) not, to be convinced by "Berosus" and company.

Unlike Josephus and Renaissance Adamologists such as Steuco and Leone Ebreo, Annius had no interest at all in defending the Bible. Instead, he used Pseudo-Berosus as a crowbar to make room in biblical history for his own very secular preoccupations. Pretending to champion the Bible, he exploited it to glorify the papacy, imagining a kind of apostolic succession of *pontefices maximi* (high priests) descending from Noah through the Etruscan and Roman priesthoods, by whom perennial religious truths were transmitted to the Christian popes.

Annius's ambitions were only marginally connected to his pretense of defending the Bible or Christian theology. Instead, he defended Italic civilization and the cultural politics of the Latin Church by proving their descent from a supposed autochthonous Etruscan culture founded by Noah, predating any Greek influence by many centuries. Using his forgeries as supplements to the Book of Genesis, Annius abused Josephus's *Jewish Antiquities* to rewrite the Bible and all of human history. Pseudo-Berosus's chronicle hustled the Hebrews off the stage of the ancient world, "proving" that the real stars of Old Testament times were the Babylonians, the Egyptians, and, even earlier, the Noachian Etruscans. Annius argued that the Etruscan "ancient theologians" not only foreshadowed papal Christianity but, coincidentally, produced some of his own ancestors.

However, to complete the irony, Annius may have been at some level a sincere Christian. He made one plausible attempt to vindicate the truthfulness of Scripture by resolving the differences between Matthew's and Luke's genealogies of Christ so as to demonstrate Jesus's credentials as the prophesied Messiah. The closely argued proof shows Annius's signature mixture of traditional Christian scholarship with dubious assertions based on his falsely claimed familiarity with Hebrew; but he argued so credibly that his proof withstood scrutiny into modern times. The ironies were thick: over the following two centuries, Annius's prostitution of Scripture would be copied and amplified not only in service of political conflict but to justify the colonization and enslavement of non-Europeans.

PROTESTANT HISTORIANS INVENT A PROTO-BIBLE

Both Jews and Catholics avoided making the Protestant claim that *Scriptura* stood *sola* as the fount of monotheistic theology. Strangely, however, Annius's historical fantasies made arguments that appealed to early Protestants' idealization of the Bible. Traditional Adamology, whether oral-Kabbalistic or written-Catholic, simply could not convince them: too many intermediaries separated the original utterances or events from the moment they were written down. Yet Lutheran theologians' insistence on written documents made them susceptible to the idea that in the beginning, the core of biblical theology had existed in the form of written documents. Their fantasy that "It Was Written" already in the time of Adam betrays a wistful desire for provable certainty about the Bible that prefigures such modern fundamentalist projects as the Ark Encounter in Kentucky. Early on, and throughout the sixteenth century, Lutheran authors found it comforting to assume that the ultimate truth had been faithfully and accurately handed down because "It Was Always Written." Protestants' preoccupation with primeval writing harmonized with their promotion of lay literacy and their dependence on the printing press to standardize and disseminate biblical texts. But it also left them susceptible to Annius's antediluvian fantasies.

Since Eusebius's time, Christian authors had assiduously rewritten the world history that they inherited. Because the Bible made human history a linear process from creation to the apostles, early Christian chroniclers felt a duty to validate the Bible and domesticate its history as far as possible by interweaving it with pagan sources. We have seen that, alongside Josephus's *Jewish Antiquities* and Augustine's *City of God,* medieval historians continued the labor, in universal chronicles that included Peter Comestor's *Historia scholastica* (ca. 1173), Godfrey of Viterbo's *Pantheon* (ca. 1191), and Vincent of Beauvais's massive *Mirror of History* (*Speculum historiale*, 1260s). Still other medieval universal histories, imported from Byzantium, were eagerly translated into Latin from the sixteenth century onward.

Less than two decades after Martin Luther's rebellion, Protestants were already feeling the urge to retell world history and make it reveal the truths they cared about. Like chroniclers since Eusebius, Augustine, and Paulus Orosius (d. ca. 418), Protestant polemicists deployed world history to explain God's unfolding providential plan as a blueprint for their own religious doctrines. In June 1531 a court astrologer signing himself Johannes Carion (his real surname was probably Nägelin) sent an epitome of world history in German to his longtime friend, Luther's colleague Philipp Melanchthon, the humanist "educator of the Germans." According to his son-in-law Caspar Peucer, Melanchthon found Carion's work unsatisfactory, quickly rewrote it entirely, and published it in 1532, leaving Carion's name in the title but otherwise not crediting his authorial role.

One notable task of the *Chronicon Carionis* was to corroborate the Protestant complaint that Catholicism had corrupted the virtuous "primitive" Church of the Apostles with decadent human traditions. This entailed a strong, though largely implicit, defense of Luther's principle of *sola scriptura*. Ironically, to effect this, the *Chronicon* resuscitated Catholic traditions about the textual history of the Bible and even invented new elements, creating a more radical definition of "primitive." In effect, the *Chronicon* transformed the Bible's account of early human history into a prehistory of Lutheranism. There was nothing particularly new in this emphasis: Christians had always had an ambivalent relation to Judaism, since they adopted the Hebrew Bible only to reinterpret (and often misinterpret) it. But the *Chronicon*'s more radical solution was to define Judaism itself as the continuation or reinterpretation of an even older, more Christian-seeming church, somewhat as Annius had already done with his bogus Etruscan-centric history. The obvious problem for Protestants was that, unless they imitated Annius's outright forgeries, proof of this primal religion would have to come from the Bible itself, which inconveniently gave scant evidence of formal religion before Moses. Moreover, confirming such pre-Mosaic religion would necessarily depend on invoking the history of writing, yet such proof was also absent from Genesis. Even Josephus's Sethian columns, the closest approximation to documentary scriptural authority, regarded astronomy rather than religion and only alluded to the origin of writing.

Melanchthon well understood Annius's mendacious pretense that his *Antiquities* were a defense of the Bible as history. Like many of his contemporaries, he ostentatiously rejected Annius yet sometimes echoed his forgeries, whether unwittingly or surreptitiously, to defend the Bible. As we saw, Annius's pretended vindication of Genesis depended on identifying a common historical source for both the "Chaldean" and the "Jewish versions" of primeval history. Like Annius, Melanchthon recruited Josephus to supply the historical proof that Genesis lacked but for the opposite reason. The Viterbese forger had defended papal authority over politics by inventing a spurious "pontifical" continuity from Noah through the Etruscans to Pope Alexander VI. Melanchthon commandeered Annius's ploy for the inverse reason, to disprove papal authority over both politics and religion.

Melanchthon's redaction made the *Chronicon Carionis* accessible to ordinary literate Lutherans of 1532 and recounted a simple story about the origin of religion and learning in general.

> Josephus writes that Adam and Seth made two tables, one of terra cotta [*jrden*] and one of stone, and wrote thereon God's word and the prophecies through which God's word would be preserved. Josephus also writes that they were the first to divide the year into twelve months, and to observe and teach the course of the stars. And truly, if God had not given them a special revelation about this, it would not have been possible for the unaided human mind to discover such a wondrous thing. So we have inherited God's word, along with letters and the highest [i.e., liberal] arts from Adam and Seth; and we find many references even among the Greeks that writing and all the arts come to us from the Jewish patriarchs, for in his fifth book Herodotus says that the Greeks received writing and all the arts from the Phoenicians.[18]

Melanchthon reinterpreted Josephus's antediluvian columns even more radically than Annius had, attributing them not to Seth or Enoch but to Seth's active collaboration with Adam. He added other more far-reaching claims: along with astrological data, writing, and the liberal arts, Adam and Seth recorded "God's Word and Prophecies," an implicit reference to Christ. *Pace* Josephus, in the time of Adam God prophesied not only the Flood but the entire history of salvation. For good measure, readers were informed that "all the arts" were invented by the "Jewish Fathers" and that Herodotus admitted the Greeks had received "the arts and writing" from the Phoenicians (although Josephus had carefully distinguished the Hebrew people from Phoenicians).[19] Unlike Josephus, Melanchthon specified that discoveries such as Adam and Seth's could not be made through human ingenuity and longevity; instead God had "opened their minds" through revelation.[20]

Thus the *Chronicon* repeated and modified hoary medieval claims about the Sethian monuments as if they were a known source for the early chapters of Genesis. Regarding the issue of historical authority, it skirted the question of whether records passed from Adam to Moses by human means or were revealed anew to Moses by God. Appropriately, Melanchthon mentioned neither the *Life of Adam and Eve* nor medieval Catholic versions of Adamic writings. Nor did he claim, like Godfrey of Viterbo, to have accessed "more ample" historical accounts that Moses rejected when compiling Genesis.

The *Chronicon* omitted long-familiar episodes such as the antediluvian giants' outrages and the Cainites' invention of music and metalworking. These episodes could be seen as fantastic or ambivalent—or even as perilously reminiscent of Annius's forgeries, which had strongly emphasized them. Melanchthon agreed with tradition that, as Matthew and Luke would later do, Moses mentioned the names of Christ's antediluvian ancestors so as to document his legitimacy as the Messiah. The sole exception was Enoch. Unlike Annius, Melanchthon omitted Enoch's role as writer and traced both prophecy and writing to Adam. So he declared that Genesis recorded "nothing memorable" until the Flood, except for Enoch's momentous translation to heaven (Genesis 5:24): Moses mentioned it briefly "so that the world would know and believe that there is immortality after this life, and that God will judge everyone, saving the pious and punishing the impious."[21]

It is unlikely that Melanchthon misquoted Josephus deliberately. As we saw in previous chapters, authors habitually quoted the *Jewish Antiquities* from memory, their recall influenced by Josephus's claim to be defending Moses's account of primeval history. What mattered most to Christian historians was not accurate quotation but appropriateness. Given the importance of *scriptura* to Scripture, it seemed obvious that writing began with Adam; as Steuco queried, "How could we suppose otherwise?" Although both Josephus and Annius had made defending the Bible subordinate to glorifying an ethnic or political group, the Lutheran chroniclers were focused on defending the Bible as *scriptura*.

Yet the Lutherans' objective was equally partisan and ultimately political. A Latin translation of *Chronicon Carionis* by Hermann Bonnus appeared in 1538, and in 1558 Melanchthon revised and expanded Bonnus's version.[22] Melanchthon's foreword remained silent about Carion, but praised Bonnus's erudition; thenceforth, the *Chronicon* was oriented toward a learned, Latin-reading audience. After Melanchthon's death, Caspar Peucer took over the chronicle, and in 1572, revised it again, laying greater emphasis on the perceived abuses of Catholicism. Expanded even further from the original—now comprising 746 large format pages—the *Chronicon Carionis* became the canonical Lutheran account of world history.

In his 1558 revisions, Melanchthon took greater liberties with Josephus's old anecdote and transformed it into an unmistakable foreshadowing of the Lutheran Reformation. He made explicit that the *primitiva Ecclesia* was not the one founded by the Apostles and corrupted by the medieval church; rather, the Apostolic church reincarnated the *original* church founded by Adam, and both churches were resurrected through Luther's reforms. The tablets now became a physical church rather than an abstraction.

> Josephus writes that Adam set up two stone tablets, on which he wrote the beginning of creation, the Fall of Man, and the promise

> [of the Redemption]. I think those tablets were like a sort of temple, and the sign of a certain place where Adam was wont to convoke his Church, where sacrifices were made and doctrines recited. There the voice of the promise [of the Messiah] was a testimony distinguishing the true Church from the assembly of Cain, who broke away from his father and created his own rites and sect. Thus right from the beginning a part of the human race deserted the true Church and forgot the promise; and yet, when they founded their city [Enochia, Genesis 4:17], even they had to retain some parts of the Law.[23]

To increase the authority of the tablets as the first draft of Genesis, Melanchthon completely eliminated Seth from Josephus's anecdote and modified the biblical account by casting Cain as the original apostate and the founder of a heretical church rather than a solitary, outcast fratricide. Read typologically, Cain's apostasy could be seen to foreshadow the errors of both Judaism and Catholicism. Melanchthon's emphasis on Adam's church is strongly reminiscent of the medieval appendix to the *Life of Adam and Eve*, where Solomon discovers the tablets of Seth, revealing "where Adam and Eve used to worship the Lord God," which inspire him to build the Jerusalem Temple on that spot.[24]

Melanchthon now found the Bible's neglect of events between Adam and the Flood even more significant than he had in 1532. So he explicated God's intention for translating Enoch to heaven. "Afterwards, Moses briefly records the sequence of years down to the Flood. For, as was said previously, God wants the beginnings of things and the times of his covenants to be known. Yet there is only one wonderful thing narrated for that period, and it is about Enoch, who did not die but was transported alive into Heaven." Understanding God's purpose entirely, Melanchthon simply knows that there must have been witnesses, so he imagines the precise circumstances that would make that story and the rest of Hebrew patriarchal history worthy of credence.

> Now let us not suppose that [Enoch's] rapture was accomplished secretly, with no onlookers, but rather that it happened either

during a meeting of Adam's Church, or else with some trustworthy Patriarchs as witnesses. For God wanted that translation to be a proof of eternal life and salvation, to be given to those who pray righteously to Him in the knowledge of the promised Seed [Christ]. And in fact two similar proofs were shown to humanity before the coming of the Messiah, one through Enoch before the Flood, and the other through Elijah [2 Kings 2:1–12] after it. These two prophets testified that the Messiah would return, would abolish death, would bring eternal life, and thereby trample the head of the Serpent [cf. Genesis 3:15].

The *Chronicon*'s conjectures now made Enoch's life and prophecy into explicit foreshadowings of Christ. "Enoch's prophecy about the Last Judgment is also quoted in the Epistle of Jude [vv. 14–15]. . . . Without a doubt this prophecy contained a fuller teaching about God's promise, about the Messiah, and about the future vicissitudes of the human race."[25] Reasoning almost novelistically, Melanchthon assured readers that Enoch's prophecy contained considerably more detail than he dared reconstruct. Moreover, he now maintained, "it is reasonable to believe" that Enoch's entire prophecy was actually "written down," since the Epistle of Jude quotes it.[26] Ironically, Melanchthon could not know that the actual source of Jude's reference was the infamous 1 Enoch or that it overflows with details of exactly the wrong sort for his pious purpose.

He boldly adapted Steuco's "kabbalistic" history of the biblical text but, like Annius, applied it to the history of writing: "Now the letters of the alphabet [*figura literarum*] must have been invented at the beginning of the world. For letters were necessary in order to record the genealogies of the patriarchs." As before, like Matthew and Luke, Melanchthon wished to establish Jesus as Messiah by demonstrating his descent from the royal House of David. True, Genesis recorded too few events for anyone defending Scripture to characterize it as comprehensive history. Still, Melanchthon professed himself unperturbed. "The narration of this first book is brief, although without a doubt God produced many illustrious testimonies about Himself, and there were various struggles between the Church and the empires, and many things worthy of our wonder were discovered and transmitted by the arts that were invented then."[27] Repeating "without a doubt" (*haud dubie*) in these passages foreclosed questions that could seem all too logical about Genesis. The figure of Moses as a severe editor with a sharp blue pencil had appealed to Josephus, Godfrey of Viterbo, and, spectacularly, to Annius; it seems clear that Melanchthon was of the same mind.

Indeed, he chafed perceptibly at the need for speculative restraint. In a foldout table, he notes that "Nimrod founded the first empire after the Flood, that of the Assyrians and Chaldeans. Before the Flood all government was domestic and paternal. Few writers about this period have survived; in fact, few things are known about it. The principal surviving writers are the [authors of the] Bible, Berosus, Manetho, Diodorus of Sicily, Herodotus, Justin, and Eusebius in his *Chronology*."[28] The desire for news of the antediluvian world inspired Melanchthon to accept chronicles from all periods as authoritative, and other sources "can be added without difficulty" to this list, according to the *Chronicon*. But like many of his contemporaries, he understood the need to be selective in his references to Berosus and Manetho. Repeating Josephus's quotations from the two old Gentiles was safe, but the same was not true of the ridiculous "fables that Berosus relates about Tuiscon," Annius's imaginary founder of German peoples.[29] However, despite his ostentatious repudiation of Annius's forgery, Melanchthon was far from immune to secular patriotic impulses. Already in 1532, a little map illustrating the postdiluvian world had pushed the Germanic lineage back beyond the discredited "Tuiscon" to Noah's grandson Gomer (Genesis 10:2–3), giving German history an indisputably biblical beginning.[30]

Melanchthon's new Latin *Chronicon* imagined a primeval world that was damaged by Adam's fall, yet nonetheless enjoyed a golden age of piety and humanly attainable wisdom. It reprises the traditional medieval identification of Noah's son Shem with Melchizedech, the King of Salem who offered bread and wine to Abraham after he defeated Gentile kings who had abducted Lot (Genesis 14:18–20). Since the Book of Hebrews (5:6–10; 7:1–28) identified Melchizedech as a typological forerunner of Christ, the *Chronicon* imagined the astoundingly old Shem/Melchizedech forming a holy "academy" with his descendants Abraham and Lot: "they had an assembly [*coetus*] of disciples who worshiped God with them. Now think how righteous such an academy [*collegium*] was, presided over by Shem and Abraham, who both witnessed God's revelations, and whom He adorned with the highest gifts!" Like its putative Lutheran reincarnation, Shem's Church was surrounded by a larger, corrupt and impious society: "And if their Church [*Ecclesia*] was not much frequented, this was because their region had a great multitude of the worst sort of men, contumacious and inimical to these Teachers of Divinity [*doctoribus*] and their assembly."[31] The few virtuous "doctors" of Shem's Church could hardly avoid prefiguring Luther, Melanchthon, and other Lutheran divines at the University of Wittenberg.

The evil majority who flourished after the Flood, during the Babelic confusion, and in Abraham's time provides a dark foil to the church of the three patriarchs: "Let us think [*cogitemus*] what a beautiful academy [*collegium*] was that of Shem, who witnessed the Flood, and the other two patriarchs." Here the true utility of all those genealogies finally reveals itself: this proto-Lutheran church is conjured from nothing more than the list of "begats" in Genesis.

> We should think [*cogitandum*] what a sweet association there was between Shem and Abraham and their kinsmen, when Shem saw the eighth generation of his descendants called to the Church and a definite seat promised to him,[32] along with the renewal of the promise of the Messiah, by Whom death would be abolished and eternal life restored. So let us not think that the Church of those times was an anonymous and useless mob; rather, let us know [*sciamus*] that the guiding lights of the human race were there, and that they saw many proofs of the presence of God.[33]

Whether we choose to see bad faith in these proto-Lutheran churches is beside the point. More significant is the testimony these fantasies provide to the mystique of writing and the romance of Scripture's antiquity. Given the premises of the Carion chroniclers' fantasy, who could doubt that *It was written* "in the beginning?" Who would want to?

For about a century, the *Chronicon* was a best seller, translated into the major vernaculars and read by Catholics as well as Protestants; vernacular poets, including the Huguenot Guillaume Salluste Du Bartas, a favorite of Milton, adapted its version of biblical history. Du Bartas (d. 1590) reimagined the Lutherans' antediluvian "academy" and its posterity in his narrative poem the *Sepmaines*, or *Divine Weeks*, describing how Adam and Seth's wisdom was preserved in a "hieroglyphic" format. Du Bartas imagined Josephus's two columns as temple-like towers, one of brick and one of stone, which housed allegorical "portraits" of the arts of arithmetic, geometry, astronomy, and music. The section on *Les colomnes* (The Columns) opens by describing how Heber, the mythical ancestor of the Hebrews, initiates his son Peleg into antediluvian wisdom after the Babelic confusion, explicating the portraits of the arts in the surviving stone tower to expound how the heavenly constellations form the hieroglyphics of God's own primeval book and how the music of the spheres turns the cosmos into a vast pipe organ of praise to the creator. Thus did Seth's wisdom journey from Josephus's few lines about astronomy to a substantial poem exploring the most suggestive intersections between God's two mutually referential books, the Bible and the cosmic Book of Nature.[34]

Other French Protestant narrative poets of the time found similar inspiration in the myth of an antediluvian science that was also the Ur-text of the Bible and of Protestantism itself. Interestingly, late in the next century John Milton was familiar with much of this parascriptural literature—especially Du Bartas—but declined to reference it in his own major poem. *Paradise Lost* (1667–1674) eschews any pretense of a primordial, proto-Mosaic text of the Bible, opting for the epic and biblical premise of divine inspiration:

> Of man's first disobedience, and the fruit
> Of that forbidden tree, whose mortal taste
> Brought death into the world, and all our woe,
> With loss of Eden, till one greater man
> Restore us, and regain the blissful seat,
> *Sing heavenly Muse*, that on the secret top
> Of Oreb, or of Sinai, didst inspire
> That shepherd [Moses], who first taught the chosen seed,

> In the beginning how the heavens and earth
> Rose out of chaos.

Blind like Homer, the Miltonic poetic "I" implicitly dismisses his Protestant predecessors' scriptural "histories" and "epics" as mere romances, opting for a repetition of Moses's divine inspiration. He creates a new "Life of Adam and Eve" as a gripping versified novel of psychological depth and plausibility. As Milton may have seen, the Lutherans' conjectural Ur-texts were far too reminiscent of the "rediscovered manuscripts" of forgers like Annius (and, we might add, of jocular novelists like Rabelais and Cervantes).

CATHOLIC VIEWS

In 1575 Giovanni Battista Camozzi published at the Vatican printshop a tiny *Oration on the Antiquity of Writing.*[35] Though only ten pages long, Camozzi's *Oration* was reminiscent of Bibliander's and Sixtus of Siena's encyclopedic studies, managing to cite most of the antediluvian and primeval writings familiar to the sixteenth century. Camozzi makes abundant use of classical, Jewish, and Christian sources but also devotes positive attention to Annius of Viterbo's pseudo-authors, treating them as genuine documents in the history of writing. Although brief and unoriginal, Camozzi's pamphlet is an intriguing witness to interest in the history of writing among scholars in the Roman Curia during the decade before Pope Sixtus V rebuilt the Vatican Library and reerected the neglected Egyptian obelisks that ancient emperors had brought to Rome. Camozzi also foreshadows the brief academic essays on the history of writing that would be common at northern European universities in later decades (see chapter twelve).

For the new library, Sixtus commissioned several series of frescoes to illustrate the history of writing, and his prefect of the Vatican Publishing House, Angelo Rocca, provided the program. In 1591, after the frescoes were completed, Rocca published *The Apostolic Vatican Library, Transferred by Sixtus V into More Splendid and Commodious Quarters.* Among the many sumptuous depictions of the history of writing (including the history of censorship and book burning) that adorn Sixtus's new library, those arrayed in the Salone Sistino are particularly relevant to our discussion.[36] The series depicts not one or two, but twenty-six inventors of writing known to the late sixteenth century, beginning with Adam, whose caption paraphrases Steuco's assertion that "reason itself" shows he must have invented writing. Others include Abraham, Moses, and Hermes Trismegistus, as well as the time-honored Cadmus and Carmenta. There are some historical figures, such as Ulfila or Wulfila, the creator of the Gothic alphabet, as well as the semi-historical Cyril to whom the old Slavic alphabet was formerly attributed. The most curious figure is Jesus Christ himself, who as *Logos*, holding the letters Alpha and Omega, crowns the entire series. Likewise, the full-length portraits of all these heroes are adorned with conjectural or fanciful examples of their alphabets. It is probably no accident that all but two of the frescoes are executed on pilasters, perhaps to remind us that the primeval "columns" of Hermes and the Sethians were often considered the first libraries. But in the Salone Sistino, the Sethian columns are no longer the primordial book, an honor that belongs to Adam; nor are they a source of Moses's Book of Genesis. Instead, Moses is pictured with his iconic decalogue tablets, having just descended the mountain.

Rocca's treatise in some ways resembles the elaborately illustrated catalogs of *Wunderkammern*, or chambers of wonders, that began a few years later and endured a long time afterwards. But *Wunderkammer* books discussed wonders of all kinds, from fossil shells to unicorn horns (narwhal tusks), whereas Rocca's program and his guidebook suggest something more limited: that writing and everything connected with it is a multifarious complex of wonders. Although the book contains no woodcuts or engravings, it describes all the frescos in substantial detail over the space of a hundred pages, dedicating about four pages of now-familiar lore to each inventor.

When discussing the fresco depicting the Sons of Seth, Rocca asserts, as Carion and his continuators had, that the Sethians included Adam's prophecy about the coming of Christ among the writings on one of the columns. In fact, Rocca anthologizes numerous distortions of Josephus's *columnae Sethianorum*, ranging from Peter Comestor to the Lutherans. Recalling Genesis (4:26), Rocca also relates that some writers say that the formal, ritual worship of God began in the time of Seth's son Enosh, or even that Enosh invented the use of "certain images to stimulate the prayers of worshippers."[37] Rocca thereby appropriates the Lutheran myth of "Adam's Church" and turns it against Protestant propagandists, by claiming it as the origin of one of the Catholic customs most abominated by reformers. Moreover, Rocca cites certain "manuscript and printed copies" of Josephus's *Jewish Antiquities* "in Hebrew," which, he says, quote compatible evidence from "a letter written by Alexander the Great to his tutor Aristotle about the tomb of Cainan the son of Enosh (who lived before the Flood), which was found in a city of Persia; Alexander testifies that its epitaph, written in the most ancient Hebrew alphabet, prophesied Noah's Flood."[38] The spurious letter of Alexander to Aristotle appeared in the tenth-century Latin translation of the Greek *Alexander Romance*, one of the most widely read and influential travelogues of the Middle Ages.[39] This independent pagan corroboration of Adam's prophecy on the *columnae Sethianorum* seemed valuable indeed.

Rocca asserts that "Alexander's story of Cainan" was repeated in a Hebrew text of Josephus; this claim points to the so-called *Josippon*, a ninth- or tenth-century Hebrew text falsely attributed to Josephus, first printed in the late 1470s, and translated into Latin by the German Hebraist Sebastian Münster in 1541. Münster's contemporary, the Italian Jew Azariah dei Rossi, already recognized that *Josippon* contained passages translated from the Alexander romance.[40] (Rossi also worked in the other direction, giving a Hebrew translation of the pseudepigraphic Letter of Aristeas, concerning the Septuagint translation and its inclusion in the Library of Alexandria.) As we have seen previously, all such lore reveals how strongly Jews and Christians were invested in human *scriptura* and its connection to Holy Writ. That the papal library itself displayed the legends so prominently constitutes a Catholic declaration of ownership over the entire body of lore.

Pope Sixtus himself exerted control over Catholic discussions about the history of writing through literary and historical works written by scholars he commissioned. In 1590, his engineer Domenico Fontana published *On the Transportation of the Vatican Obelisk and on the Construction Projects of Sixtus V*, in which he described his own feat of moving the 327-ton obelisk across Rome and reerecting it, using nearly a thousand men and seventy-five horses[41] (figure 11.1). The four

massive obelisks that Fontana resurrected and crowned with crosses were the only symbol necessary to celebrate the victory of Catholic biblical culture over the best ancient Egypt could offer.

Another of Sixtus's employees, the protonotary Michele Mercati, also wrote a book *On the Obelisks of Rome*. Not only was it published the year before Fontana's *Transportation*, it also contained the most complete history and description yet published of Egyptian obelisks, their symbolism, and their hieroglyphic inscriptions. As Fontana would soon do, Mercati published *Gli obelischi di Roma* in Italian rather than Latin, to celebrate for a broader cultured audience the Egyptian obelisks brought to the capital in ancient times. Along with Rocca's *Bibliotheca Apostolica*, which mentions Mercati's book favorably, *Gli obelischi di Roma* represents the apex of late sixteenth-century Catholic lore about the history of human and divine *scriptura*. Quoting Josephus more chastely than other early moderns, Mercati simply noted that the sons of Seth "wrote their knowledge on

FIGURE 11.1. Transport of the Vatican Obelisk from its ancient site alongside Old Saint Peter's in Rome and its re-erection in front of New Saint Peter's in 1586. Natale Bonifacio and Giovanni Guerra, *Disegno, nel quale si rappresenta l'ordine tenuto in alzar la Guglia li dì ultimo Aprile M.D.L.XXXVI*. Print. Rome: Apud Bartholomeum Grassium, August 1586. Sheridan Libraries Special Collections, Johns Hopkins University, Bib# 7591237.

two columns, one of which was of bricks, and the other of stone, so that if the column of bricks were ruined by the flood, the one of stone would at least remain, through which the science of astrology and the memory of the columns' inventors could be reborn for posterity."[42] Like many Christian writers since the ninth century, Mercati argued that the Sethian columns demonstrated that letters had passed hereditarily from Adam to Noah, thence to Abraham and finally to Moses. He reasoned that "even if the Book of Enoch is apocryphal, it demonstrates that the ancient Jews believed that the patriarchs had possessed writing; therefore it is reasonable to attribute its invention to Adam. And just as God gave language to Adam to explain his thoughts to anyone present, so he gave him letters to do the same for absent persons and for his posterity, facilitating and preserving human memory with them."[43] Mercati's declaration synthesizes a millennium of Christian tradition about *scriptura*, from Isidore of Seville's praise to Agostino Steuco's "reasonable" conjectures.

As for the hieroglyphics, it was true that the reerected obelisks of Rome were impressively massive and their picture-writing aesthetically pleasing, but it was not *true* writing.

> Understanding these natural figures does not presuppose any language or other medium; the unaided imagination of things is sufficient, which follows concepts immediately and is common to all peoples, and in this way is represented simply by pictures. Thus, if we had not had the abovementioned information, which proves that our custom of writing with letters is extremely ancient and was practiced by the earliest patriarchs, we would have to assume, in accord with nature, that the first manner of writing introduced among men was the one that represents the things themselves that we wish to consider, and that it had been universally used until, by some supernatural instinct, humans found a more perfect means of writing—which is that of our letters, the invention of which is of such subtlety and excellence that we can see no better way of attributing its origin than to a special gift of God. . . . Even Plato recognized that the invention of letters was divine, but if God infused it into anyone, it is more logical to attribute it to Adam. . . . It is clear that Abraham wrote with letters, since God added a letter to his name [Genesis 17:5].[44]

The earliest postdiluvian patriarchs took the use of alphabetic letters into Egypt, thus the Egyptians did not have to invent them or imitate the pictographic writing of the Ethiopians, says Mercati. Instead, the Egyptian priests invented the hieroglyphics for the express purpose of concealing "their noblest and most recondite sciences, and, in short, all holy knowledge," from the vulgar herd, who could have appropriated it if the priests had used "our letters, which are understood by everyone"; such negligence would have caused the downfall of the priestly order.[45] Unperturbed by the contradiction in his two descriptions of hieroglyphics, Mercati neatly summarizes centuries of orthodox thinking about alphabetic and hieroglyphic writing, concepts which dated back to *Horapollo* and that many would still defend in the eighteenth century.

Milton's favorite modern poet was Torquato Tasso (1544–1595), whose masterpiece was the epic *Gerusalemme liberata* (*Jerusalem Delivered*, 1575–1581), a fictionalized account of the First Crusade based on medieval chronicles. One of Tasso's last works, *Il mondo creato* (finished 1594 but published several years later), resembles Du Bartas's *Sepmaines* and Milton's *Paradise Lost* by retelling the biblical creation story. Like Milton, Tasso claimed divine inspiration for his divergences from biblical narrative and refused to claim proto-biblical sources, thereby avoiding any problematic Protestant resonances of *sola scriptura*.

Accordingly, Tasso does not simply begrudge wonderment at the art of writing but actively ridicules it. In fact, book six of *Il mondo creato*, on the creation of man and woman, refuses to express marvel at any human accomplishments whatever. Instead of celebrating the human genius behind the invention of writing, Tasso belittles it by exalting the instincts of brute beasts like the dog. Unlike humans, he says, who take inordinate pride in their use of logic and their ability to solve problems and transmit knowledge through writing, dogs have an innate and infallible knowledge. When faced with the question, "Which way did my prey go from here?" a dog has no need of syllogisms or writing: his *pronto senso* takes the place of the *lunga arte* accumulated by humans.[46] Tasso's comparison, though earnest, seems almost worthy of Rabelais, who at his most seriocomic called the dog the most philosophical animal because of its patient, "studious" extraction of marrow from bones.[47]

Tasso repeatedly disparages the activity of writing in terms that reflect but systematically invert ancient, medieval, and early modern expressions of wonder at the human accumulation of knowledge through letters. True, he says, humans invented and explored logic and the other branches of philosophy through writing, initially scratching their syllogisms into sand or dirt; yet we fail to understand that, unless we are enlightened and inspired by God, whatever truths we discover are as impermanent as those rudimentary inscriptions, which the wind or sea obliterated in a day. More to the point, Tasso rebukes the ancients for taking pride in the supposed permanence of certain primeval inscriptions. The ancients, he says, celebrated these epigraphic monuments as the foundation of all subsequent knowledge and a triumphant symbol of human permanence through writing. Yet such celebrations are vain and delusional, for all writing eventually dies.

> And though the ancients praised themselves and boasted
> of their sacred glyphs and lofty columns,
> whereon the noble arts were engraved
> in Mercury's sacred ornate temple,
> and th'other famous columns where they kept
> a thousand olden memories of the world
> secure from floods and from combustion,
> no dust or other vestige remains of these nor those,
> so deep a night engulfs their names and works.[48]

This passage provides an important key to Tasso's contrarian dialogue with praisers of writing. The lofty stelae, *colonne eccelse,* displaying texts of the liberal arts, indicate that Tasso was thinking of Mercury as Hermes Trismegistus, the legendary sage whose epithet "Thrice-Great" commemorated his status as philosopher, king, and magus. Tasso read Iamblichus's *On the Sacred Mysteries of the Egyptians* in Marsilio Ficino's Latin translation; as we saw in an earlier chapter, the ancient Neoplatonist averred that "Egyptian writers attributed their own books to Mercurius since they believed that he had discovered all things. . . . The columns of Mercurius are filled with learning."[49]

Tasso's verses disparage not one but two sets of columns; the other columns were intended to preserve *mille antiche memorie,* that is, centuries of history, from floods and fires. Tasso referred very precisely to both sets of columns in *Il Conte,* a dialogue on writing and symbolism published in 1594, the year in which he had wanted to publish *Il mondo creato*. He begins the dialogue with a description of his wondering contemplation of the obelisk that Fontana had recently erected in front of the Lateran palace; but rather than marvel at writing, he celebrates the obelisk as one of the wonders of Pope Sixtus's benefactions. Writing itself is profoundly unimpressive.

> The columns of the sons of Seth, one of which was made of [bricks and] mortar against the flood, the other of stone so as to be proof against fire, and those of Mercury, in which the sciences of the pagans were later written down, as Iamblichus writes at the beginning of his *Mysteries*, and the epitaphs of Semiramis or Jacob,[50] and the pyramids and the obelisks were written upon [*riscritti*] in letters that are less ancient than those inscribed on our souls, if it is indeed true that they do not resemble a *tabula rasa*, empty of writing.[51]

Tasso finds a way to deprecate and diminish the importance of all human writing, whether alphabetic or hieroglyphic. Interestingly, he reversed the traditional materials of the Sethians' columns, leaving the brick one vulnerable to flooding and the one of stone to fire. Was this a mistake or sabotage? In either case, the detail illustrates Tasso's hostility to the idea of anything resembling a proto-Bible. As in the *Mondo creato*, he refuses to acknowledge the wondrousness of a feat which, he says, antiquity wrongly celebrated for conferring permanence on human knowledge. Time long ago swept both sets of columns into the night of oblivion, Tasso claims, down to their last vestiges. Thus he does not merely dispute the antiquity of writing and the arts, but denies them any homage, since the original tablet is the human soul and God the eternal, original writer.

Tasso relied on Mercati's *Obelischi* for much of this writing lore, although, as several previous chapters have shown, a scholar of his time would have had many alternative sources. In fact, there was no need at all for Tasso to read a book on the subject, thanks to Pope Sixtus V: he could have discovered the columns of the Sethians, Hermes Trismegistus, and other illustrious pioneers of writing among the frescoes of the Salone Sistino (1588) in the new Vatican Library. But whether or not Tasso ever saw the frescoes, his writings intimate that he concluded they were, beyond their association with a pope that he courted, hubristic and banal. In his own hyper-Catholic and Neoplatonic way, Tasso rejected *sola scriptura* by refusing to countenance proto-Bibles or, indeed, to dignify the art of writing at all.

CHAPTER TWELVE

The Age of Grand Collections, 1600–1800

> The word *bibliotheca* refers to three things: a *library*, a *bookcase*, and a *collection of books*. The word is Greek but came into use in Latin too. Latin does have the word *libraria*, but that actually refers to a shop with books for sale.
>
> —JUSTUS LIPSIUS, *De bibliothecis* (1602)

In early modern Europe, organization and institutionalization allowed commerce and politics to flourish as never before, inspiring and underwriting dramatic new intellectual developments. Improvements in navigational technologies; the foundation of the great Dutch and English trading companies; Portuguese, Spanish and French colonial ventures; and Catholic missions to Latin America and China gradually opened European eyes to the existence, and ultimately the realities, of writing systems and lore as different from the alphabetical culture of Latin, Greek, and Hebrew as dawn from midday.

Since the late fourteenth century, Europeans' knowledge and interest in the history of writing had been deepening. It had also been expanding into areas that medieval writers had known little about, and often cared less. The sixteenth century had hosted a concentration and specialization of studies about writing; rather than being ensconced in universal histories and studies of individual authors such as Pliny the Elder, Livy, and the epic poets, extensive lore about the history of the art could now be found in works such as Sixtus of Siena's *Bibliotheca Sancta* and Theodor Bibliander's *Common Principles of All Languages and Letters.*

In the seventeenth and eighteenth centuries, even more extensive histories of writing were produced, some of them much larger than anything previously available. The histories of books and of libraries were not only developing within the pages of the new treatises on writing, but would be explored separately after 1600. The advent of printing had disseminated lore about writing, books, and libraries far more widely than in the manuscript ages and had ensured its survival through hundredfold reproduction of texts. Increased availability of books had important consequences for the study of both secular and sacred literature. The reform of religious practice and doctrine led to sectarian institutionalization at a level not seen over the previous centuries. Lutherans, Calvinists, and Tridentine Catholics fortified or completely transformed their doctrines of worship and their organizational structures. Universities long established continued to grow, and new universities were founded, particularly in Protestant lands.

Private, municipal, and national libraries were growing larger, more magnificent, and more numerous throughout the seventeenth and eighteenth centuries, when secular authorities and wealthy private citizens were competing with monastic foundations and popes by collecting manuscripts and printed books on an ever-larger scale.

BRICKS, MORTAR, AND BOOKS

Intellectuals interested in the history and lore of writing involved themselves in the work of institutionalization through associations and edifices that ranged far beyond the churches, monasteries, and universities that had traditionally preserved writing and books. Scholars carried their activity of collection, classification, and study to previously undreamed levels, thanks to increasing support and funding by powerful patrons, whose interest in written culture was as much political as intellectual. Institutions proliferated for the purpose of expanding all forms of knowledge about writing, and the ideal of a transnational "Republic of Letters" was beginning to affirm itself.

Museums and libraries made these activities concrete, giving them a visitable, physical home. Moreover, numerous publications illustrated in words and images the material and intellectual greatness of the Vatican Library and its princely rivals.[1] As we saw in chapter ten, the contemporaries of Pope Sixtus V were already engaged in propagandizing his reconstruction of the papal library, both in its rich pictorial decoration concerning the history of writing and in a variety of encomiastic books.

The ideal of a "universal" library containing every book had existed since Ashurbanipal. Hellenistic institutions came even closer: though it had been rivaled by Pergamum, the archetypal library for Europeans was still the Alexandrian Mouseion, or "house of the Muses," which originally housed the Ptolemies' book collection. Over subsequent centuries, however, libraries, while more numerous, had shrunk in relation to the magnitude of their surrounding cultures (remember the hardships and frustration endured by fifteenth-century bookhunters). In medieval Christendom, the idea of a universal library was partially supplanted by collections of relics and reliquaries maintained by churches and monasteries, which preserved the memory of saints by enshrining their physical remains as conduits of charismatic power. Secular princes of the fifteenth century increasingly owned libraries, along with assortments of relics, artifacts, treasures, oddities, and curiosities. By 1600, *Wunderkammern*, or chambers of wonders, were housing collections of natural and human-made marvels. As the name indicates, *Wunderkammern* were often founded on no other criterion than beauty, strangeness, or inexplicability.[2] These princely and bourgeois proto-museums developed into such public institutions as London's British Museum, founded around the collection of Sir Hans Sloane in the year of his death (1753), and Oxford's even older Ashmolean, based on the collections of John Tradescant (d. 1638) and his namesake son (d. 1662).

Aside from these brick-and-mortar institutions, the age's interest in writing also nurtured more abstract learned collections in the predominantly secular associations of scholars interested in a common topic. Already in the fifteenth century, academies—learned societies based outside of universities that were originally more like salons or private clubs—had surrounded such philosophically inclined intellectuals as Pomponio Leto in Rome and Marsilio Ficino in Florence.

By 1500, modern languages and literatures were coming to be seen as worthy to compete with their ancient models, and academies were formed to explore and propagate ideas and examples of vernacular excellence. Patriotism still had a role to play in such efforts: when the patriotic ideology of an academy was not explicit, it was implicitly or intrinsically connected. Two of the greatest early examples are the Accademia della Crusca (1583) and the Académie Française (1635), each formed to regulate or "purify" a modern language, canonize its most cherished literary classics, and collect their riches in dictionaries. Nor were books the only source of knowledge and pride. The Académie des Inscriptions et Belles-Lettres (1663) and the Society of Antiquaries in London (1707) also undertook the study of artifacts other than books, whether produced in antiquity or in their own cultures. On a looser and more inclusive pattern resembling that of a gentleman's club, the Society of Dilettanti (1743) welcomed antiquarians, archaeologists, travelers and artists; its members had a hand in founding the Royal Academy of Arts (1768), and it is still active.[3]

During these two centuries, books that embodied the idea of a "grand collection" of writings flourished as never before. These included the linguistic dictionary, the library catalog, the encyclopedia, and, most explicitly, the treatise on the history and characteristics of writing. None of these institutional genres was brand new in 1600. Each had previously existed in some form, often for centuries. The *catalog* listing the holdings of any sort of material collection testified to its existence and locale. Thanks to the spectacular talents of engravers, the Age of Grand Collections perfected several varieties of grand, illustrated catalogs, including the luxurious tomes that displayed the holdings of chambers of wonder. *Wunderkammer* books were museums on the page—visual, verbal, virtual surrogates for visiting the collections in situ. The earliest was *Dell'historia naturale*, published by the Neapolitan apothecary Ferrante Imperato in 1599 (figure 12.1). Other, more specialized books explored particular categories of marvelous items, from minerals and fossils to exotic animals and distant peoples. The "monstrous" exerted a powerful lure: anomalous births and such portentous astral events as Kepler's supernova of 1604, as well as disquieting cultural practices such as cannibalism,

FIGURE 12.1. A *Wunderkammer*. *Historia naturale di Ferrante Imperato*. Naples: Costantino Vitale, 1599.
Sheridan Libraries Special Collections, Johns Hopkins University, 508.I m73 1599 QUARTO c. 1.

polygamy, and human sacrifice, were reported and often embellished, both anecdotally and in catalogs of varying specificity.

BIBLIOGRAPHIES

As we should expect, a particularly important species of cataloging in this age was the bibliography. Finding inspiration in earlier works like the legendary inventories of the Alexandrian Library, and even more in such virtual or ideal collections as those contained in Conrad Gesner's *Bibliotheca universalis* (1545, 1574) and Michael Neander's *Erotemata* (1565), scholars used every available resource—including abundant money and time, large collections of their own, and extensive networks of patrons and correspondents—to compile encyclopedic bibliographies. They also explored subtopics, including ancient historians who wrote in Greek (Gerhard Johann Voss, d. 1649), Greek literature from pre-Homeric times to the fall of Constantinople, biblical pseudepigrapha, medieval Latin literature (all by Johann Albert Fabricius,

FIGURE 12.2. The Imperial Library and *Wunderkammer* at Vienna.
Edward Brown: *Durch Niederland, Teutschland, Hungarn [. . .] Reisen.*
Nuremberg: Verlegts Johann Zieger, 1686.
Sheridan Libraries Special Collections, Johns Hopkins University.

d. 1736), a panoramic history of philosophy from the pre-Socratics to modern times (Johann Jakob Brucker, d. 1770), and even histories of literature in the modern vernaculars.[4]

The bibliographies of Vincent Placcius (d. 1699; *Theatrum anonymorum et pseudonymorum*, 1708) and Christoph August Heumann (*De libris anonymis ac pseudonymis*, 1711) confronted the problem of authenticity by compiling encyclopedic lists and analyses of forged, plagiarized, apocryphal, and pseudepigraphic works. The treatise *On Learnèd Impostors* (*De doctis impostoribus*, 1703) by the German philologist Burchard Gotthelf Struve (d. 1738) was a lively study that dissected both forgery and related forms of literary fakery, foreshadowing modern scholarship.

The dream of grand collections is nowhere better expressed than in an engraving of the Imperial Library at Vienna from 1686, which accompanies the account of a grand tour taken by Edward Brown. The panoramic view concentrates on the towering stacks of the library's holdings in the foreground, but a series of open archways leads the eye successively to two rooms of a *Wunderkammer*, a courtyard, and what appears to be a street scene beyond the building. Altogether, the engraving seems to suggest a mental journey from reading to the contemplation of rare or precious objects, to the everyday world (figure 12.2).

ENCYCLOPEDIAS

The most fortunate vehicle for propagating knowledge in the Age of Grand Collections was the universal encyclopedia in a vernacular language.[5] Few of these lacked something substantial to say about writing; at a minimum they declared how fundamental letters were to the perpetuation of human memory. Extensive articles on writing, books, and libraries appeared in Ephraim Chambers's *Cyclopaedia, or a Universal Dictionary of Arts and Sciences* (first ed. 1728, 2 vols.), the first truly modern encyclopedia. But the iconic work of this age was D'Alembert and Diderot's *Encyclopédie*, with its dozens of volumes and thousands of pages, compiled by an international team of scholars between 1751 and 1772. It was based in part on expanding the articles in smaller, previous encyclopedias, especially Chambers's *Cyclopaedia*. It would take an entire chapter to discuss the numerous long and highly informative articles on writing, books, and libraries in volumes 2, 5, and 9 of the *Encyclopédie*.

Encyclopedias centered on aspects of religion were numerous. An illustrious example of an encyclopedic biblical scholar was the Benedictine Antoine Augustin Calmet (d. 1757), who published in French several variants of his commentaries on and translations from the Bible, containing a number of dissertations or compact analyses of topics mentioned in or relevant to the Bible, including everything from Old Testament giants to modern Hungarian vampires. Nor did he forget to include a dissertation "Concerning the Material and Form of Ancient Books, and the Divers Manners of Writing."[6]

The Age of Grand Collections reveled in its ability to mass-produce copies of even the most monumental earlier efforts, including the four huge tomes of Vincent of Beauvais's thirteenth-century *Speculum Maius* (*Greater Mirror*, printed 1624). Vincent's several thousand pages exemplify the long-standing, nebulous borderlines between biblical commentary, universal history, and the encyclopedia.[7] Thirteenth-century Scholastic culture had produced other, more specialized works akin to catalogs and encyclopedias, such as the *Legenda aurea* (*Golden Legend*) and Aquinas's *Systematic Theology* (*Summa theologiae*). The *Legenda* was a *Who's Who* of Christian history and mythology: assembled from sources written over many centuries, it was filled with pious hagiographies showcasing wondrous divine confirmations of the saints' holiness and charisma, and was a "best seller" even before the age of printing. After 1615 an entire confraternity, the Bollandists, took over the task of collecting saints' lives, creating the gargantuan *Acta Sanctorum* (68 volumes, 1643–1940). Aside from Aquinas's, other summas of Catholic theology attempted to condense hundreds of years of commentary and speculation about articles of faith (what to believe) and morals (how to behave). Catechisms, both Catholic and Protestant, have been called the dominant literary form of the Reformation and early modern era; they distilled the contents of summas into formulas that could be more readily understood by parish priests and Protestant pastors, and more easily taught to their congregations.

Histories were becoming ever more specialized and, to a considerable extent, secularized. Histories of writing, books, and libraries were no exception and in fact were often substantially free of religious bias, even toward Jews and Muslims. At the beginning of the seventeenth century, book-collections received their due when the renowned Flemish humanist Justus Lipsius (Joest Lips, d. 1606) brought out his *On Libraries*. *De bibliothecis syntagma* (1602) was the first specialized, self-standing treatise on the subject, but until recently, the book was not considered among Lipsius's major works, perhaps because in size it was practically a pamphlet. However, its example was massively important, inspiring much larger, more thorough works throughout the seventeenth and eighteenth centuries. Lipsius began by conjecturing that the origins of writing and libraries were entwined: "The library is an ancient institution, and—if I am not mistaken—one as old as writing itself. Once knowledge and wisdom came into the world, writing followed soon after. And writing could not have been useful unless books were preserved and organized for present and future use."[8] We notice right away that, in contrast to the Adamological speculations of previous centuries, Lipsius considers that knowledge and wisdom preceded the advent of writing, despite the implications of his first sentence. Because he intended to write history rather than myth, Lipsius quoted Diodorus Siculus, the pagan universal historian who best defined the genre in human, empirical terms, rather than follow the theocratic, top-down model of Christian historiography: "The first king to have a famous library (so far as memory preserves) was Ozymandias of Egypt. He built, among other famous works, 'a sacred library, and on its façade had it written . . . *psuches iatreion* a hospital for the soul.'" Admitting that the pharaoh's library had perished in antiquity, Lipsius concluded, "Although Ozymandias was among the most ancient of kings, there is no doubt that the precedent of building libraries survived . . . and that in Egypt, some libraries always existed from then on, particularly in temples under the care of priests." For 250 years, until Shelley's "Ozymandias," Lipsius's remarks were to make the ancient king's name synonymous with the ideal of preserving human culture by means of writing rather than infamous for destroying it through tyranny and hubris.

Lipsius added that "there is much evidence" about ancient Egyptian libraries, and, like Herodotus and Diodorus, he connected Greek culture to a barbarian predecessor, precisely through its oldest literary monument. He quoted a twelfth-century Byzantine commentary recording that "a certain Naucrates accused Homer of plagiarism, saying that he had come to Egypt and had discovered the books of a woman named Phantasia, who had written the *Iliad* and *Odyssey*, and had deposited them at Memphis in the temple of Vulcan. He alleged that Homer found these books, pretended they were his own, and published them." We have seen that accusations about Homer's dishonesty were as old as Pseudo-Dictys of Crete, if not Herodotus, but Lipsius cautioned that "I believe that [Naucrates's] story is wrong about Homer." Nevertheless, he said, "it confirms the existence and tradition of libraries in temples." Lipsius understood that even slanders, deliberate or accidental, have documented emotional attitudes toward writing in different eras.

Joachim Johann Mader's *On Libraries and Archives*, published in 1666, collected evidence about the history of writing. Rather than write a unified treatise like Lipsius's, Mader compiled a sourcebook of histories of libraries, creating a state-of-the-art archive that collected knowledge and speculation about the history of writing from antiquity to his own time. Lipsius's little book was the second item in Mader's collection, preceded only by Isidore of Seville's chapter on libraries. *De bibliothecis* was edited and reprinted in 1702 and then quickly supplemented by two further volumes in 1703 and 1705.[9] The three tomes gathered accounts of writing and libraries, including Richard of Bury's complete *Philobiblon* and parts of Neander's *Erotemata* (and numerous other accounts mentioned above), making *De bibliothecis* a valuable resource for historians of writing and libraries in the eighteenth century.

Already in Mader's time, other scholars, more in the tradition of Lipsius, were disentangling the history of libraries from two millennia of myth and legend. Three years after Mader, in 1669, another treatise bearing the title *On Libraries* was published by Johannes Lomeier, a pastor and teacher in the Dutch town of Zutphen. *De bibliothecis liber singularis* was, as its subtitle proclaims, "a treatise in one book" rather than a sourcebook like Mader's (see figure 12.3).

SCHOLARSHIP

Among many histories of writing in this period, perhaps the most comprehensive was Daniel Georg Morhof's *Polyhistor*, its title meaning "writer in many fields of knowledge."[10] Even more completely than Bibliander in the previous century, Morhof (d. 1691) attempted to survey everything relating to the nature and history of *litteratura* in its broadest sense, that is, the cultures of writing. First published in 1688, by 1747 the enormous work had undergone three further editions curated by renowned scholars. The first book, *Polyhistor litterarius*, opens with chapters on *polymathia* (mastery of many disciplines), literary history, library science, manuscripts, condemned books, pseudonymous and anonymous works, mystical and secret books, and "secret physical-science books, especially [al]chemical." These 120-plus pages are followed by supplementary essays concern-

FIGURE 12.3. King Solomon in his study. Title page, Johannes Lomeier, *De bibliothecis liber singularis*, 2nd ed. Utrecht: Johannes Ribbius, 1680. Sheridan Libraries Special Collections, Johns Hopkins University, Z721.L65 1680 c. 1.

ing "what is divine about studies," and discuss learned societies (both secret and not), learned conversation, writers about libraries, catalogs of libraries, biographies of authors, the usefulness of library history, writers' commonplaces, and polygraphers (writers on multiple topics). Six further books follow: *On Method, On Excerpting, On Grammar* (beginning with writing, its excellence and inventors, ways of writing, and artificial, "universal" languages), *On Criticism* (including a chapter on antiquarians), *On Oratory*, and *On Poetry* (which includes material on mythographers, those scholars who collected ancient myths or studied them systematically).

Morhof's ambition and enthusiasm dwarf those of his predecessors: he begins by comparing the task of writing his huge book to a transoceanic voyage of discovery, casting himself as a kind of Columbus, an admiral of the literary Ocean Sea. "The word *polymathia* embraces great extents of territory, although it has definable spaces and borders. But clearly, the history of writing [*historia litteraria*] is even broader: it comprises not merely *polymathia*, but the very birth and progress of individual sciences down to our own time. No one has been able to cross this Ocean, though not a few have cruised along its shores. Partial histories of study [*philosophiae*], whether ancient or modern, have been attempted by many—the entire history of writing since its birth, by no one!"[11] Rarely has a scholar made so good on such a titanic boast. It may be possible to find an important work of scholarship that Morhof or his continuators neglected, but the task would be unenviable. Burchard Gotthelf Struve, one of the great early eighteenth-century scholars, published his own influential *Introduction to the Knowledge of Literature and the Use of Libraries* for the guidance of students. Struve's handbook was republished for decades after his death, and by 1754 several additional layers of notes and commentary by other experts had swelled the *Introduction* to over a thousand pages.

Throughout the seventeenth and eighteenth centuries, scholarship expanded its historical and geographic reach, so histories of literature and philosophy included ever more detailed histories of writing and books, as in the examples of Morhof and Struve. Bible scholarship and the study of the classics were far more closely linked than today, for both relied to a remarkable degree on continual reexamination of ancient authors. The scientific archaeological study of Egyptian, Assyrian, and Chinese antiquity was still in the future, but attempts to write the history of writing did more than simply mark time. Spanning the late sixteenth and early seventeenth centuries, Joseph Juste Scaliger's attempts to reconstitute Eusebius's original Greek *Chronicle* yielded much fuller knowledge of Chaldean and Egyptian antiquity,[12] and recovered large swatches of both the real Berossos's *Chaldaica* and the infamous Book of Enoch, each almost erased by early Christian theologians. Discoveries by scholars like Scaliger revolutionized the history of writing even before systematic archaeology became possible around 1800.

Three-quarters of a century after the solitary printing of Bibliander's book *Common Principles*, the most famous and long-lived general treatise on the history of writing in the early modern period was published by the Jesuit Hermann Hugo. Hugo says he was casting about for a way to make his reputation in the world of letters when he noticed a curious fact. There was no "history of writing," for no one had yet tried to trace its complete evolution; it had always been discussed unsystematically, piecemeal and in parentheses, as it were. So in 1617, Hugo published *On the First Origin of Writing and the Antiquity of Literary Endeavor (De prima scribendi origine et universa rei literariae antiquitate)*.[13] Being careful to defend his "few little pages" as not "only about letters" and as "a treatise not of Grammar, but of Philology," Hugo stipulated that it examined "everything relative to the antiquity of the whole literary endeavor." As he intended, his book became an obligatory reference for any discussion of the history of writing, particularly after being republished with extensive annotations in 1738. His claim to be the first true historian of writing was exaggerated: as we saw in chapter ten, generalized histories of writing and collections of alphabets had begun in the mid-sixteenth century, with writers like Bibliander and Postel. But over the next two centuries, numbers of writers emulated Hugo's effort, slowly but steadily increasing and disseminating historical knowledge about writing. Thomas Bang's *Caelum Orientis*, or *Eastern Sky* (1657), began by opposing Pliny's arguments that writing had "always" existed, and provided an exhaustive summary of ancient and modern speculative histories of the subject.[14]

Despite ongoing fascination with traditional myths and legends, scholarship on writing was becoming recognizably modern. The Benedictine monastic order, which had a thousand-year history of preserving knowledge by copying manuscripts, produced two thoroughgoing treatises on the history of handwriting, which are still touchstones for paleographers, the scholars who classify and date scripts from antiquity to early modern times. Jean Mabillon's *De re diplomatica* (*On Manuscripts*, 1681) and Bernard de Montfaucon's *Palaeographia graeca* (1708) built on centuries of studying and thinking about the material history of writing, from Pliny to Bibliander and Hugo; they produced veritable encyclopedias of manuscript culture and its practical engagement with society.

The bread and butter of paleography in this period was furnished by the massive increase of forgeries over the previous several centuries. Mabillon was particularly concerned with the creation of forged Latin charters in the Middle Ages and their influence on monastic and secular politics; the history of handwriting held the key to dating manuscripts and thus to judging the validity of claims to valuable properties. Other forgeries and their debunkers depended less on the characteristics of handwriting than on general historical and textual considerations. We saw that Lorenzo Valla neither had nor needed an original charter to see through the "Donation of Constantine" but relied instead on principles ranging from historical linguistics to political psychology. Such criteria were continually refined in the three following centuries.

For over two hundred years, the forgeries of Annius of Viterbo, and the imitative reforgeries they spawned, stimulated research into Greek and Roman archaeology, biblical history, Egyptology, and ancient Near Eastern languages. Annius also inspired—and tainted—evolving attempts to understand the mysterious Etruscan civilization, ranging from the *De Etruria regali* of the Scotsman Thomas Dempster (d. 1625) to Athanasius Kircher's *Latium* and *Arca Noë* a half-century later. Annius had attributed "lost and rediscovered" *texts* to famous ancient authors but for whatever reason had not produced any spurious *manuscripts*, even of the supposed Latin translations he published, although he had forged several inscriptions on stone, including one in anachronistic Greek. Collections of inscriptions would evolve during and after the sixteenth century into the science of epigraphy.[15] Annius's forgeries of Etruscan history also inspired a thorough winnowing of ancient writers for clues to the material history of writing and books.[16]

A future Vatican librarian, the Greek expatriate Leo Allatius (Leone Allacci), was so outraged over pseudo-Etruscan forgeries that he provided an important breakthrough in the history of writing materials. In 1634, Curzio Inghirami, a teenage prodigy inspired by Annius, began "discovering" strange buried capsules containing "ancient manuscripts" on his father's estate near Volterra.[17] These supposed eyewitness accounts described the destruction of Etruscan Volterra by Roman armies in the first century BCE. (The parallel with the "eyewitness accounts" of the Trojan War by "Dares" and "Dictys" is suggestive.) Inghirami could not have counterfeited the Etruscan language, which survives as a mere few dozen words even today, so although his fictitious eyewitness was a patriotic Etruscan priest, Inghirami composed his texts in Latin, the language of his supposed conquerors. Bitter controversy quickly arose between Inghirami's supporters, who saw his "discoveries" as confirmation of Tuscan and Volterran antiquity, and the learned doubters who raised historical and philological objections to the capsules and their contents. Foremost among the skeptics, Allacci enumerated similarities between Inghirami's "discoveries," Annius's notorious Etruscan forgeries, and the ancient "Books of Numa."

Inghirami and his defenders held out stubbornly, but the callow scholar's unfamiliarity with the history of materials doomed his arguments. His elaborate hoax finally collapsed, said Allaci, when a sharp-eyed critic noticed that one of his manuscript "originals" showed part of a watermark.[18] Knowing that watermarks were modern, Inghirami had tried to eliminate them from the paper he used, by trimming it into sheets of various sizes; but he fatally overlooked one watermark. Worse, as a very young man who read a lot, he understood Latin words but connected them to anachronistic things. He had misunderstood the Latin word *charta* (designating both papyrus and parchment), thinking it identical to the modern Italian *carta*—rag paper—on which he wrote. He was further confused by the famous linen books of Etruscan priests—which were written on cloth, not paper made from linen rags.[19] To annihilate the credibility of Inghirami's manuscripts, Allacci dedicated a hundred pages to the history of writing materials, quoting extensively from Pliny's description of papyrus and later historical sources. Undeterred, in 1645 Inghirami published a vindication of his "discoveries" in a large-format illustrated book of over a thousand pages, but his defense fell heavily on deaf ears.[20]

The scandal was so notorious that when Mabillon set criteria for identifying forged manuscripts in *De re diplomatica*, he used Allacci's caustic remarks on Inghirami's forgeries to alert posterity to the difference between paper, parchment, and papyrus.[21] In the seventeenth century, philological debunkers in the mold of Allaci proliferated. Aside from contesting the authenticity of individual works, their researches were often vital to advancing the historiography of writing. Forgeries of ancient manuscripts such as Inghirami's were an especially important laboratory.

The seventeenth and eighteenth centuries were a golden age for pre-archaeological speculation about the history of writing materials and the morphology of ancient books. Classically trained scholars, including professors of law and medicine, wrote numerous dissertations on the chronology of writing. These works, usually only a few dozen pages, served as a professorial prelude to freewheeling oral examinations for university doctoral degrees; they included items like Christian Gottlieb Schwarz's *First Disputation* on the ancients' construction and ornamentation of their books (1705) and Christian Saalbach's *On the Books of the Ancients* (1725).[22] These brief treatises quoted, combined, and recombined much of the lore we have discussed in previous chapters, and they were usually explicit or implicit defenses of Christian scriptural culture, but they also contained information on the material history of writing that has stood the test of time.

IN PRAISE OF WRITING

Emotional, explicitly poetic tributes to the art were not lacking. At London in 1711, William Nicols published *On the Invention of Writing* (*De literis inventis*), an anachronistic six-book Latin epic poem on this unwarlike topic.[23] Feigning modesty, Nicols said he wrote the epic in his youth, and his unusual merging of genres does sound like an appropriate project for a confident and enterprising adolescent. (Having died in 1655, Inghirami could not avail himself of Nicols's sensible alternative to forgery.) Nicols's choice of verse rather than prose casts *On the Invention of Writing* as an *Aeneid* of the pen rather than the sword, starring the art of writing itself rather than a heroic soldier like Aeneas or Achilles. A more likely model than the *Aeneid* was the somewhat older six-book poem *De rerum natura* (*On the Nature of Things*) by the Epicurean philosopher-poet Lucretius (d. 55 BCE). Early modern writers deplored Lucretius's materialistic denial of human immortality and a creator-god but celebrated the sensual appeal of his verses, and often cited his dictum that a beneficial poet resembled physicians who, "to coax children to take / Foul wormwood" or other bitter medicines, "first brush the rim of the cups with the sweet and golden juice of honey."[24] Though earlier, Lucretius's medicinal metaphor was a step beyond Horace's decree in the *Art of Poetry* that a good poem provides delight (*dulce*) as well as instruction (*utile*).

It was therefore appropriate that Nicols allusively dedicated *De literis inventis* to Ozymandias, that old pharmacist of the soul who had intrigued everyone since Diodorus. Nicols's woodcut frontispiece shows a scholar—presumably himself—hard at work in an elegant manorial library, well lit by lofty windows (figure 12.4). Over a door leading to the stacks, Ozymandias's motto *psuches iatreion* is inscribed in Greek capitals. Nicols's homage was timely. In Sweden, the Royal University Library of Uppsala had the Greek motto stamped on the covers of its books in 1710; when the Swiss monastery of Saint Gall built a new library in 1760, the monks, like Nicols, inscribed the phrase over the library's entrance, where it remains. *Psuches iatreion*, "house of healing for the soul," was especially appropriate for this collection, situated next to the infirmary where the monks' physical ailments were cared for.[25] Nicols's choice of frontispiece and motto reminded his readers that any praise of libraries was a tribute to writing and vice-versa, as the twin conservators of memory and of humanity itself, as Pliny had declared. As well-read as most writers of prose treatises on the subject, Nicols chronicled how writing had been praised over the centuries.

Medieval authors such as Boethius and Dante often mixed prose and verse in the same work, a genre known as *prosimetrum*. Similarly, Nicols's volume included his own prose commentary on his quirky epic, couched in his era's increasingly popular genre, the footnote.[26] His documentation was so meticulous that it threatened to submerge Horatian *dulce* in a flood of *utile*, by adding 71 pages of notes, and 68 more of erudite addenda, to scarcely 271 of poetry. Unsurprisingly, Nicols's opus was stillborn, never again printed. But it opens with sentiments that had a familiar ring to learned audiences:

> I've often wondered whom God did first inspire
> To paint the voice, allow us to see words;
> And by rude strokes and magic figures made it so
> That things believed impossible to catch
> Should be at last forever fixed in stone.
> Was it Moses or one more ancient still
> Who taught the strange new rite of painting words?[27]

Isidore's idea that letters are fetters is prominent here, along with his description of writing as an artful synesthesia that bypasses the ears and "speaks" to the eyes.

Nicols knew that a recent French poet had improved Isidore's praises enormously. In his 1682 translation of Lucan's epic *Civil War* (1st c. CE), Georges de Brébeuf transformed the Roman poet's brief aside about the Phoenicians' invention of the alphabet into verses that remained famous for centuries. In the 1820s, a French bibliographer recounted that Pierre Corneille (d. 1684), one of three consummate playwrights in Louis XIV's court, opined that "he would trade two of his best tragedies to have written Brébeuf's four famous verses."[28] True or not, the anecdote confirms how widely Brébeuf's brief tribute to writing was prized in the seventeenth and eighteenth centuries. Brébeuf added such charisma to Lucan's throwaway line that Nicols retranslated the Frenchman's verses into Latin as a compliment to both poets, the ancient and the modern. Nicols's "strange new rite of painting words" (*ritu pingere verba*

ΨΥΧΗΣ ΙΑΤΡΕΙΟΝ.
ΨΥΧΗΣ ΙΑΤΡΕΙΟΝ
animusque vicissim
Aut curam impendit populis, aut otia Musis.
S. Gribelin sculps.

novo) unmistakably recalls Brébeuf: "It's from him [i.e., Cadmus] we receive th' ingenious art / Of painting speech and speaking to the eyes, / And, penning sundry strokes of letters, / Giving color and body to our thoughts."[29]

Nicols sanctifies Brébeuf's *ingenuity*, recasting it as divine inspiration, simultaneously transporting readers back to Homer's Bellerophon and the almost tangible *power* of writing. Dethroning the legendary Cadmus, Nicols made the inventor of letters anonymous but cast him as a magus or a minor god, a greater hero than Achilles or Aeneas, perhaps echoing Horace's pledge to immortalize Lollius in his verses.

> Whoe'er he was—for time destroyed his name,
> He gave life after death to other men;
> None perhaps was ever worthier
> Of enduring statue or Grecian marble column.
> With godlike genius founding this art,
> He conjoined the peoples of both poles in speech.

Indeed, the anonymous wonderworking inventor conquered time as well as space, for

> He made the living and the dead converse,
> Allowed us to consult the dead from long ago.
> He brought them from their tombs, and gives us too
> The chance to speak to every future time;
> Joined age to age, so any learnèd man
> Might have the chance to live in every age—

FIGURE 12.4. "[No part of life is lost: all that is withdrawn from the law courts is devoted to the study, and] *thy mind in turn either bestows its efforts on the State or its leisure on the Muses.*" Claudian, "Panegyric on the Consulship of Fl. Manlius Theodorus" (399 CE) vv. 65–66 (Loeb Classical Library, trans. Maurice Platnauer). Frontispiece, William Nicols, *De litteris inventis libri sex*. London: Henry Clement, 1711.
Sheridan Libraries Special Collections, Johns Hopkins University, P252 .N53 1711 C c. 1.

> Both long ago, and in the future too—
> Yet ne'er be absent from his own.
> See, no distance now forbids men's voices
> To be heard, nor yet can fierce death steal them![30]

By writing in Latin, Nicols and earlier praisers of writing had limited their direct influence to the highly educated. But after Nicols, authors increasingly explained the history of writing to vernacular audiences unable or unwilling to read Latin.

To realize how the history of writing was transformed by vernacular writers during the eighteenth century, consider Daniel Defoe, the author of *Robinson Crusoe* (1719). Best known as a pioneer of the European novel, Defoe was by trade a journalist and pamphleteer, who earned his living by penning works on many subjects. In 1726, he published *An Essay upon Literature, or, An Inquiry into the Antiquity and Original of Letters*. Although no professional scholar, Defoe had originally aspired to join the clergy, so his ideas about the history of writing resembled those of contemporary experts, who were still dependent on ancients like Pliny and Josephus—except that he expounded them in English.[31]

Defoe's long subtitle proclaims that "there was no Knowledge of Letters, much more of Writing, before that of the two Tables of Stone written by the Finger of God in Mount Sinai."[32] Like others we have seen, Defoe preferred to believe God had obviously invented writing. A few dissenters reasoned that unless writing had existed before Moses, no one could read the Commandments (an idea that anachronistically assumes a large audience of readers, such as existed in 1726); but such quibblers were a minority.[33] To exclude or minimize God's role in creating the art seemed both improbable and impious; as yet no right-thinking Europeans or Americans openly doubted the Bible's assertion that it was the Word of God who had personally inspired or even dictated Moses's account in Genesis. The "Five Books of Moses" were obviously the oldest surviving document, and Moses was the oldest human author they mentioned—QED. The circular reasoning, reminiscent of sixteenth-century Adamologists, still felt like a solid exercise in logic.

True, there was no direct claim in the Bible that God invented the Hebrew alphabet; but Defoe agreed with experts that no evidence *outside* the Bible proved that writing predated it. Material evidence seemed concordant. Not even the ruins of Babel (or Babylon, Genesis 11:8), disproved the historical primacy of the Hebrew alphabet. "If there was anything known before this, 'tis more than we have any Account of in History or Monument among the Antiquities of the most antient Buildings. The ruins of the most antient cities show no Inscriptions. The old Babel, part of which remains to this Day, has no appearance of anything Written." Likewise, the Bible failed to mention writing in Nimrod's Babel or Cain's much earlier city (Genesis 10:10, 4:17). Lack of evidence was evidence of lack.

Defoe conceded that in Egypt, hieroglyphics could still be seen (whether he had actually seen them or only read about them is another question). Although ancient and well-preserved, hieroglyphics could not possibly be writing: "The Ægyptian Pyramids, the next piece of Antiquity to Babel, at least that we know of, which are Fair, and preserv'd entire, have yet no figures or Semblance of Letters left upon them."[34] The Bible shows the Egyptians had a highly sophisticated society, but "we yet know of no knowledge of Letters among them." Egyptians wrote, but "by a way particular to themselves . . . by *Hieroglyphicks* or paintings of Creatures and Figures." These were pictograms, not true writing: "However Ingenious the Ægyptians were in suiting those *Hieroglyphicks* to their own Understanding, it must be allow'd that it was but a poor Shift, compared to the Present Improvement of Letters." For over a millennium, Europeans had supposed that the hieroglyphics were symbols that concealed stupendous truths from all but a philosophical or religious elite. Learning to interpret them required personal initiation of the sort Du Bartas had described in *Les columnes* a century and a half earlier. Some of Defoe's predecessors pursued the chimaera of "translating" the hieroglyphic symbols, but, aside from Aldo Manuzio's "make haste slowly," few agreed what any particular hieroglyphic meant. Their attempts at interpretation, like analyses of symbolism by amateur Freudians in modern times, depended on assumptions about the "obvious" or "universal" meaning of symbols. Defoe merely assumed that hieroglyphic pictograms were a primitive and unsatisfactory expedient, a "poor shift." He remained uninterested in *how* hieroglyphics transmitted ideas: they were self-evidently not writing but "reminders," as Plato's Socrates would have scoffed.

In 1763, William Massey, who styled himself the "Master of a boarding school for many years at Wandsworth in Surry [*sic*]," published his *Origin and Progress of Letters*. He named Defoe's *Essay*, along with Nicols's epic and Hermann Hugo's *First Origin*, as among the "few writers, who have [expressly] treated of the invention of letters, and their various alterations and gradual improvements to the present time." Massey began by declaring that "the invention of letters, and their various combinations, in the forming of words in any language, has something so ingenious and wonderful in it, that most who have *ex professo* treated thereof, can hardly forbear attributing it to a divine original, and speaking of it with a kind of rapture. Indeed, if we consider of what vast, and even daily service it is to mankind, I think it must be allowed to be one of the greatest and most surprizing discoveries, that ever was made in the world."[35] Massey echoes Defoe's declaration that "the use of Letters is of divine Original" because "there was no knowledge of Letters, much more of Writing, before that of the two Tables of Stone written by the finger of God in Mount Sinai."[36] Massey might have also remembered Nicols's title page, which displayed a little woodcut of God's finger emerging from a cloud to inscribe the Decalogue tablets (see figure 5.1 in chapter five).

Schoolmasters were Massey's intended audience, so alongside the obligatory classical authors, he quoted English praisers of writing, rather unexpectedly including Alexander Pope's love poem "Eloisa to Abelard" (1717), on the infamous twelfth-century sexual scandal. The radical theologian Peter Abélard (d. 1142) had clashed with religious conservatives, including Saint Bernard of Clairvaux. Worse, around 1115 he seduced and impregnated his brilliant pupil, Héloïse of Argenteuil. In revenge, Héloïse's outraged uncle—Abélard's erstwhile benefactor—had his retainers castrate the priest. Rather than sympathize with the mutilated seducer, Pope imagined Héloïse as more engaging, a sorrowful cloistered nun, writing to Abélard that letters could overcome their enforced separation: "Heav'n first taught letters for some wretch's aid, / Some banished, or some captive maid; / They [letters] live, they speak, they breathe what love inspires / Warm from the soul, and faithful to its fires . . . / Speed the soft intercourse from soul to soul, / And waft a sigh from *Indus* to the *Pole*." What were nuns, if not captive maids, imprisoned against their will?

Pope's insight that women prized writing as a means of abolishing distance between loving individuals moved praise of the art from the "masculine" public sphere—preserving civilization over time—to the private domain of emotion. By overlooking Héloïse's Latin erudition, his poem exemplified the eighteenth- and nineteenth-century cult of Abélard and Héloïse as tragic lovers, a notion that hinged directly on the publication of their correspondence in 1616. After their letters were disseminated in vernacular translations, Jean-Jacques Rousseau allusively updated and reenacted their dramatic affair in *Julie, ou la Nouvelle Héloïse* (1761). As Pope might have imagined, such passionate epistolary novels were a staple of sentimental fiction throughout the century.

Massey, however, gave evidence that real women shared the respect for writing that Pope and Rousseau imagined for their heroines. Not that women's praise of writing was unheard-of: we recall Athenaeus's anecdote about Sappho's riddle, and Homer's many alleged female sources. But Massey took Pope's hint and also reprinted "verses said to be wrote by a lady" from a contemporary periodical: "Blest be the man! his memory at least / Who found the art thus to unfold his breast, / And taught succeeding times an easy way, / Their secret thoughts by letters to convey; / To baffle absence, and secure delight, / Which, till that time, was limited to sight."

In addition to Nicols's Latin, Brébeuf's four French verses inspired a whole dynasty of writing-praisers in the modern vernaculars. Massey translated them thus: "Whence did the wondrous, mystic art arise, / Of painting speech, and speaking to the eyes? / That we, by tracing magic lines are taught / How to embody, and to colour thought?"[37] He quoted another ecstatic rhapsody that combined Brébeuf's abstract praise with the private concerns of Pope's Eloisa and Athenaeus's Sappho.

On the Art of Writing.

QUERY.

Tell me what genius did the art invent,
The lively image of a voice to paint?
Who first the secret how to colour sound,
And to give shape to reason wisely found?
With bodies how to cloath [clothe] ideas taught,
And how to draw the picture of a thought?
Who taught the hand to speak, the eye to hear,
A silent language, roving far and near,
Whose softer notes outstrip the thunder's sound,
And spread their accents thro' the world's vast round;
Yet with kind secrecy securely rowl [roll],
Whispers of absent friends, from pole to pole.
A voice heard by the deaf, spoke by the dumb,
Whose echo reaches long, long time to come;
Which dead men speak, as well as those alive,
Tell me, what genius did this art contrive?[38]

Massey saw women's interest in the art of writing as a path to independence, confessing that he "often wondered that women, who have a genius" for penmanship, didn't "qualify themselves to be teachers of writing and accounts to the youth of their own sex." Skipping over the entrepreneurial difficulties most women faced, he concluded, "I leave this hint however for future consideration" by women, "either to be teachers in girls' boarding-schools, if they be single women, or [whether] married, or single to set up female academies for themselves."[39] The lives and fictions of the Brontë sisters and other nineteenth-century women bore him out, but frequently fell far short of the happy independence Massey evidently imagined.

Young girls were influenced by Massey as well, most likely through their mothers or other private teachers, some of whom may have been just the sort of instructors Massey envisioned for "female academies." A needlework sampler, now in the Johns Hopkins University libraries' special collections, is signed, "Mary Godfrey finished her Sampler in the / Year of our Lord one Thousand seven Hundred / seventy five" (see plate 10). Mary embroidered verses that echo the wording and rhymes in Massey's praiser of writing:

> Great was that Genious [*sic*], most sublime that thought
> That first the curious Art of Writing taught.
> This image of the Voice did Man Invent,
> To make thought lasting, reason permanent.
> Whose softest Notes with secresy [*sic*] can roll
> And Mysteries proclaim from Pole to Pole.[40]

There may be numbers of other surviving samplers proclaiming such thoughts. But even if not, all samplers document that literacy ranked with sewing and other more domestic accomplishments of girls and women in Britain and North America.

Meanwhile, highly educated men continued to praise writing's public benefits. Thomas Astle was keeper of the records in the Tower of London for King George III. This lofty appointment gave him privileged access to centuries of official documents, which he could scrutinize for their use of handwriting, just as others studied them for their legal, political, and intellectual content. In 1784, and again in 1803, Astle published his results in English. But *The Origin and Progress of Letters* was not concerned only with historical documents. Like Defoe and William Massey, Astle realized that once discussions of the history of writing were no longer conducted solely in Latin, they became interesting to readers who were not professional scholars. Astle, too, wanted to reveal the invention and development of writing to a public that, since the turn of the century, had been avidly consuming newspapers, as well as best-selling novels like Samuel Richardson's *Pamela* (1740), in English.

Astle's book was larger, more detailed, and more erudite than Defoe's and Massey's; it was thickly illustrated with engravings showing historical samples of handwriting and, accordingly, more expensive. But despite its wealth of medieval and modern information, Astle's exposition of ancient writing showed little progress over his predecessors'. He lamented that finding the inventors of writing was a thankless, even hopeless, task and, as if updating Polydore Vergil's prediction, he complained that writing of all sorts "has been so long known and used, that few men think upon the subject; so inattentive are we to the greatest benefits from their having been long enjoyed." This was a paradox, for "the uses of writing are too varied to be enumerated, and at the same time too obvious to need enumeration."[41] How much truer his words are in the twenty-first century! Aside from specialized historians, how many users of writing bother to wonder about its history and emotional significance, much less investigate them? The centuries since Astle have roundly confirmed his and Polydore's complaints.

Since writing and alphabets were still synonymous in the eighteenth century, Astle, like Defoe before him, confidently asserted that the latest researches failed to show common ground between alphabets and hieroglyphics. "Monsieur Fourmont, Bishop Warburton, and Monsieur Gebelin, have endeavoured to shew, that alphabets were originally made up of hieroglyphic characters; but . . . the letters of an alphabet were essentially different, from the characteristic marks deduced from hieroglyphics, which last are marks for things and ideas, in the same manner as the ancient and modern characters of the Chinese; whereas [letters] are only marks for sounds."[42] But literary reality was changing slowly in the seventeenth and eighteenth centuries. As the foregoing anecdotes illustrate, progress in understanding the history of writing was still incremental, given the authoritative backlog of writing-lore sanctioned by literary tradition.

ANTEDILUVIANISM

Antediluvianism, the notion that writing was "eternal," was far from dead, though its health was slowly declining. There were two persistent forms of antediluvianism, the literal sort, still concerned with biblical patriarchs, and a more figurative kind, interested in archaic figures such as Ozymandias and Homer, who were attested by pagan, nonbiblical sources. The writings of the stupendously productive Johann Albert Fabricius (d. 1736) are some of the most precious resources concerning antediluvianism, although his reach was much broader. He published editions of Morhof's *Polyhistor* in 1731 and 1747, and further investigated several related subjects. As previously mentioned, he compiled enormous multivolume histories of classical, biblical, and early Christian literature, notably his immense *Bibliotheca graeca*, a history of Greek literature from its origins to the Muslim conquest of Byzantium in 1453 (4th ed., 14 vols., 1790–1809). His lifetime mission was (to paraphrase Sixtus of Siena) winnowing extant works from mythical and inauthentic ones; but to debunk the myths, he had first to collect them. The first volume of *Bibliotheca graeca* devoted over three hundred pages to ancient descriptions of writings attributed to Greeks who lived before Homer.[43] His much smaller *Codex Pseudepigraphus Veteris Testamenti* (2 vols., 1722–1741), performed a similar triage of books ascribed to biblical patriarchs.[44] Some, like the Book of Enoch, were still considered lost, destroyed by censorious ecclesiastics; others were truly lost, or simply mythical. Fabricius dedicated the first 94 pages of the *Codex* exclusively to books attributed to Adam; he only reached Methuselah on page 224, while pages 283–91 treat Noah's son Shem, also known as Melchizedech, whom the Carion chroniclers had celebrated as rector of the proto-Lutheran postdiluvian church. Not until page 334 did Fabricius leave Noah's Flood behind him. Here, as in the *Bibliotheca graeca*, he was thorough to a fault: his sources included the Talmud and Kabbalah in addition to the Hebrew Bible, plus pagan and Christian discussions from the Alexandrians to his own time; and he compiled similar studies of the New Testament and of Patristic literature. Since Fabricius's time, his sort of erudition has sometimes been considered obsolete, not to mention frivolous or merely obsessive, all too similar to the bibliomania that collects books without reading them. But from an emotional rather than technical perspective, such apparent eccentricities are central to humanity's long involvement with writing.

Until the late seventeenth century, and even later for more tradition-bound writers, the topic of antediluvian books and libraries continued commanding attention, since it documented the will of divine providence to preserve human civilization, as well as theological doctrine, through writing. Perhaps the most persistent characteristic of all the Bible-related stories is their emphasis on writing as wonderful, a God-given marvel. Because of the centrality of Noah's Flood as a radical recommencement of humanity, biblical literature—not only the texts of the Bible but also sources and predecessors that were presumed or imagined for it—received special attention throughout this period. The last half of the seventeenth century saw the peak of enthusiasm for these utopian bibliographies.

Ironically, considering his rather traditional encyclopedic ambitions, Hermann Hugo witnesses that fatigue and skepticism about dating the invention of writing were already beginning in his time. By 1617 theories were so numerous that he wrote them off en masse:

> Philo of Alexandria maintains that Abraham first invented writing; Flavius Josephus says it was thought up by Enosh, the son of Seth, long before Abraham's time, and he finds agreement in Polydore Vergil [d. 1555], Pieter van Opmeer [d. 1595], Pierre Grégoire [d. 1617], Juan Luís Vives [d. 1540], and Annius of Viterbo [d. 1502]. Bibliander [d. 1564] attributes it to Adam. Eupolemus, Eusebius, Clement of Rome, Cornelius Agrippa [d. 1535], Crinitus [d. 1507], Ravisius Textor [d. 1524], and Lilio Gregorio Giraldi [d. 1552], credit Moses. Clement and Cyril of Alexandria, Herodotus, Pomponius Mela, Herodian, Rufus, Festus, Zopyrion, Phornutus, Pliny the Elder, Lucan, and Joseph Scaliger [d. 1609], assign its invention to the Phoenicians, St. Cyprian to Saturn, Tacitus to the Egyptians. Several writers mentioned by Diodorus of Sicily credit the Ethiopians; other writers mention other inventors, ad nauseam.[45]

Hugo's speculators range from Herodotus in the fifth century BCE to his own contemporary Joseph Scaliger, the rediscoverer of Berossos, but the majority lived either in the first centuries of Christianity or between 1500 and 1700—two ages of Scripture and periods of acute cultural rivalries and sectarian turmoil. Summarizing two millennia of conflict, Hugo deplored "so many brawls among writers, so many warring opinions" on the subject, "a conflict that will wear out every tribunal." He concluded, "What will *you* believe, what disbelieve?"[46] Still, Hugo had more patience for the topic than this passage implies; he weathered the strain for fifteen pages, in the course of which he examined all the sources he mentions here and more besides. That he persevered suggests that the emotional significance of writing had not actually rendered a complete listing of theories stomach-churning.

Jacques Boulduc revealed the ultimate aim of such speculations in his treatise on *The Church before Moses* (1626), which expands on the Adamic Church of the *Chronicon Carionis*. Boulduc concluded a chapter titled "On the Progress of the Church in the Time of the Patriarch Seth, and on His Wondrous Sanctity" by asserting, "I easily convince myself" that Seth "conjoined supernatural, divine, natural and human knowledge" and that "he received many divine mysteries pertaining to our redemption, either from his father Adam or through God's unmediated inspiration, and in some excellent way accepted the mystery [*arcanum*] of the Incarnation of the Word of God, which, along with many other things, he took care to inscribe on those two columns."[47] Appropriateness is all. . . . Nearly a century later, even Thomas Bartholin's sober *De libris legendis* suggested that the Sethians' column of brick perhaps inspired the Babylonians' and Phoenicians' use of "baked bricks"—Pliny's term for clay tablets—to transmit their "customs, laws, and teachings."[48] The second and third essays of Thomas Bang's 1659 *Eastern Sky* surveyed opinions about the "literature of the Patriarchs," Adam, the Sethians, Enoch, Noah, and the Annian Etruscans. Bang took issue with Angelo Rocca and the Sistine murals about the multiple inventors of alphabets, and "modestly" contradicted Saint Jerome's idea that Ezra had to reinvent the Hebrew alphabet after the Jews returned from their Babylonian captivity.

Joachim Johann Mader's preface, titled "On Antediluvian Writings and Libraries," resembled Giovanni Battista Camozzi's little *Oration* of 1575: in two dozen pages it summarized generations of European scholars' Bible-centered bibliographic daydreams. Concerning the invention of writing, Mader refuted Diodorus Siculus, whom he had already attacked for denying providence by asserting that human language evolved rather than being bestowed by God. Reasoning like Steuco or Bibliander, he declared, "What Diodorus makes up about his Mercury [Hermes] is much more appropriate for Adam: 'He was the first who set forth the common speech in an articulate manner, and bestowed names on many things that had hitherto been without names: he also discovered the use of letters.'"[49]

As Christians had done since antiquity, Mader assumed that pagan mythology must be a faint, distorted memory of Moses's "history," so he declared that the myth of Hermes demonstrated that Adam invented writing. To corroborate this reasoning, Mader appended a hodgepodge of medieval Christian authorities.

> Which things ... Gobelinus Persona [d. 1421] included in his *Cosmodrome* so as to establish firmly that Adam was the inventor, or rather the architect of writing: "It is clear," he says, "that the invention of letters was made by the first man. Writing was discovered on account of memory. ..." Suidas [the *Suda*, ca. 1000] says the same thing most authoritatively under the heading "Adam, to whom arts and letters belong." And although Suidas also bestows the glory of this invention on Adam's son Seth ... Freculf of Lisieux, who is a much older author than Suidas (since Freculf flourished about 830 years after Christ), says this: "It is clear as day that the invention of writing took place in the time of the first man, who was still alive when his children and grandchildren studied wisdom so fervently."[50]

True, Mader admitted, Augustine was ambivalent:

> Although the excellent Bishop of Hippo, Saint Augustine, says in one place that he does not know "how it can be proved, as some people claim, that writing began with the first men, and endured until Noah's time, and passed thence to Abraham's ancestors, and on to the people of Israel," still, in another work he says that "We must not believe that Heber [or Eber, Genesis 10:24–25, 11:14–16], from whose name the word Hebrew is derived, preserved and transmitted the Hebrew language to Abraham only as a spoken language, so that the Hebrew letters would have begun with the giving of the law through Moses, but rather that this language, along with its letters, was preserved by that succession of patriarchs."[51]

Unsurprisingly, grist for Mader's mill also came from Godfrey of Viterbo, "who was familiar with holy libraries, and who gained great knowledge of both the world and its languages. [He] declares explicitly in the first part of his chronicle that 'Adam,'tis said, formed great columns of brick, / And decreed all events be recorded on them; / From them we copy all our ancient history.'"[52] These, said Mader, were none other than "the columns of Seth's sons mentioned by Josephus." So, "why indeed couldn't Father Adam, after whose pattern the whole world was arranged, have been strong-minded enough to perform such an endeavor before his sons and grandsons?"[53] Did Mader understand that Godfrey falsified Josephus's archetype? Perhaps not, since writers quoting from memory had distorted the columns for a millennium and a half. But no matter: like others before him, Mader found it *appropriate*, and therefore all but certain, that before his fall, the first man, created by the very hand of God, should have possessed writing and all other humanly available knowledge.

Throughout his preface, Mader's logic is poetic, not historical. "Whither, indeed," he asks, "would words go, those most excellent of all things, quickly traveling through the beaten air, never outlasting their own sound, if there were no signs for them in the letters . . . , by means of which the eyes can show words anew to the mind, even perhaps to the person who once uttered them?" Still mixing ancients with moderns, he quoted Maximus of Tyre (d. 185), that "human weakness itself has invented letters so that, by hastening to meet our mind's feebleness, their signs could remind it of our activities," and collated it with Isidore's comparison of letters to fetters: "by letters memories are tied down, lest they escape into oblivion. In such a great variety of all things, we could neither learn everything by hearing, nor contain it in memory." He recalled that Isidore added that "the signs of words" contained "so much power that . . . absent people speak to us without a voice: for they bring words in through the eyes, not through the ears."[54]

The doyen of such erudite speculators was the well-connected Jesuit polymath Athanasius Kircher (d. 1680), the most famous scholar of his era. Kircher was too much of an intellectual omnivore to devote entire works to the topic of loss, but his immense, lavishly funded and well-illustrated treatises on hieroglyphics and obelisks, on Noah's Ark, and on the Tower of Babel contained catalogs of lost books that would have made Richard of Bury blink. His *Arca Noe* (1675) encompassed not only the complete history of politics and science before

Noah's Flood, with diagrams of the exact dimensions and internal layout of the Ark, but also the literary history of the antediluvian patriarchs.[55]

Writers almost as erudite as Kircher often focused their attention quite narrowly on literary masterpieces before Moses. As we have seen, Mader was prominent among these fantasists. He commended Kircher for attempting "to prove that ancient tradition of the Jews and Muslims, according to which Adam was not only the inventor of letters but also a writer of books, citing many testimonies of those peoples in book one, chapter one of his *Obeliscus Pamphilii*." Kircher was both a connoisseur of written objects and the creator of a famous *Wunderkammer* in Rome, containing an obelisk among other marvels (figure 12.5). Several of Kircher's publications were likewise calculated to evoke wonder about the history of writing, and in the work that Mader mentions, Kircher was particularly enthusiastic about deciphering the Egyptian hieroglyphics and connecting them to antediluvian literature. Mader continued:

> According to Kircher, Abulhessam Mahumed Abn Abdalla Elhessadi says this in the Saracen History of things done since the world's beginning (a work still in manuscript), as Kircher translates it into Latin: "Thereafter God gave Adam twenty-one pages written and inscribed with his own letters.[56] This was the first book, and it was written in the first language, and it contained precepts and traditions that would come to pass in future generations; and it showed the translation of letters, pacts, statutes, and prophecies from the entire future world. And Almighty God portrayed therein the generations of individual men and their likenesses, and their organization, along with their kings, and everything that would be done on earth, right down to men's food and drink. Then when Adam saw in these pages what would happen to his offspring, he wept and mourned greatly. Then God commanded that he write it all down with a pen. And he took the skins of sheep and prepared them until they were perfectly white, and wrote on them with twenty-nine letters," and so on.[57]

So God not only foresaw all human history but wrote it down in a kind of hypertext and providentially made Adam copy it out (somewhat as he later made Moses recopy the tablets of the Decalogue in Exodus 34:27). Mader cherishes the evidence so much that he warns readers not to think that Kircher referred to

> those books concerning Adam, and written under his name, that heretics once forged and propagated, such as *The Revelation of Adam, When God Put Him to Sleep* [i.e., while creating Eve], *The Book of Genealogy, or, On the Sons and Daughters of Adam,* and that other one, on *The Penitence of Adam,* which Sixtus of Siena mentions, drawing on the works of Pope Gelasius [d. 494].[58] These are works that were long ago rejected as spurious and listed with other condemned books, including the works that *Sethite* or *Sethian* heretics dreamed up, claiming they were given by Adam to his son Seth, as you can read in Epiphanius's catalogue of heresies [ca. 377] under number thirty-nine.[59]

Kircher and Mader inspired entire monographs about innumerable writings of unrepeatable wisdom from the beginning of time. Gottfried Vockerodt's *History of Antediluvian Literature and Literary Societies* (1704) extrapolated the concept of learned academies into the world before the Flood: if writing existed at that early date, then scholars must have formed academies where they held learned discussions.[60]

In 1728, the pioneering historian of German literature, Jacob Friderich Reimmann, would damn Mader with faint praise, calling him more notable for his peculiar topic and the number of authors he excerpted than for critical acumen or careful analysis.[61] Despite Reimmann's greater learning and systematic approach, we might wonder whether his dismissive judgment about Mader's preface "On Antediluvian Writings and Libraries" was entirely serene. He himself had published *An Investigation Toward an Introduction to the History of Antediluvian Literature* only twenty years earlier.[62] Was he perhaps a bit sheepish about his own foray into the thicket of mythical bibliography? He had quoted Mader approvingly on "certain stones and columns to be found" in the Near East, "showing ancient and

FIGURE 12.5. "The theater of Nature and Arts in the residence of [Athanasius] Kircher, the likes of which can hardly be seen elsewhere." *Romani Collegii Societatis Jesu Musaeum celeberrimum.* (Museum of the Collegium Romanum of the Jesuit Order). Amsterdam: Ex Officina Janssonio-Waesbergiana, 1678. Sheridan Libraries Special Collections, Johns Hopkins University, AM101.R665 1678 FOLIO C. 1.

unreadable writings about which some would have it that they had already been described before the Flood," though Mader had merely referred to the Sethian columns and Pliny's Babylonian baked bricks. In this earlier book Reimmann had declared that "anyone who takes Adam for the first author of speech and yet wonders about the first inventor of letters does the same as a man who would search for a horse while sitting on it." Indeed, "the origin of speech is much harder to determine than the origin of writing," perhaps because spoken words "fly away," while writing "remains."[63]

Differently from Mader and the young Reimmann, Johannes Lomeier took a somewhat skeptical view of antediluvian literature. He reviewed traditional lore about writing and libraries in his second chapter, "On the Preservation of Historical Memory before the Time of Moses." As Melanchthon had been, he was intrigued by tales based on the Epistle of Jude's reference to Enoch's apocalyptic prophecy. "Because memory of [the prophecy] was preserved for such a long time, it can hardly be doubted that Enoch collected his prophecies in books. . . . The bolder [speculators] would have it that Enoch's book was large enough to contain 4082 lines." However, "that those books, written before the Flood, could survive it is in no way arguable [*verisimile*], since it devastated the entire earth." Referring to Tertullian's anomalous conjecture, Lomeier demanded to know "how anyone can presume to teach that Noah took [Enoch's] books into the Ark, since everything brought on board was carefully enumerated [in Genesis], and yet concerning those books the silence is absolute?"[64] It was much simpler to think that Noah taught his descendants viva voce, particularly since they were few and lived a long time. Nor was it necessary for God to reveal every detail about the creation of the cosmos or of subsequent history directly to Moses, who "could have learned those things from tradition." But Lomeier denied, with implicit contradiction, that Moses's "holy ancestors" had been "so stupid or mute as not to expound in their books whatever pertained to world history and redemptive doctrine, and entrust it to their descendants for faithful transmission." At any rate, God made a new beginning with Moses, "so that the purity of celestial doctrine, confirmed by divine authority and testimony, would be propagated and preserved, so that questions and controversies about the ancient, genuine, and pure doctrine of the patriarchs should not arise, nor new and detailed revelations be constantly sought after and expected."[65]

Despite mentioning the Sethian "columns" with respect, he opined that neither Josephus's authority nor his supposed eyewitness testimony to the survival of the stone stele were decisive. Surprisingly, however, he still concluded that Adam had the best claim on the invention of writing. Although some had attributed the invention to other biblical patriarchs, to Saturn, the Egyptians, the Ethiopians, or other ancients, Lomeier repeated Melanchthon's long-ago certainty "that [letters] took their origin from Adam. Who could believe that that most wise of all men lacked such necessary instruments [*organis*] by which to transmit his wisdom to his descendants? Remember that the first mention of letters comes from the time of Seth, and since he was the son of Adam, who could doubt that Seth learned his knowledge of them from his father?"[66] So earlier was still better, and Lomeier questioned Josephus's authority only insofar as the old historian had not explicitly named Adam as the inventor of writing.

The columns of Seth's sons remained a favorite topic of northern European doctoral dissertations: one was published in 1668, two in 1669, one in 1733, another in 1759, and some were reprinted. But they now accommodated broader considerations of the history of writing and its materials, which predisposed them to doubt the story. Times and tastes had changed since Melanchthon and Peucer: far from making the columns the source of Genesis, the doctoral treatises expressed skepticism that they ever existed. They dismissed the many variants in authorship and materials attributed to the columns as evidence of myth or imposture; and as the first author to mention the columns, Josephus came in for a fair amount of skepticism, some of it baldly anti-Semitic. The authors all agreed that no one could find evidence of writers older than Moses, whose authority they considered firmly established in Exodus.

CHAPTER THIRTEEN

Skepticism and Imagination, 1600–1800

> One of them, brand-new and nicely bound, was hit so hard that its innards spilled out and its pages were scattered. One devil said to another:
>
> "See what book that is."
>
> And the other devil responded: "This is the second part of the history of Don Quixote of La Mancha, composed not by Cide Hamete, its first author, but by an Aragonese who is, he says, a native of Tordesillas."
>
> "Take it away from here," responded the other devil, "and throw it into the pit of hell so that my eyes never see it again."
>
> "Is it so bad?" responded the other one.
>
> "So bad," replied the first, "that if I myself set out to make it worse, I would fail." And they continued with their [tennis] game, hitting other books.
>
> —CERVANTES, *Don Quixote*

Because legendary texts remained central to premodern cultural-religious identities, the stories about them went on being discussed exclusively in terms of their historical truth or falsehood. Nowhere do stories and studies of antediluvian literature make even a momentary concession to the idea that the Sethian stelae and other primeval texts might have a merely figurative significance. The antediluvian classics were discussed straightforwardly, as either historically true or untrue, that is, consonant with Genesis or not; otherwise, they were described with merely tacit approval or disapproval, treated implicitly as plausible or questionable. The absence of symbolic and allegorical interpretations shows how strongly Christian scholars remained invested in the literalist tradition dating back to Berossos and the *Timaeus*. No sooner had philology, paleography, historical linguistics, and even logic chipped away the final vestige of a story's plausibility than it metamorphosed directly from a wondrous testimony of truth to a symptom of naivety or fraud.

All through the seventeenth and eighteenth centuries, however, there was another, less emphatic but more ironic motive for disbelieving these stories: when read skeptically, they argued by default for the credibility of Genesis—usually the one extant book discussed—as a historical document. If Moses was not the oldest author, or at least an heir to Adam's or Abraham's invention of writing, where did that leave the Bible? It was safer and simpler to believe that God dictated the Pentateuch to Moses, making his five books the earliest writings.

A major source of erosion about antediluvian writing was the exaggerated hope that the sixteenth century had invested in Josephus's anecdote of the Sethian columns as evidence of pre-Mosaic scriptures. Treatments of this story showed more sudden and dramatic changes in attitude than did other writings on antediluvian classics. There could be several reasons. The increasingly fanciful Protestant elaborations on the myth of Adamic *Scriptura* held no appeal for Catholic writers and artists such as Angelo Rocca and the painters who frescoed his multiple inventors of writing on the pilasters of the Vatican Library's Salone Sistino. Perhaps Catholics welcomed a broader and more ecumenical history of writing because their theology made abundant room for postbiblical doctrinal traditions. But both tendencies created ideological expectations that could not bear the strain imposed on them by philology. Increased awareness of religiously and politically motivated counterfeits, both ancient and modern, especially in the wake of Annius of Viterbo's forgeries, clearly added to the skepticism. Finally, Flavius Josephus lost value as a witness. Although Godfrey of Viterbo's reference to him as an author of Holy Writ reflected the attitude of Christian exegetes after Isidore of Seville, by 1600 Josephus was beginning to be seen as a "mere" Jew:

Christians doubted his trustworthiness as an exegete, while his cozy relations with Roman conquerors tempted Jews to consider him a traitor to their culture and religion.

In the eighteenth century, the Sethian myth encountered outright mockery, even among writers in the vernacular languages. William Massey called it "silly," "highly improbable," and "a lame story."[1] In the second edition of his *Scienza nuova* (*Principles of a New Science*, 1730), the Neapolitan philosopher Giambattista Vico declared it was time to relegate Josephus's story to the *Museo dell'impostura*, the Museum of Hoaxes; in the third and final edition (1744), he softened the reference somewhat to the *Museo della credulità*, or Museum of Gullibility.[2]

There were rare, incongruous moments of modernity amid all the speculations about antediluvian writings and libraries. Several chapters in Lomeier's *De bibliothecis* give a history of the library as an institution. Aside from traditional mythical lore about the libraries of the Hebrews, Chaldeans, Arabs, Phoenicians, Egyptians, Greeks, Romans, and Christians before, during, and after the "Dark Ages (*barbaries*)," Lomeier discussed the libraries of "various other peoples" and those of famous individuals. He also examined the "siting, arrangement, and decoration" of libraries, their directors, and, like Richard of Bury, he enumerated the natural and human "enemies of libraries." But in contrast to Richard, to Mader, and to parts of his own book, Lomeier's chapter "On the Most Famous Libraries of Europe" is anything but mythical. He gave abundant historical and practical information about modern Christian libraries; for that reason, his book was translated, massively annotated, and published by the University of California in 1962, in a series of publications on library history. The editor praised Lomeier's chapter as rare and valuable testimony about late seventeenth-century European libraries. In fact, Lomeier's book was apparently seen as practical in his own time; it was reprinted eight times, wholly or in part, before 1962. Aside from Lomeier's own second edition (1680), *De bibliothecis* appeared in Johann Andreas Schmidt's second supplement (1705) to Mader's *De bibliothecis*, three printings of a pirated or unattributed edition in French (1680, 1685, 1697), and an English plagiarism of the French plagiarism (1739).[3]

It little mattered that Egyptians had no tradition of a Great Flood; Plato's *Timaeus* had made clear that far remoter historical origins could be imagined for writing than the cramped chronology of Genesis allowed. We saw that Herodotus and other Greek writers conceded that their culture derived from Egyptian civilization as early as the time of Homer. Perhaps for this reason, seventeenth- and eighteenth-century scholars mentioned Naucrates's scandalous accusation that Homer had stolen the stories of his epics from manuscripts by a woman named Phantasia in an Egyptian library, but didn't reveal whether they were puzzled or irritated by it. Conceivably, like the many other legends scholars collected, it was just a good story about the antiquity of libraries. But others besides Lipsius must have suspected it was false or wondered about Naucrates: Was he a cynical joker or perhaps an unimaginative literalist who misunderstood when a wittier scholar personified Homer's genius by inventing a female source named "Imagination"?

At any rate, the legend of Phantasia strengthened early modern scholars' conviction that libraries—the ultimate sources of reliable proof—had been invented eons previously, probably by the Egyptians, possibly by the very king that Diodorus mentioned as founding the library with the interesting motto. Certainly, Homer was the oldest extant Greek author, so even if Naucrates was a liar, the story still corroborated the antiquity of libraries in terms that Greeks could respect. Even if Phantasia never existed, she symbolized what Greeks considered a historical fact: that thanks to ancient Egypt, Greek culture had roots in deep antiquity, as Herodotus and Plato had claimed.

Photius, a ninth-century Byzantine who has been called the first book reviewer,[4] came to prominence among Latin readers in the seventeenth century. He cataloged and summarized 280 books, of which only about half have survived; among them he found other candidates for the honor of being Homer's source; tellingly, most of them were women. But unlike Phantasia, several were Greeks; indeed, Photius counted twenty-four "celebrated Helens" from the Trojan War era. Was it coincidental that this is the exact number of letters in the Greek alphabet, and of the books into which Alexandrian scholars divided the *Iliad* and the *Odyssey*? Photius records that one of the Helens, "the daughter of Musaeus of Athens," supposedly wrote a history of the war that inspired Homer.[5] This idea, like Naucrates's tale of Phantasia, was probably inspired by a scene in the *Iliad*. When Homer first introduces Helen of Troy, she is creating a pictorial history of the conflict over possession of herself: "She was weaving a great web, / a red folding robe, and working into it the numerous struggles / of Trojans, breakers of horses, and bronze-armoured Achaians, / struggles they endured for her sake at the hands of the war god."[6] Although the tablets of Bellerophon are the only writing the *Iliad* mentions, here Homer could be read as intimating that Helen's woven narrative was somehow his historical source, a kind of Bayeux tapestry without the captions. Homer's implicit comparison between textiles and texts remained a commonplace pun in Latin and the Romance languages (*textus* can be an adjective meaning "woven" or a noun for written "text").[7]

Such speculations could have been fortified by Book Four of the *Odyssey*, which presents Helen even more explicitly as a charismatic narrator. She tells Telemachus that when his father, Odysseus, entered Troy in disguise to spy, she alone recognized him, and she claims to have detained, entertained, bathed, and clothed him. Her story is implausible, as it would have neutralized "crafty Odysseus's" disguise and endangered his mission. But Helen's husband corroborates that she was even craftier than Odysseus: Menelaus claims that before Helen's reunion with him, she understood that the Trojan horse was a trick and almost exposed it by imitating the voices of the wives of the Greek soldiers inside it, tempting them to betray their presence.[8] Might Photius or some other early commentator have assumed that Homer had implicated Helen as his source, a kind of first-person human muse?

More modestly than Photius, the late seventeenth-century writer Theodor Ryck listed only sixteen female authors, including Helen and Phantasia, as possible written sources of Homer. It may not be coincidental that Pliny the Elder says the Greek alphabet was born when "Cadmus imported an alphabet of *sixteen* letters into Greece from Phoenicia."[9] Like the proto-biblical memoirs of Adam, Enoch, and Noah, these legendary women's originals of the *Iliad* and *Odyssey* all shared an intriguing, provocative failure to survive. So the broader tradition about Homer's inspiration may have had far more to do with nostalgia for a foundational moment of culture than with Naucrates's accusations of literary dishonesty.

At any rate, Phantasia's story was often repeated in the eighteenth century and frequently linked with the legend that Ozymandias invented libraries. Lipsius had written that "we cannot be certain who invented libraries, nor can the question be decided, since it is of such ancient date. If we said that the ancient Egyptians invented purely human wisdom, perhaps we would not be in error. This appears from the wealth of traditions about the Egyptians, and from their hieroglyphic figures and such things, from which we infer that a not inconsiderable use of libraries was introduced by them."[10] So far, so reasonable. But 150 years later, Struve noticed that some ancients had misattributed the invention to Aristotle, a venerated but much later author. "Strabo [d. ca. 19 CE] says in book seven of his *Geography* that Aristotle was the first who convinced Egypt to take such care of its books; but clearly Strabo was in thrall to some error, for Diodorus Siculus is our witness that the most ancient king of the Egyptians, Osymanduas [*sic*], built a library, and placed the inscription *Psuches Iatreion* [Healing-Place of the Soul] on its façade; and Osymanduas lived long before the time of Aristotle." Having thus defended both Ozymandias's historicity and the great antiquity of libraries, Struve turned to the even more speculative relation between Homer and Egyptian libraries. Paraphrasing Lipsius, Struve felt obligated to include Homer's alleged source, Phantasia.

> It can be easily conjectured that Osymanduas's concern for books was carried on by his successors, especially in their temples, through the attentions of the priests. One reason for thinking this is that a certain Naucrates accused Homer of plagiarism, saying that when Homer traveled to Egypt, he discovered the books of a woman named Phantasia, who had written her own *Iliad* and *Ulyssead* and deposited them in the temple of Vulcan at Memphis. Eustathius [d. 1198] tells this in his preface to the *Odyssey*, but he exonerates Homer of the charge [of plagiarism].[11]

The old stories were still holding out; whatever their inconsistencies, nothing convincing had yet come along to replace the charm of oft-repeated stories about the antiquity of writing.

No one needed to read Latin or Greek to make the connection between libraries and the ancient pharaoh: even the moderately educated could find it in D'Alembert and Diderot's French *Encyclopédie*, a work that had enormous circulation.[12] Cora Lutz once observed that the eighteenth century consistently associated Ozymandias with the phrase "healing for the soul,"[13] so Diodorus's mention of the optimistic motto may explain the old king's appeal for scholars. Being the founder of libraries is certainly a more inspirational claim to fame than Shelley's later, "I am Ozymandias, King of kings," which cast him as the obtuse, hubristic despot of a vanished desert kingdom.

DESTROYED LIBRARIES

Some scholars took a more personal interest in the fate of libraries. The renowned Danish anatomist Thomas Bartholin was haunted by the destruction of libraries, not the history of their foundation. He composed two books on the subject to mourn the incineration of his large, personal library, which included numerous manuscripts of his own works. The first, published in 1670, described his loss in a long letter to his sons.[14] To compound his sorrow, his library burned to the ground while he was attending the funeral of his mentor.[15] Two years later, in 1672 , he wrote *On the Necessity of Reading Books*, publishing it in 1676. It is difficult to imagine the emotional magnitude of disasters such as Bartholin's, but strangely, he did not dramatize it further in *De libris legendis*. Although still too frequent, such disasters are rare in our time, compared to Bartholin's day. Nowadays, accidental bibliocausts seem to have little impact on public opinion, whereas deliberate book burnings continue to fascinate average readers.

After losing his own collection, Bartholin soon received another library, or at least the custody of it, when he was appointed librarian of the University of Copenhagen. In celebration, he wrote *De libris legendis* to commemorate the fact that "the loss of my personal library opened the way to this public one."[16] Proud of his knowledge of the humanities as well as medicine and anatomy, Bartholin dared to one-up Ozymandias: "On the façade of the sacred library of Egypt, Ozymandias wrote that it was the healing-house of the soul; we shall call ours the healing-house of both soul and body."[17] Bartholin was so grateful to the University's academic council for this honor that he claimed he couldn't have been prouder if he had been an ancient prefect of the Alexandrian Library.[18] Bartholin's relief and gratitude stimulated him to recall more positive bibliographic legends than he previously had listed. His homage to primordial libraries inspired him to mention Homer's discovery of Phantasia's *Iliad* and *Odyssey* not once but twice.[19]

But like Michael Neander in the previous century, Bartholin was at heart a poet of loss and erasure, for even better reasons. Bibliocausts ran in his family; discussing the burning of his library, he averred, "I bear the calamity more bravely because it has long been customary for Vulcan [the Roman god of fire] to rage in our family. The very famous medical and mathematical books of my grandfather, Magister Thomas Finck, were consumed by fire, a loss that the venerable old man never recalled without tears, although he was otherwise unaffected by emotions."[20] Bartholin's chief consolation, if it can be called that, was the miserable company of great biblioclasms in antiquity. He mentions the Library of Alexandria, of course, but without neglecting misfortunes that befell smaller collections.

He recounts the famous story of how the "books of the Cumaean Sibyl," the most famous of the pagan Roman scriptures, fell victim to a disastrous business negotiation. "The same fortune awaited the Sibylline books, for when Tarquinius Superbus [d. 496 BCE] did not pay that price which a woman asked for the nine Sibylline books, six of them are said to have been consumed by fire, and the same price had to be paid for the remaining three."[21] Bartholin saw no need to pursue the tale, since his readers would have known that after the woman destroyed the first six, the three unburned "books of the Sibyl" became fundamental instruments of ancient Roman divination, only to be erased about 80 BCE, in the fire that destroyed the Temple of Jupiter Capitoline. Afterwards, they were supposedly reconstituted from fragments scattered all over the empire, but the new collection then burned definitively in 405 CE, during the barbarian invasions that led to the fall of Rome. The Roman general Stilicho, himself the Christian son of a Vandal cavalryman, may have considered the pagan oracles impious but more likely destroyed them because they were being exploited to undermine the Christian Roman government. (In fact, Rome fell to Alaric I and his Visigoths in 410, due to the vacuum of military leadership worsened by Stilicho's execution in a rebellion two years earlier.) Another collection of Sibylline oracles was compiled soon after the burning of the "restored" Sibylline books; but these prophecies, still preserved among the biblical pseudepigrapha, exploited the fame of the pagan collection to lend authority to Jewish and Christian religious propaganda.[22]

That Tarquin allowed six of the nine Sibylline books to be burned reminds Bartholin that "there has been no people, no age, in which the best books have not suffered harm from Vulcan the destroyer."[23] He dedicates ten further pages to anecdotes about the fire-god's depredations, interspersed with forlorn attempts to see the bibliocausts as providential dispensations of the Christian deity. Aside from the Alexandrian library, Bartholin recalls the destruction of the library at Constantinople, among whose holdings was "the intestine of a dragon twenty feet long on which the *Iliad* and *Odyssey* of Homer had been written in letters of gold."[24] Among deliberate book burnings, Bartholin pays close attention to religious bigotry: he cites the burning of Pietro Pomponazzi's *On the Immortality of the Soul* (1516) and of works by that philosopher's ancient forerunner Protagoras, who also "denied the immortality of the soul." He recalls as well the Roman Senate's order, in 181 BCE, to cremate the "Greek books of Numa," supposedly rediscovered in his grave, because they "seemed in some degree to oppose the religion" that Numa himself had imposed on Rome.[25]

The vulnerability of libraries lent them an explicitly erotic charm in Bartholin's eyes. As a kind of temple, they are the home of Pallas Athena, he says. She is the most beautiful of the goddesses, and men should not neglect her, "lest we resemble those eunuchs who guard their mistresses but never taste the delights of Venus." Despite salacious appearances, Bartholin has not confused Athena with Aphrodite. "[Pallas] brooks no repulse, so let us throw ourselves into her embrace [*amplexus*], lest we deprive ourselves of the best part of her proffered gifts, for she graciously offers her entire self to us, so that we can see her completely nude. When our minds are admitted into the voluptuous delights of intimate contact with such a goddess, no repentance is required of us. Although such a chaste goddess would be ashamed to strip off her tunic, she kindly suffers us to untie her girdle."[26] To a true scholar, says Bartholin, Athena behaves as a wife, not a trollop. His lubricious allegory can perhaps be forgiven as the ultimate proof of his love for books and *litterae*; certainly the specter of Vulcan's pyromania intensified his lust for learning. The old doctor's memories of his lost library must have been as intense as any graybeard's nostalgia for youthful lovemaking.

ENCYCLOPEDIAS, OR PORTABLE LIBRARIES

The general encyclopedia was in its way a library in miniature but one that could be reproduced on a grand scale as a hedge against the kind of universal erasure that antediluvianists had deplored since the times of Berossos and Plato. D'Alembert's idealistic *Preliminary Discourse* (1751) proposed the *Encyclopédie* as "a sanctuary where human knowledge is sheltered from time and turmoil," and he set out its objective in terms that would have stirred the enthusiasm of Flavius Josephus, Petrarch, or Poggio Bracciolini.

> What a benefit it would have been for our ancestors and ourselves, if the accomplishments of the ancient peoples, the Egyptians, Chaldeans, Greeks, Romans, and others, had been transmitted in an encyclopedic work, one that also explained the true principles of their languages! Let us therefore do for future centuries what we regret that centuries past did not do for our own. We dare affirm that if the Ancients had completed an Encyclopedia . . . and if this manuscript were the only one that had survived from the famous Library of Alexandria, it would have been able to console us for the loss of all the others.[27]

Richard Yeo explicates this grand dream, noting that Diderot's entry in volume five for *Encyclopédie* shows that "he regarded the work in which he was then engaged as a time capsule of the Enlightenment. In the event of a catastrophe, it would be a summary of intellectual accomplishments to be reactivated by a later age. This is the most powerful encyclopaedic vision: a work containing the collective knowledge of a community which might be put together again if all other books were lost."[28] D'Alembert's vision was, in sum, a practical, eighteenth-century relaunch of the kinds of bookish imagination about the powers of writing that had animated fantasies about antediluvian encyclopedias since the time of Berossos.

But the *Encyclopédie* was not uniformly celebratory of writing: as it documented, the Age of Grand Collections suffered more dramatically than its predecessors from information overload.[29] Even Vincent of Beauvais, who died in 1264, had lamented the *librorum multitudo* and *temporis brevitas* on the first page of his massive *Speculum maius*, and we saw Erasmus complaining of the same scourge in the sixteenth century.[30] After discussing strategies for managing the flow of books, the *Encyclopédie* concludes that the number of books is now so huge that "it is impossible not only to read them all, but even to know how many of them there are, or keep track of their titles," so that "one can only judge a very small number of books by reading them, given . . . the extreme brevity of human life. Besides, it is too late to judge a book after one has read it cover to cover. How much time would one risk losing by having such patience?" The *Encyclopédie* featured a number of choice anecdotes, including the paradoxical story of an implausible bibliomaniac who, to save space in his extensive library, tore out and saved only the worthwhile pages [*qui méritent d'être lûes*] of the books he read, consigning the mutilated remainders to his fireplace—even if that meant saving only "six pages from a work in twelve volumes."[31] Books had progressed from being a defense against the Deluge to being a flood in their own right.

COUNTERCURRENTS

Despite a common desire to find the prototypical creators of writing and the overwhelming religious conformity that undergirded most efforts in that direction, all was not sweetness and enlightenment in the histories of writing that scholars composed in this period. They were beginning to devote deeper critical attention to literary falsification, discovering along the way that not all forgeries were deliberate, malicious, or modern. Many were consecrated by traditions that were relatively ancient, rendering moot the question of "original" manuscripts. One of the more famous mythical examples is the *Tabula Smaragdina*, an alchemical text supposedly discovered in the tomb of Hermes Trismegistus. An early modern defender of its authenticity claimed that Sarah, the wife of Abraham, had discovered the *Tabula* "in the Valley of Hebron, in the tomb and in the very hands of the cadaver of Hermes"; this was an unsubtle attempt to annex it to the canon of proto-Christian "ancient theology."[32] But already in antiquity, Egyptian writers had depicted Hermes himself as a finder of primeval books. In fact, the ancient tales depict Hermes as an actual hunter of books, prefiguring the fifteenth-century humanist Poggio Bracciolini.[33] But serendipitous, accidental, or unintentional discoveries have frequently excited more wonder and imagination than books recovered after an active search. A book discovered accidentally implies a kind of grace, as if history or the gods had bestowed a precious gift.

However, by the seventeenth century, texts that peddled dreams of antique wisdom, even for nonpatriotic motives, were coming under suspicion. Reasoning along the lines of Lorenzo Valla, Isaac Casaubon exposed the inauthenticity of the *Corpus Hermeticum* in 1614, showing it could not have been composed at the dawn of time and that its doctrines smacked of Christianity. Although Casaubon's attack had little impact at the time, in 1661 Johann Heinrich Ursin assailed the *Corpus* anew, expressing similar skepticism about the "primordial" writings attributed to Zoroaster and Sanchuniathon.[34]

But the sword of skepticism was double-edged, and scholarly lovers of the Bible were as susceptible to the law of unintended consequences as defenders of Hermes. A major crisis came in 1655, when the Calvinist Isaac de La Peyrère completely upended Moses's claim to the invention of writing—and with it the authority of the Pentateuch's version of universal history. In his *Preadamites, or an Essay on Chapter Five, Verses Twelve, Thirteen, and Fourteen of Paul's Epistle to the Romans, Which Imply the Creation of Men before Adam*, La Peyrère advanced exactly the hypothesis that his title proclaims. Don Cameron Allen neatly paraphrased *Preadamites* in a manner that demonstrates the book's continuity with concerns explored in my previous chapters. "There is too much in the Pentateuch . . . that Moses could not have written. The great patriarch may have written *a* Pentateuch, but the one we have is only a bad copy of his manuscript. Moreover, how are we to know that there were not other writers before Moses? It is absurd to think that he was the first to write history. Moses probably derived from others, then his account was copied by successors, who epitomized the first 1600 years of Jewish history into five meager chapters [i.e., Genesis 1–5]."[35] Ironically, La Peyrère was not attempting to undermine the authority of the Bible but to preserve it against erosion. For anyone who thought as he did, the encounter with an entire unrecorded hemisphere of the world, peopled by infinite tribes whose existence Jews and Christians had never suspected, along with recent discoveries in pagan mythology and chronology, had rendered indefensible the idea that Genesis was objective history. If the Bible was going to retain its authority as revealed *doctrine,* a rational reconsideration of its historiography was necessary. Or so La Peyrère thought. So he based his book not on advances in scholarship or on Josephan legends but on a novel interpretation of suggestive verses in the Bible itself. However, instead of earning gratitude, he "gave Christian Europe a tremendous jolt. A flood of books and pamphlets appeared at once" to refute and revile him, "and throughout the remainder of the seventeenth century, his name is usually joined with that of Spinoza and Hobbes to make a triumvirate of [atheistic] devils incarnate." And yet, Allen contended, "this miserable sinner was a product of the theologians' labors; they had invented the questions and stirred up the doubts that caused him to rack his brain for a comfortable answer. La Peyrère's chief offense was that he was a rationalist; he tried only to get a reasonable answer."[36]

The seventeenth and eighteenth centuries discussed even stranger ideas relating to the Bible. Burchard Gotthelf Struve's pantheon of fraudsters in *De doctis impostoribus* (*On Learnèd Impostors,* 1703) included a radical gambit by the unsavory Pope Gregory IX (d. 1241). In the medieval struggle between Church and Empire for political domination, Gregory scurrilously libeled his arch-enemy the Holy Roman

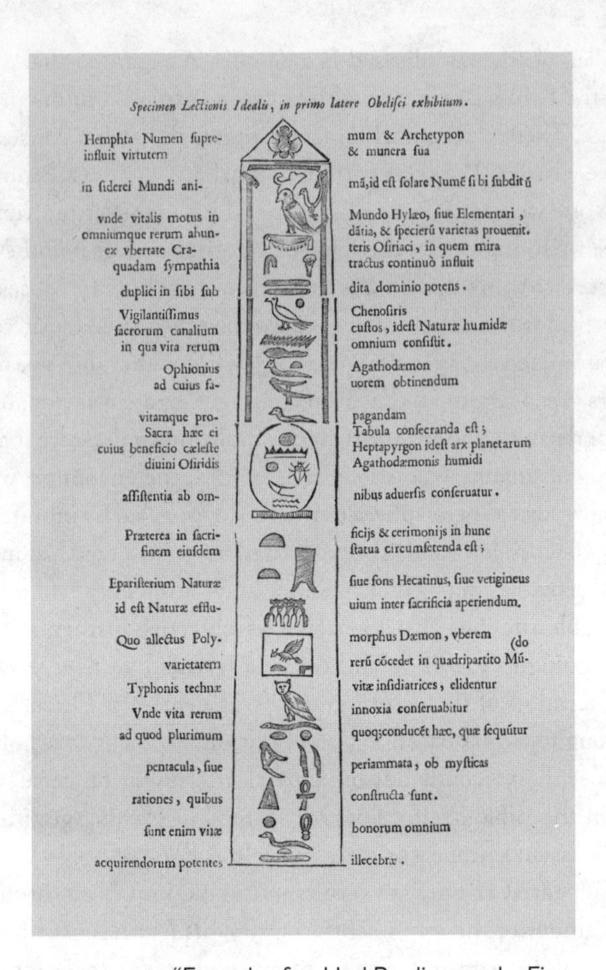

FIGURE 13.1. "Example of an Ideal Reading, on the First Face of the Exhibited Obelisk." Athanasius Kircher, *Obeliscus Pamphilius*. Rome: Ludovico Grignani, 1650. Sheridan Libraries Special Collections, Johns Hopkins University, PT1531.P3 K5 1650QC c. 1.

Emperor. The Pontiff accused Frederick II of declaring that Moses, Jesus, and Muhammad were all charlatans, and the three Abrahamic religions a swindle. When Frederick's chancellor, Pier delle Vigne, attempted to answer the false charge, both were accused of having enshrined their allegations in *The Book of the Three Great Impostors* (*Liber de tribus magnis impostoribus*). In fact, this book was nothing more than a shocking title, but unlike other orphan titles cataloged by early modern scholars, it had pernicious consequences. Gregory's slander ricocheted off Frederick but then rattled around Europe for half a millennium, pinging off every adventurous intellectual from Boccaccio to Spinoza and tainting them with atheism. Given its repellent reputation, "the atheist's Bible" was finally written—twice—in the eighteenth century. But the horrendous "secret" that monotheistic scriptures were forgeries had gone stale, and it became a mere "banal, outdated pamphlet of purely historical interest."[37] Struve, who lovingly tracked Gregory's imaginary book through the centuries, parodied its title for his history of imposture.[38]

Giambattista Vico was consumed for over twenty years by his project of reconciling the historical claims made about Genesis with his extensive readings in classical and modern historians. Vico intended, like so many before him, to vindicate the superior antiquity and truth of the Bible. His *Scienza nuova* (three editions between 1725 and 1744) was in fact not all that new in what he claimed for the Bible itself.[39] But he provided a radically new vision of the ancient world. Vico complained of the precocious excellence that ancient pagan writers, most of them rediscovered since 1450, had claimed for their writing systems and civilizations. His objectivity did not extend to the Bible, however, and in this he was typical of a surviving majority of Christian writers. For them the ultimate stakes of the writing question remained the antiquity, uniqueness, and authority of the Bible, which had been menaced, and in Vico's eyes still were, by rediscoveries of archaic Egyptian and Chaldean texts and assertions about Chinese antiquity made by Jesuit missionaries and other Western enthusiasts.[40] Had key doctrines of Moses and Christ been anticipated by ancient pagans? The problem of *sterminate antichità*, or the "infinite antiquity," of pagan literary culture had assumed increasing prominence ever since Joseph Juste Scaliger's publication (1606) of hitherto unknown Greek passages of Berossos and Manetho (including Oannes's antediluvian pedagogy) gleaned from Eusebius's quotations.[41] Vico dismissed the claims that such ancient texts made for pagan cultural precedence as a symptom of *la boria delle nazioni*, the ethnocentric boastfulness of ancient cultures, and considered their return to prominence after 1500 as expressions of *la boria de' dotti*, the vainglory of modern scholars who made anachronistic, fantastical claims about the matchless achievements of archaic civilizations. Vico opposed all claims of antediluvian wisdom and proposed that after Babel all peoples except the Hebrews devolved into bestial primitives; except for the Bible's inspired truth, wisdom was a late development, anachronistically projected on the ancient world. The true wisdom of the early pagans was typified by Homer's poems: it was poetic, imaginative, and childlike. And far from being repositories of arcane philosophical and religious wisdom as Athanasius Kircher and other scholars still claimed, the Egyptian hieroglyphics were the visual equivalent of pagan myths: inchoate relics of a primitive mentality forever lost (figure 13.1.)

FICTIONAL AND SATIRICAL CONTESTATIONS

Vico's contemporary Voltaire was another brilliant, adventurous intellectual who wrote in the vernacular, but he left himself open to charges of atheism. He savagely mocked legends of Adam's writings in his *Philosophical Dictionary*, though he attributed them only to "the rabbis."[42] His novel *Zadig* (1747) went much further and staged Vico's concept of *la boria delle nazioni* as a ridiculous dinnertime quarrel about literary precedence among ancient patriots from Babylonia, Egypt, India, China, and Gaul. Voltaire's Chaldean reprimands both the Egyptian and the Indian for their claims:

> "You're both quite wrong," said a Chaldaean sitting next to the Indian. "It is the fish Oannes to whom humans owe such blessings, and it is proper to thank him alone. Besides, you both come from an origin too plebean and too recent to argue with me over anything. Egyptian civilization is only a hundred and thirty-five thousand years old, and the Indians can only boast of eighty thousand; yet we have astrological almanacs dating from four thousand centuries ago. Believe me, give up your silly dispute, and I'll give you each a nice portrait of Oannes."[43]

In the "Orientalist" ambiance of the story and the succession of interlocutors, Voltaire entertainingly chronicles the dreads and infatua-

FIGURE 13.2. "Hermes Carves the Rudiments of the Arts and Sciences on Columns." Jean-Jacques Rousseau, *Emile, ou De l'éducation*. Leipzig: heirs of Weidmann and Reich, 1762. Sheridan Libraries Special Collections, Johns Hopkins University, LB511 1762 t. 2 c. 1.

tions of European scholars from the time of Ficino down to his own day, as he did in several of his other fictions.[44]

Confirmation that legends exalting the wisdom and writings of the ancients were losing their hold on Europeans was provided in 1762 by a passage in Jean-Jacques Rousseau's *Émile*. In this radical treatise on the education of children, Rousseau declared that books are worthless and claimed to hate them all, dismissing the information they contain as dead. Interestingly, Rousseau was unwilling to imagine a humanity entirely deprived of books, although he thought one would suffice: not the Bible but *Robinson Crusoe*. The extent to which myths of antediluvian wisdom had lost their fascination is indicated not only by Rousseau's cavalier allusion to *some* columns and *a* flood but also by the engraving that illustrates the passage in the 1762 edition of *Emile* (figure 13.2). It shows a classical Greek Hermes—naked, wearing winged anklets and a winged headgear reminiscent of a fedora, holding his caduceus while engraving geometric symbols (rather than words or hieroglyphics) onto one of two very Doric-looking columns. The only hints that this might be an *Egyptian* Hermes are provided by a small statue of a crouching sphinx in the foreground and an obelisk (or a very acute pyramid) in the distance. Rousseau's source for "Mercury's columns," if it was not scholarly conversation, was perhaps some modern treatise that mentioned Iamblichus's reference to the columns of Hermes Trismegistus.

FICTIONS OF WRITING IN THE NOVEL

By the eighteenth century, increasing numbers of servants and other common folk were able to read and write in their everyday vernacular languages. Over the course of the century, the fear spread that reading could become dangerous to morals as well as religious orthodoxy, another path to perdition rather than salvation. The runaway popularity of Samuel Richardson's *Pamela* (1740) and other early novels among readers across social classes suggested to moralists that such works glamorized moral turpitude while pretending to condemn it. At the other end of the reading spectrum, serious works of history and natural science, including the *Encyclopédie*, were increasingly published in French, English, Dutch, and other languages of the political and commercial powers, rather than Latin.

While the wealth of eighteenth-century novels is outside the range of this book, one more should be added to the few mentioned thus far, for its unprecedented commentary on the art and practice of writing.

Laurence Sterne's *Life and Opinions of Tristram Shandy*, published in installments between 1759 and 1767, purports to be an autobiography. However, Shandy finds it impossible to compose a straightforward narrative of his life ab ovo. The novel is thus a near-infinite regress of digressions peppered with neurotic, "necessary" authorial explanations, so that Tristram's very birth is postponed thoughout the first third of the narrative. Tristram's name, owed to his father's inadvertent conflation of Hermes Trismegistus and the Tristan of medieval romance, reflects the hybridity of Sterne's bookish narrative; *Shandy*'s frequent plagiarisms of Rabelais, Robert Burton's *Anatomy of Melancholy* (1621), and other rambling, encyclopedic vernacular works, are punctuated by innovative typographical jokes, crowned by a "blank" page that belies the word's etymology by being entirely black rather than white.[45] The novel plays with the mystique of writing, exposing its limitations and contradictions while also celebrating its power to deploy an organized imagination beneath a deceptively chaotic appearance.

ASPIRATIONAL ADAPTATIONS

We have seen that religions from Babylon to Protestantism tended to accumulate bookish myths about their own foundation. A new religion, or at least the facsimile of one, arose in the early eighteenth century. Despite the Freemasons' pretensions to a "rational" theology, they were just as enamored of secrecy, elitist initiatory rites, bookish legends, and the other bric-a-brac of esoteric philosophy as early modern Christians from Ficino to Kircher. Masonic myths center on the brotherhood's supposed role as the preserver of ancient wisdom and *prisca theologia*, and they show a distinctly religious pedigree. Yet one scholar remarks that "for all their talk of ancient origins, [Freemasons] are a product of the early modern age, and the traditional accounts of their origins were first settled in the early decades of the eighteenth century."[46] In 1723 James Anderson, a Presbyterian minister, published at London *The Constitutions of the Free-Masons. Containing the History, Charges, Regulations, &c. of that most Ancient and Worshipful Fraternity*.[47] Although late medieval sources had imaginatively traced a brotherhood of masons to Euclid, Anderson took advantage of recent centuries' fascination with scripturality and related issues. The *Constitutions* were naive compared to Defoe's exactly contemporary *Essay Upon Literature*, but the same biblical culture animates the Worshipful Fraternity's mythology, though its seriousness is uncertain. Its antediluvian pretensions appear immediately on the title page: it is published "In the Year of Masonry 5723 / Anno Domini 1723." In the initiation rite that opens the *Constitutions*, the first sentences echo a refrain familiar from centuries of repetition: "Adam, our first Parent, created after the Image of God, the great Architect of the Universe, must have had the Liberal Sciences, particularly Geometry, written on his heart; for even since the Fall, we find the Principles of it in the Hearts of his Offspring."[48] The "liberal sciences" are the seven liberal arts familiar from medieval myths such as the fourteen columns of Peter Comestor's Ham/Zoroaster, but the Masonic rite stresses that "Adam taught his sons geometry" in particular. This is certain, because "Cain, we find, builded a City, which he call'd, consecrated, or dedicated after the name of his son Enoch [i.e., Enosh]." The eulogistic characterization of Cain and his progeny is novel and something of a shock. The rite blithely asserts that Cain became "the Prince of the one Half of Mankind" and expresses the conviction that "his Posterity would imitate his royal Example in improving" geometry.[49] (In fact, traditional biblical commentaries insisted that all Cain's descendants were sinful and had died in Noah's Flood.) Anderson's positive interpretation of the technological history in Genesis 4:17–22 also upends previous centuries' negative judgments of the Cainites' character; he declares that "other Arts were also improved by them, viz., working in Metal by Tubal Cain, Music by Jubal, Pastorage and Tent-Making by Jabal, which last is good Architecture."

J. M. Roberts wrote, "Masonic scholars have regretted Anderson's 'fertile imagination' and find his work an excellent illustration of 'learned credulity.'"[50] Anderson's account is both strange and unique; it gives precedence to Genesis before dutifully but grudgingly alluding to Josephus's *Antiquities*, yet declining to mention author and book by name. Scholars have traced some of the account's eccentricity to the late fourteenth-century Christian chronicle of Ranulf Higden, but it echoes several features of the Sethian columns tradition.[51] Like

Annius of Viterbo, Anderson attributes the columns to "godly Enoch" so as to quote the canonical Book of Jude on Enoch's prophecy, but belatedly admits that "some ascribe" the columns to Seth, whom he describes dead-pan as not "less instructed" than Cain, and as "Prince of the other Half of Mankind." (Given their notion of Cain's heretical "church," one wonders what the Lutheran Carion chroniclers would have thought of this parity between Cain and Seth!) Anderson calls Seth "the prime Cultivator of Astronomy," who "would take equal care to teach Geometry and Masonry to his Offspring, who had also the mighty Advantage of Adam's living among them." These are precise allusions to the account in Josephus's *Antiquities*, but the Jew's name and his book's title continue unmentioned. By sacrificing the traditional outlines of the invention narrative and constantly alluding to Genesis, Anderson emphasized the supposed protoscriptural source of Freemasonry but actively hid its dependence on Josephus.[52]

Freemasonry soon traversed the Atlantic, and the *Constitutions* were reprinted at Philadelphia in 1734 by none other than Benjamin Franklin. Since that time, enthusiasts have made many additions and revisions to the mythology. As the first great monument of Judeo-Christian architecture, Solomon's temple has remained an important feature in Masonic myth, so Anderson's two columns of Enoch were eventually doubled by two columns that the Bible described as the work of Hiram of Tyre and located at the entrance to the Temple erected by the "wisest man who ever lived"[53] (figure 13.3). In Masonic myth, Hiram is supposed to have died to preserve the secrets of the brotherhood, and the mythology expanded rapidly to claim Masonic origins for any and all secret doctrines, whether associated with biblical personages or with others.

How seriously most Masons took the mythology is questionable; it certainly seems to have been more a recreational daydream, a sort of role-play for adults, than a recognizable religion. Indeed, recreation was a primary goal of the organization; "lodges" of stonemasons had begun admitting "non-operative" members during the late seventeenth century from "the desire for improved social arrangements and greater opportunities for conviviality." These essentially honorary members were known as "speculative" Masons to distinguish them from the de-

FIGURE 13.3. The Columns of Hiram. James William Bowers, *A Manual of the Three Degrees of Ancient Craft Masonry, for the Use of Master Masons*. Baltimore: J. H. Medairy, 1888. Sheridan Libraries Special Collections, Johns Hopkins University, HS457 .B78 1888 c. 1.

clining "operative" membership of working masons. In 1717 "masters" of several lodges met at a London pub and formalized the first "Grand Lodge." The Masonic emphasis on secrecy, ritual, and a written quasi-religious mythology doubtless explains the rapid growth of membership throughout the eighteenth century, but it also inspired the strong suspicion, even paranoia, that Catholic authorities have continually expressed toward Freemasonry and the subsequent growth of conspiracy theories concerning "Illuminati," Jews, and other groups.[54]

IMAGINARY BOOKS

The Sethian columns were an archetype of imaginary books, textless titles or bibliographic ghosts created because of the emotional need that they *should* exist. Sixtus of Siena had already cataloged whole categories of dubious books in the Bible and its commentators. As we saw in chapter ten, the sixteenth century was also the golden age of imaginary books as a scholarly joke or pastime, when authors like

Folengo and Rabelais burlesqued the imaginary source-books of forgeries and romances.

But the seventeenth and eighteenth centuries awoke to more sophisticated and inclusive ideas concerning the implications of nonexistent books. In its entry for "Book," Ephraim Chambers's *Cyclopaedia*, the first truly modern encyclopedia, classified even nonexistent books as if they were physical objects, "according to [their] circumstances and accidents." Chambers divided them into books "*lost*, those which have perished by the injuries of time or the malice or zeal of enemies. . . . *Books promised*, those which authors have given expectations of, which they have never accomplished. . . . *Books fictitious*, those which never existed: to which may be added divers feigned [imaginary] titles of books."[55] To varying degrees, all putative books in these categories are fictitious, since they cannot be found or read, but Chambers's classification gave them a shadowy reality.

In 1765, D'Alembert and Diderot's *Encyclopédie* incorporated Chambers's article into its entry on *Livre*, and further developed his categories.[56] Its most notable modification was to translate Chambers's "fictitious" books as *livres imaginaires*, books "having no foundation in reality." Both Chambers and the *Encyclopédie* cite scholarly compilations that amount to catalogs of imaginary books. Not all these were supposed histories or Books of Memory; there were also lists of Books of Mastery—important books on practical, technological, or magical subjects that *should* be written but had never yet been undertaken. Unlike the satirical imaginary books of Rabelais or the fantastic ones of Jorge Luis Borges, most of the titles mentioned by the eighteenth-century encyclopedias appear idealistic and probably reflect the massive expansion in printed books and readers capable of appreciating them. Prominent among such handlists are Theodore Janson's *Bibliotheca promissa et latens* (*The Hidden Library of Promised Books*), Adrien Baillet's *Jugemens des sçavans* (*Scholars' Verdicts*), and Valentin Ernst Löscher's *Arcana literaria* (*Mysteries of Literature*), all published between 1688 and 1700. According to Chambers, "M. Dugono has a whole volume of *schemes*, or *projects of books*, containing no less than 3000."[57] The lists of Janson and Löscher, in particular, reflect enthusiasm for physical science and practical knowledge, along with confidence that books, and literature in the broadest sense, can transmit them successfully. Titles are presented as prophecies of positive development—of progress in something like our sense—rather than as entertainment or satire.

SLAVE NARRATIVES

Hermann Hugo repeated a story about the powerful mystique of writing that had been told in 1572 by a European missionary. It presents a Brazilian slave's wonder and bewilderment at first encountering writing, and is something of a modern archetype for European pretensions about the superiority of their written culture over "primitive" oral societies. The slave's experience with the unfamiliar medium, like that of Bellerophon, was far from benign. His European master sent him on a solitary errand, to deliver a basket of figs and a letter to a European friend nearby. On the way, the hungry slave ate many of the figs; seeing the small quantity of fruit remaining, his master's friend berated him for theft, exhibiting the letter as testimony, but the slave denounced the letter as a liar and an untrustworthy witness, as if it were a real person. By his boldness, the slave escaped punishment, but a few days later, his master sent him on precisely the same errand.

This time, the slave thought he had learned from his previous mistake, so when he stopped to consume some of the figs, he first extracted the letter from the basket and hid it under a rock; then he sat down on the rock while enjoying his snack. To ensure that his readers understood the slave's anthropomorphizing blunder, Hugo had him taunt the hidden missive as if it were a human sentry: "Now, letters, you will not witness my treat." But yet again the slave's confidence was misplaced: in the second letter, his master had craftily specified the exact number of figs he was sending to his friend. When caught out a second time, the slave was even more astonished, convinced that the letter had been able to observe his transgression superhumanly, from underneath the rock.[58] And so, says Hugo, the slave was able to understand the power of writing only when "his master thoroughly bescribbled his back, using elm-switches for a pen (*ab hero suo tergum eius totum conscriberetur stylis ulmeis*)."[59] Presumably, the master's perverse "writing" produced long-lasting memory for the slave. Hugo's anecdote has the hallmarks of a "just-so story," a historical fiction explaining the superiority of written culture. It was repurposed in 1641 by John Wilkins in an essay on cryptography, *Mercury, or the Secret and Swift Messenger.*[60]

Daniel Defoe quoted a variant of the "marveling savage" commonplace concerning Captain John Smith, the early Virginia adventurer.

> Captain Smith . . . happening to be taken Prisoner among the Indians, had leave granted him to send a message to the Governor of the English Fort at James Town, about his Ransome; the Messenger being an Indian, was surpriz'd, when he came to the Governor, and was for kneeling down and Worshipping him as a God, for that [because] the Governor could tell him all his Errand before he spoke one Word of it to him, and that he had only given him a piece of Paper: After which, when they let him know that the Paper which he had given the Governor had told him all the Business, then he fell in a Rapture the other Way, and then Captain Smith was a Deity and to be Worshipp'd, for that he had Power to make the Paper speak.[61]

Although more benign than Hugo's story in its attitude to the individual "savage," Defoe's anecdote is also more hyperbolic and, if anything, somewhat more racist, in its explicit comparison of Europeans to gods. Defoe projected the attitude deep into the past, doubting that the builders of the Tower of Babel could have had "so glorious an Improvement as this of writing down their Speech by the Help of Letters, and as the poor Indian said, making the Paper or the Tables [tablets] they wrote upon, to Speak." He implies that the Babelians' lack of writing reflected an "immense Dulness in the People of that Time," so their foolish project was "begun upon the most ignorant Notions of Things, that could be supposed to come into the Heads of rational Creatures."[62] Of course it was, since this was ages before God delivered writing to Moses.

The stories Hugo and Defoe repeated were first told before the legalized enslavement of Africans in English North America. However, one of the most notable literary developments during the eighteenth century was found in the autobiographies written by slaves of African origin who, by one means or another, achieved literacy and freedom. Contrary to the poor Brazilian's and Virginian's experiences, the more fortunate among enslaved Black people presented their discovery of reading and writing as a major episode in their liberation. Nonetheless, the initially mystified and astonished reactions of some Black autobiographers resemble the tales related by Hugo and Defoe, and hint that such stories were not always a European fabrication. Slaves' wonder at the "talking book" is a less melodramatic attitude than the Brazilian's putative astonishment that the letter could observe him from underneath the rock and relate his misdeed to his master.[63] But by learning to read, some North American slaves escaped the Brazilian's fate, which, we can only assume, was lifelong bondage and continued savage whippings.

Olaudah Equiano, who was born in 1745, was kidnapped from what is now Nigeria and enslaved at the age of eleven. His vivid account (1789) of his early impressions about white Europeans includes a gradual awakening to the importance of their books. "I had often seen my master and Dick employed in reading; and I had a great curiosity to talk to the books, as I thought they did; and so to learn how all things had a beginning. For that purpose I have often taken up a book, and talked to it, and then put my ears to it, when alone, in hopes it would answer me; and I have been very much concerned when it remained silent."[64] Equiano's apparent allusion to Genesis implies that the book in question was the Bible, which he names explicitly in later passages. He had acquired one around the time he was baptized, but a tyrannical master had sold him suddenly, making him abandon his possessions, so he was jubilant when eventually able to purchase another Bible.[65] With help from several friends and benefactors, Equiano learned to read and write fluently, and earned enough money to purchase his freedom. Luckily, he was sold early on to shipowners, thus learning a trade that enabled him to travel and, more importantly, to avoid the homicidal treatment visited on agricultural slaves.

Frederick Douglass, now better-known than Equiano, escaped to freedom several decades later, in 1838. (Though his book falls outside the chronology of this chapter, it was explicitly influenced by Enlightenment ideals.) Douglass's autobiographical *Narrative* (1845) paints a somber picture of the relation between enslavement and illiteracy. Transferred to Baltimore from the brutal misery of plantation childhood on Maryland's Eastern Shore, he initially received a little instruction from his master's wife, but once the master learned of it, Douglass's tutoring abruptly terminated. "Mr. Auld found out what was going on, and at once forbade Mrs. Auld to instruct me further, telling her, among other things, that it was unlawful, as well as unsafe, to teach a slave to read. . . . 'Now,' said he, 'if you teach that nigger (speaking of myself) how to read, there would be no keeping him. It would forever unfit him to be a slave. He would at once become unmanageable, and of no value to his master.'"[66] According to Douglass, overhearing these remarks was for him "a new and special revelation, explaining dark and mysterious things" that he had "struggled in vain" to comprehend.

> I now understood what had been to me a most perplexing difficulty—to wit, the white man's power to enslave the black man. . . . From that moment, I understood the pathway from slavery to freedom. . . . [T]he very decided manner with which he spoke, and strove to impress his wife with the evil consequences of giving me instruction . . . gave me the best assurance that I might rely with the utmost confidence on the results which, he said, would flow from teaching me to read. . . . [T]he argument which he so warmly urged, against my learning to read, only served to inspire me with a desire and determination to read.[67]

The value of reading for Douglass appears powerfully in his acquisition somewhat later of "a book called 'The Columbian Orator,'"[68] which, in keeping with the values of the Enlightenment, contained eloquent indictments of slavery. "These were choice documents to me. I read them over and over again with unabated interest. They gave tongue to interesting thoughts of my own soul, which had frequently flashed through my mind, and died away for want of utterance. . . . The reading of these documents enabled me to utter my thoughts, and to meet the arguments brought forward to sustain slavery."[69]

Without the ability to read, Douglass realized, he could never have articulated his thoughts, much less memorialized them through writing. The effects of literacy were not all positive, however: "I would at times feel that learning to read had been a curse rather than a blessing. It had given me a view of my wretched condition, without the remedy. It opened my eyes to the horrible pit, but to no ladder upon which to get out. In moments of agony, I envied my fellow-slaves for their stupidity."[70] Literacy was a means to power, but not of the automatic or magical sort.

Like the ability to articulate thoughts of freedom, the mechanics of writing required imitating models: "By this time, my little Master Thomas had gone to school, and learned how to write, and had written over a number of copy-books. . . . When left [alone], I used to spend the time in writing in the spaces left in Master Thomas's copy-book, copying what he had written. I continued to do this until I could write a hand very similar to that of Master Thomas."[71] Later, sent back from Baltimore to plantations on the Eastern Shore, Douglass was forced to become an agricultural field hand under several cruel masters and overseers. Unlike Equiano, Douglass never had the opportunity to purchase his freedom and had to "steal himself" by a daring escape to the northeastern states.

HOW WRITING MAKES US HUMAN

Both Equiano and Douglass became well-known advocates for the abolition of slavery, Equiano in Britain, Douglass in the United States (though he lectured widely, including in Britain). A constant theme in their and others' narratives is the liberation—if not of the body, at least of the mind—effected by becoming literate, and its importance to their transition from "beast" of burden to full humanity. Equiano writes that he initially suspected white Europeans of being "spirits" because their superior technology seemed magical; but eventually "I no longer looked upon them as spirits . . . and therefore I had the stronger desire to resemble them, to imbibe their spirit and imitate their manners. . . . I had long wished to be able to read and write."[72] Like Douglass, Equiano describes the acquisition of literacy as a long and discontinuous process, requiring tenacity and the benevolence of impromptu teachers. But Douglass's case also required he develop Odyssean powers of stealth and strategy. Prefiguring Mark Twain's Tom Sawyer, Douglass says he tricked other boys into donating their help. Before finding Master Thomas's copy-books, "when I met with any boy who I knew could write, I would tell him I could write as well as he. The next word would be, 'I don't believe you. Let me see you try it.' I would then make the letters which I had been so fortunate as to learn, and ask him to beat that. In this way I got a good many lessons in writing."[73] By this trick, Douglass provoked free (presumably white) children to break the first law of American slave culture and "forever unfit him to be a slave."

Skeptical critics have occasionally disputed both these accounts.[74] However, they are no more improbable or obviously fictionalized than Saint Augustine's or Jean-Jacques Rousseau's autobiographies. Nor do they strain credibility any more than the tales recounted about indigenous Brazilians and Virginians by self-congratulatory Europeans. Whether Equiano and Douglass experienced or imagined these episodes, they communicate an emotional truth about the value of writing. In all such accounts an educative process transports the illiterate person from wonder at the mysterious, "magical" technology to an understanding of writing's role in society, for both good and ill. In this, they hark back thousands of years to the tension between Enmerkar's effortless invention of "painting speech, and speaking to the eyes," and the Lord of Aratta's astonishment at the contrast between the art's visual banality and its mental power.

The narratives of these newly literate, newly freed slaves provide, for perhaps the first time, and against a backdrop of appalling racist cruelty, passionate first-person testimony of how writing can and should make us human.

CHAPTER FOURTEEN

The Age of Decipherment, 1800–1950

> And on the pedestal these words appear:
> "My name is Ozymandias, king of kings,
> Look on my works, ye mighty, and despair!"
> Nothing beside remains. Round the decay
> Of that colossal wreck, boundless and bare
> The lone and level sands stretch far away.
>
> —PERCY BYSSHE SHELLEY

To understand how strong the mystique of writing remained among ordinary people in the nineteenth century, consider the needlework sampler of a young schoolgirl from 1828. Even more explicitly than Mary Godfrey's sampler from 1775 (seen in chapter twelve), the newer artifact shows the influence of William Massey's *Origin and Progress of Letters*, since it quotes his familiar quatrain: "Whence did the wondrous mystic art arise / Of painting speech, and speaking to the eyes / That we by tracing magic lines are taught / How both to color and embody thought?"[1] As we have seen, Massey's emphasis on the *wonder* and *magic* of writing, its *mysterious* ability to work an *incarnation* of thought and speech, has a pedigree stretching back four thousand years to "Enmerkar and the Lord of Aratta" in ancient Mesopotamia. Although Massey's verses may now seem oddly exaggerated, they describe an emotion that was still real to his contemporaries.

That emotion was even more intense, as we saw in chapter thirteen, for newly literate slaves such as Olaudah Equiano and Frederick Douglass, especially for the latter's epiphany regarding "the white man's power to enslave the black man." But like Equiano and Douglass, the Cherokee inventor Sequoyah illustrates the limits of literacy to effect liberation from oppression. Sequoyah singlehandedly created a syllabic writing system in 1821, which the Cherokee nation formally adopted in 1825. Unfortunately, Cherokee literacy was unavailing against their forcible "removal" with other Native peoples from their Southeastern homelands during the 1830s, and their resettlement in the "Indian territory" that became the state of Oklahoma. The legalized violence of the Indian Removal Act (1838) and the horrors of the Trail of Tears provide an epitome of an ongoing genocide (see plate 11).[2]

Conversely, access to literacy worked dramatic improvements in the life of a white woman, Helen Keller. She first encountered writing at the age of nearly seven years, in 1887. Rendered deaf and blind—and consequently mute—by illness at nineteen months, she was all but entirely imprisoned within herself until the arrival of her tutor Anne Sullivan. Sullivan relates (and Keller herself corroborates) the moment when the profoundly isolated child *suddenly understood language because she was already familiar with writing*: "As the cold water gushed forth, filling the mug, I spelled 'w-a-t-e-r' in Helen's free hand. The word coming so close upon the sensation of cold water rushing over her hand seemed to startle her. She dropped the mug and stood as one transfixed. A new light came into her face. She spelled 'water' several times. Then she dropped on the ground and asked for *its* name and pointed to the pump and the trellis and suddenly turning round she asked for *my* name."[3] Deprived of all communication except through touch, Keller reversed millennia of cultural history, learning writing before she could speak, much less understand language as a symbolic system. In her case, writing was indeed a "wondrous mystic art," since it allowed her to *feel* speech "by tracing magic lines" that "embodied thought" literally in the flesh of her hand.

From 1800 to 1950, the wonder and mystique of writing were still most visible in the emotional appeal of the Bible felt by ordinary people as well as scholars. But that was set to change, precisely though the agency of writing. Although biblical chronology had been the norm for Jewish and Christian scholars from the Septuagint onward, developments in the sixteenth through eighteenth centuries had given renewed prominence to pagan, pre-Christian writings about the age of the world, the origin and early history of humankind, and the chronology of ancient civilizations. But the change was not only due to new narrative contents. The nineteenth century would drastically modify conceptions of writing and its history by placing new emphasis on deciphering scripts and languages that were previously unintelligible, unknown, or not even considered writing: Egyptian hieroglyphics, Mesopotamian cuneiform, and, after 1900, newly discovered scripts such as Mycenaean Linear B.[4] This process deepened Europeans' and Americans' access to nonbiblical schemes of history and chronology.

Ironically, although histories of writing had grown bulkier in the nearly two centuries between Hermann Hugo and William Astle, sources of information about its earliest developments had increased hardly at all. Scholars still depended mainly on the Bible and classical literature, retelling stories grown familiar by centuries of repetition. Soon after Astle's *Origin and Progress of Letters*, however, everything about the subject started changing rapidly. In 1822 a young Frenchman, Jean-François Champollion, announced he had discovered the key to reading Egyptian hieroglyphics. These stylized pictures had puzzled scholars for many centuries, inspiring extravagant myths about their "secret symbolism." Once hieroglyphics were revealed as encoding readable *language*, the next fifty years transformed the history of writing into a completely different discipline, based on code-breaking rather than storytelling. The very definition of writing was changing quickly and profoundly: tenacious and resourceful scholars were revealing that, contrary to centuries of received wisdom, the mysterious Egyptian picture-carvings and the "bird tracks" of Babylonian cuneiform could be—and soon were—deciphered and read. The history of writing was changing history itself; thanks to Champollion and other code-breakers, history with a capital *H* had been cracked open like a ripe watermelon, scattering seeds of knowledge every which way.

This new knowledge was both exciting and disturbing. In December 1872, another young self-educated prodigy, an Englishman with the unprepossessing name of George Smith, gave a public lecture on ancient Babylonian literature in London. Like Egypt, ancient Babylonia played an outsize role in biblical history yet its language and literature were completely unknown until the 1850s. Smith's lecture announcement sounded so important that even Queen Victoria's prime minister attended, and with good reason. Discussing *Gilgamesh*, the young scholar declared he had deciphered the "Babylonian original" of Noah's Flood, a story that Hebrew scribes, writing centuries later, had transformed into their own.

Smith's jaw-dropping announcement contradicted traditional certainties. His discovery seemed to gainsay the Bible itself, the very Word of God. For over two thousand years, devout Jews, Christians, and Muslims had assumed that Genesis was written by Moses himself and that the Hebrew Bible was the oldest surviving book. What could it mean if the Jews borrowed Noah's story from their enemies the Babylonians? Like Champollion and the Egyptologists, Smith and the Assyriologists were turning the ancient world upside-down. Their pioneering work, like that of Charles Darwin in precisely the same period, looked like an assault on divinely revealed truth.

From Champollion in 1822 to George Smith in 1872, the decipherment of ancient Egyptian and Mesopotamian writing systems culminated in a half-century of breakthroughs. The resulting expansion in nonbiblical knowledge about the pre-Hellenic world soon reclassified the history of writing as a mere subdiscipline of archaeology. Between the 1830s and the 1870s, the concept of "prehistory" emerged to designate human phenomena that had never been recorded in writing but were indisputably real. At its core, prehistory demonstrated that writing gave an incomplete account of developments in human societies.[5] Together, the concept of prehistory and the expanded practice of decipherment affected archaeology in the broad sense, that is, the study of anything ancient. From Flavius Josephus to nearly 1800, the concept of "archaeology" had been centered on the text of the Bible,[6] but now scholars began seriously to consider material culture beyond the sphere of writing. In 1816–1819, while Champollion was closing in on hieroglyphics, in Copenhagen another young prodigy, Christian Jürgensen Thomsen, was developing the "Three Age System" of human archaeology—the stone, bronze, and iron ages—to classify an antiquities collection that evolved into the National Museum of Denmark.[7] Like Champollion's work, Thomsen's built on interpretive trends that were well under way in the eighteenth century, but by attending to the circumstances of archaeological finds, Thomsen expanded the activity of decipherment from the study of written texts to the dating of nontextual artifacts through stratigraphy and typology (that is, classification by geological layers and physical forms). In effect, both trends made possible the creation of hypothetical, plausible narratives of prehistory and further weakened the dependence on biblical sources for establishing universal chronology.

Like the history of writing, geology and paleontology were detaching themselves from biblical narrative. A contemporary of Champollion and Thomsen, the French naturalist Georges Cuvier, proposed in 1813 that the creation and flood described in Genesis were not unique events but part of a series of creations, catastrophes, and mass extinctions. Cuvier aimed to account for the radical differences between living beings and extinct organisms such as the mastodon and pterodactyl, which were coming forcefully to prominence. His "Catastrophism" implied that scriptural literalism was inaccurate or inadequate, and that nonwritten criteria were more reliable for dating events in the distant past. Catastrophism did not inherently contradict the text of Genesis but could provide a broader, hypothetical context for it; this solution was adopted by some writers, despite its uncomfortable analogies to Pliny's eternal world and Lucretius's reduction of the cosmos to atoms randomly coalescing in the void. But the prehistoric "descent of man" through natural selection, as propounded by Charles Darwin and other evolutionists from 1859 onward, caused genuine panic among proponents of scriptural supremacy. Natural selection was the ultimate nontextual classification of organisms both living and extinct, human and nonhuman, constituting a refusal of the Bible's distinction between humans and God's other creatures, which gave people a separate and privileged moment of origin. Between 1829 and 1864, the discovery and classification of Neanderthal fossils in Europe inaugurated the study of paleoanthropology and provided striking corroboration of Darwin's theories as an alternative to the biblical myth of Adam and Eve as the "crown of creation."[8]

A comparable movement away from the centrality of written culture was taking place in ethnology. In 1812, the brothers Jacob and Wilhelm Grimm published the first edition of their *Children's and Household Tales,* emphasizing the collection of oral stories from illiterate people rather than the literary treatments of fairy-tales by such sixteenth- and seventeenth-century predecessors as Straparola and Basile in Italy and Perrault in France. The unrecorded culture of the illiterate subaltern classes could be studied as a kind of language-based archaeology. Folklorists, from the Grimms to twentieth-century scholars, would emphasize stories ranging from Hansel and Gretel to Rabelais's *Gargantua* as expressions of a previously unwritten culture of "the folk" or "the people," which could be (and often were) interpreted in terms of either social class or the ethnic patriotism of "a" people.[9] Despite being subject to the same sorts of exaggerations and wishful thinking as literary history, folklore broadened the scope of available thinking about human history.

ARCHAEOLOGY: THE MODERN APPROACH TO THE HISTORY OF WRITING

With his references to Babylonian and Egyptian ruins in the *Essay upon Literature,* Defoe sounded faintly like a modern archaeologist, but he had no direct experience of such relics. Almost no Europeans did. Major developments in Near Eastern archaeology did not happen until nearly 1800, when events brought epochal transformations in the history of writing. Napoleon's inclusion of numerous scholars in the armies that invaded Egypt in 1798 was by far the most dramatic. Twenty-four years later, and ninety-six years after Defoe, Jean-François Champollion's *Letter to Monsieur Dacier* (1822), delivered the coup de grâce to the centuries-old European belief that hieroglyphic inscriptions were a complex of arcane mystical *symbols,* an interpretation that had prevailed after hieroglyphic writing fell into disuse around 394 CE[10] (see figure 13.1 in chapter thirteen).

Because nothing in the history of writing happens suddenly, Champollion did not decipher the hieroglyphs all by himself: he benefited from several decades of speculation by scholars who suspected that the strange carvings were *some sort* of writing system.[11] In particular, the extent of Champollion's debt to his English rival, Thomas Young, is still somewhat controversial. The two scholars were in a race to decipher the hieroglyphic system, and in 1819, three years before Champollion's *Letter*, Young came close to the Frenchman's solution of the riddle.

At first glance, the slow and painstaking work of these early scholars looks deceptively like the activity of modern code-breakers. But there is an enormous difference between *decrypting* a spy's coded message and *deciphering* a text from thousands of years ago. Once we possess a key, we can translate the spy's message back into a well-documented modern language and culture because the message has been deliberately and systematically encrypted according to assumptions and procedures that are familiar to at least a few living people.[12]

But Champollion and his predecessors were separated from their goal by more than thirteen hundred years of historical change. Scholars' assumption that hieroglyphics were a form of writing meant they had to discover the spoken language the "pictures" expressed. This could only be accomplished by trial and error: initially researchers employed educated guesswork to confirm their hunch that clusters of hieroglyphs, isolated and repeated in oval "cartouches" within a text, recorded the names of famous rulers. From this slender evidence, the approximate sound-values of a few hieroglyphs were conjectured and tested on other inscriptions, in a slow and painstaking accumulation of tiny confirmations. Then the conjectured names required comparison to words in known languages, not only ancient tongues but also modern ones, which might be descended from as-yet unknown ancestors. Before confirming that Egyptian hieroglyphics were indeed a *system* of writing, Champollion had to identify, reconstruct, and learn the ancient Egyptian language; for this task there were no living informants. Eventually, he found the language of the hieroglyphs was recoverable by comparison to the Coptic language, which appeared to be its descendant. Coptic itself was a dead language by 1800, but luckily it was "fossilized" as the language of Christian liturgy in Egypt and nearby lands (just as Latin liturgy was preserved by Catholic culture until the 1960s). Finally, one more step remained: languages—and the objects and customs they describe—are specific to times and places, so the conjectured meanings of some ancient Egyptian inscriptions required corroboration by archaeological evidence. Champollion died in 1832, three years after his sole Egyptian expedition, so his task remained incomplete for decades.

No wonder Defoe was so confident that the Egyptian "sacred carvings" were not writing and would remain forever mute. He was necessarily unable to foresee the progress in understanding that took place in many fields during the following century. To take a major example, the famous Rosetta Stone was not discovered until 1799, and it might never have come to light if Napoleon had not invaded Egypt: it had been repurposed as a building stone and was prised out of a wall by a military demolition crew. Work on this artifact confirmed and expanded the small early victories in deciphering other hieroglyphic inscriptions. Confirmation was made substantially easier since the Rosetta Stone bore inscriptions in three scripts, one of which was a translation into ancient Greek. So it seemed likely that all three inscriptions said the same thing.

But the importance of the Rosetta Stone should not be exaggerated. It was no code-book, for unlike a modern encryption key, the Greek text contained nothing that could simply be "plugged in" to decrypt the hieroglyphic message word by word. Hence twenty-three years of painstaking educated guesses and experimentation elapsed between the discovery of the stone and Champollion's breakthrough. Ultimately, neither the efforts of previous scholars nor the stone's bilingual inscription guaranteed Champollion's success. His personality was a more crucial factor: since childhood, his ambition had been to decipher the hieroglyphs, so by 1822, he had systematically dedicated his life to learning ancient and modern languages. He had employed that knowledge during several years to make educated guesses about the relevance of Coptic to deciphering ancient Egyptian. Ironically, Champollion had been anticipated in this conjecture by Athanasius Kircher a century and a half earlier, but with inaccurate, fanciful results.[13] Nor did Champollion "close the book" on decipherment: work and discovery still continue on ancient Egyptian language and writing.

Defoe's smugness about the lack of inscriptions at "Babel" was even more understandable than his dismissiveness toward Egypt. Both

Egypt and Mesopotamia were portrayed negatively in the Bible, so that religious bias (along with racial prejudice) influenced early modern ideas of the two cultures. But the ignorance was differential: in Champollion's time, Europeans' limited information—whether accurate or legendary—about ancient Egyptian culture dwarfed their awareness of Mesopotamia. "The land between the rivers" had literally nothing memorable to show of its past; no grandiose ruins comparable to the opulent pharaonic monuments were visible. Stone suitable for building was rare in Mesopotamia, so its greatest monuments had been built of bricks, most of them merely sun-dried, like adobe, rather than fired. The erosion of unfired clay hid Mesopotamian buildings until the nineteenth century. Enormous palace complexes had weathered into mounds, or *tells,* that looked deceptively like natural features of the landscape. In striking contrast to Egypt, no grandiose statues of Mesopotamian gods or rulers survived above ground to pique anyone's curiosity. Worse, literature of other ancient cultures gave little information about Mesopotamia, amounting to a mere handful of pages in the Hebrew Bible and classical authors. Moreover, except for a few legendary personages like Queen Semiramis, most of that lore concerned times after the Persian conquest of Babylon in 550 BCE.

By that date, the art of writing had spread into the Mediterranean Basin, so that Europeans' assumptions about its history were limited to alphabetic writing, the only kind they had ever experienced. The very existence of cuneiform writing was unknown until the seventeenth century. It was first described unambiguously by the Roman traveler Pietro della Valle and the Spanish diplomat García de Silva Figueroa, who visited the sites of ancient Babylon and Persepolis in 1617 and 1618, respectively.[14] The cuneiform inscriptions were not even reproduced in a book until 1657. Defoe's contemporary Thomas Hyde, an eminent Oxford professor of Hebrew and Arabic, coined the term *cuneiform* (wedge-shaped). Nonetheless, he denied that such bizarre marks could be a form of writing.[15]

As with Egyptian hieroglyphs, the decipherment of cuneiform depended on a complex inscription, but one in three languages rather than three writing systems. Moreover, the inscription could not be easily copied and distributed to scholars for decipherment since it was not carved on a portable chunk of rock like the Rosetta Stone. The enormous autobiographical inscription by King Darius I of Persia (d. 486 BCE) at Behistun stretched across the lofty face of a nearly inaccessible cliff in remote western Iran (see plate 12). Europeans had known of (and misinterpreted) the human figures on this textual Mount Rushmore since ancient times, but the inscription was hiding just out of plain sight. Even telescopes were inadequate for visualizing and transcribing it; difficult sight-lines from below made arm's-length scrutiny essential. Thus, much of the transcription resulted from perilous feats of climbing in the absence of proper mountaineering equipment.

Until the Behistun inscription could be processed, decipherment of cuneiform was limited, as hieroglyphics had been, to presuming that distinctive sections of certain inscriptions must record the names and titles of famous ancient rulers. But since all three of the Behistun inscriptions were in the mysterious cuneiform writing, there was no help comparable to the Greek inscription on the Rosetta Stone. One Behistun inscription was clearly alphabetic, since it employed only 41 signs, but these were cuneiform characters. That text had to be transcribed, then identified as Old Persian, translated, and used as a baseline for the other two cuneiform inscriptions, in Elamite and Babylonian. The Old Persian inscription alone required over ten years of effort before translation was completed in 1846. The Elamite and Babylonian texts yielded to translation only in the 1850s.

The protagonist of cuneiform decipherment was an English army officer, Henry Rawlinson (d. 1895), whose adventures are considerably less famous than Champollion's. He was a relative latecomer, and Mesopotamian writing lacked the centuries of advance publicity Western writers had devoted to Egyptian hieroglyphs. Like Champollion, Rawlinson built on the work of other scholars, in particular two contemporaries. The German scholar Georg Friederich Grotefend (d. 1853) first described Old Persian systematically, while the Irish clergyman Edward Hincks (d. 1866) worked on Babylonian. In 1857, a controlled experiment by Hincks, Rawlinson, and two other scholars produced almost identical translations of an Assyro-Babylonian cuneiform inscription—a miniature, real-life version of the Septuagint translators' legendary feat. Within fifteen years, knowledge of cuneiform had progressed so far that George Smith could draw a textual comparison between the flood-story of *Gilgamesh* and the biblical legend of Noah.

THE ROLE OF ARCHAEOLOGY

This condensed account cannot do justice to the enormous expenditures of time, travel, study, conjecture, and cooperation required before nineteenth-century Europeans could reconstruct these two earliest stages in the history of writing. But the process that followed transcription, decipherment, and eventual publication of inscriptions would have been incomplete without material archaeology—the knowledge of *objects*, both inscribed and not. Scientific advances, ranging from Thomsen's three-age classification of tools to carbon-14 dating and other advanced technologies, have yielded a much longer and more gradual chronology of writing than Daniel Defoe could imagine. Modern archaeology strongly implies that the rudiments of writing began ten thousand years ago, a period that is nearly twice the age of the world in the Hebrew Bible's chronology.

Until the age of Champollion, there had been no choice whether to dig up objects and decipher inscriptions in unknown languages: the only comprehensible evidence was literary, and the only innovations were made by reinterpreting a few ancient texts. All Defoe's information came from reading books, some composed two thousand years before his time. Pliny the Elder had reported that the Babylonians recorded astronomical observations on "baked bricks" for hundreds of thousands of years, but he mentioned nothing about cuneiform writing.[16] Pliny's assertions were still quoted appreciatively in the eighteenth century by historians more credentialed than Defoe. Even in their original state, most such descriptions were inaccurate, owing to the isolation of societies before early modern trade and exploration. By the time Defoe read it, such bookish lore was distorted by centuries of misquotation, embellishment, and conjecture.

After Champollion's *Letter to Monsieur Dacier*, the traditional text-based history of writing no longer commanded universal belief. Not that archaeology sprang suddenly to life in Champollion's time:

scholars had studied epigraphic inscriptions in Latin and Greek with increasing rigor since before 1400, but these were mostly funerary epitaphs and official decrees, so the information they added to literary sources was limited. Few inscriptions required scholars to question the traditional Greek and Roman history of writing, which remained an accumulation of myths and legends.

NEWLY DISCOVERED WRITING SYSTEMS

Cultural preconceptions constantly distort conjectures about the history of writing. Archaeological excavation, however, is no automatic remedy, for it too is influenced by literary and cultural heritage, as exemplified by one of the most influential modern archaeologists, Sir Arthur Evans. Beginning in the 1890s, he conducted excavations on the island of Crete that seemed to reveal an advanced civilization as ancient as Egypt or Mesopotamia. Named for the legendary king Minos, Minoan culture (ca. 2600–1100 BCE) displayed a sophisticated social organization long before Athens or Sparta. Its architecture, art, and writing raised Evans's hopes that an indigenous European culture had been discovered, one far worthier of the name "Greek" than the Vikingesque homesteaders and raiders of Homer's poems. But Ilse Schoep writes that rather than base his interpretation on disinterested archaeology, Evans was "anachronistically attributing to the Minoans modern European characteristics and ideals." He and others, she writes, "not only confirmed and reinforced the distinctively European character of the Minoans but also created a past and authenticity for modern notions of Europeanness."[17] Their seductive interpretation, built on a "slim empirical and epistemological basis," has created an enduring problem. Schoep argues that Evans's interpretations are founded on racial and cultural stereotypes that were "part of his Eurocentric agenda, well before the start of his excavations at Knossos," and that his goal was "to propagate his idea of Minoan civilization as the earliest in Europe and distinct in nature from other, better-known civilizations of the East (e.g., Egypt, Mesopotamia)."[18] Similar problems are now raised by archaeologists working in Palestine, some seeking to verify the biblical stories about the magnificent empire of Kings David and Solomon, others maintaining that if these rulers actually existed in the tenth century BCE, they were nomadic Bedouin chieftains, and that ruins identified as theirs are from a much later date or another civilization altogether.[19]

A key element of Evans's program was a writing system, known as Minoan Linear A. It has never been deciphered but is presumably the precursor of the Linear B syllabic script used by the later Mycenaean culture (ca. 1600–1100 BCE). Linear B was not deciphered until 1952, a decade after Evans's death, when Michel Ventris proved it encoded an early form of Greek (figure 14.1). But the mysterious Minoan language cannot be identified, so linear A remains a mystery. Yet Evans presumed it was European; Schoep concludes that

> Evans gave Cretan writing a paramount role in the narrative of European primacy chiefly because it allowed him to dismiss the convention that the Phoenician system of writing was the oldest and most significant: "the Phoenicians did not do more than add the finishing touches" to the Cretan prehistoric writing system. Writing, Evans argued, was not invented in the East but in the West, by a European civilization that "might be regarded as in many respects the equal contemporary of those of Egypt and Babylonia."[20]

Regrettably—in Evans's view—history prevented other Europeans from experiencing a similar development, which "might naturally have been expected to have taken place" but was "cut short by the invasion of the fully equipped Phoenician system of writing."[21] The militaristic metaphor of invasion reveals Evans's polemical investment in disproving Near Eastern primacy. His history of writing updated the age-old Eurocentric, "Orientalist" cultural dichotomy between East and West, Greece and Asia Minor, "in terms such as flux vs. stagnation, inventiveness vs. conventionality, free vs. dependent, naturalistic vs. stiff."[22] As we have seen, similar chauvinistic revisions of cultural primacy based on writing stretch from Flavius Josephus to Annius of Viterbo and the two centuries of early modern cultural patriotism he inspired.

Minoan and Mycenaean archaeology motivated Evans's contemporaries to excogitate other romantic and fanciful reinterpretations

FIGURE 14.1. Linear B tablet, with inscription, list of women's names, 1300–1250 BCE. From Mycenae, West House. Archaeological Museum of Mycenae, MM 2058.
Zde, CC BY-SA 4.0, Wikimedia Commons.

of Western cultural origins. In 1897, Samuel Butler attributed the *Odyssey* to an "authoress" instead of the shadowy male bard celebrated as "Homer." Butler's archaic equivalent of Virginia Woolf was none other than Nausicaa, the adolescent princess who rescues the shipwrecked Odysseus in book six and yearns to marry him but has to settle for enabling him to return to Ithaca and Penelope. Butler's identification of an original living person behind Homer reminds us of the compulsion that drives modern amateurs to find a more "worthy"—that is, university-educated and aristocratic—author of Shakespeare's works, such as Francis Bacon or the Earl of Oxford.[23]

However, Butler's psychology was not driven by class prejudice; instead it reflected Victorian realities of gender chauvinism. In his century, "authoresses" such as Mary Shelley, the Brontë sisters, and George Eliot (Mary Ann Evans) still had to hide behind male pseudonyms until their reputations were established. Butler was clearly smitten with Homer's nubile, lovelorn, "romantic" princess, but he rightly emphasized the *Odyssey*'s consistent empathy for women, in contrast to the masculine violence of the *Iliad*.[24] Butler's "authoress of the Odyssey" had a distinguished mythical pedigree in scholarship, from Naucrates's Egyptian Phantasia to the sixteen female Homers listed by Theodor Ryck. Perhaps when known "authoresses" were as rare as hens' teeth, imagining one of them as the matriarch of writing, books, libraries, and literature appealed to Butler as strongly as the fantasies that had long swirled about Sappho. What might he have thought of Enheduanna, the Mesopotamian "authoress" discussed in chapter one?

OTHER RESURRECTIONS

Burial and disinterment, whether literal or figurative, are ideas that involve books almost as frequently as people. Inscriptions on stone and metal, though not indestructible, are durable. But papyrus and parchment are organic, not mineral, and thus cursed with the same perishability as the human body itself. Parchment is closer to the composition of the human hand that manipulates it, but in some ways, papyrus is even more fragile, and so its preservation can seem yet more wondrous, as Pliny remarked about the alleged Books of Numa.

But what if massive caches of hitherto unknown documents are suddenly encountered? The discoveries of the Dead Sea Scrolls and the Nag Hamadi cache of gnostic manuscripts discussed in chapter six are striking enough, but the paleographer Peter Parsons tells how, to quote his subtitle, "Greek Papyri Beneath the Egyptian Sand Reveal a Long-Lost World."[25] Parsons's book discusses documents recovered from a massive garbage dump accumulated in antiquity by a now-vanished Greek colonial town in the Egyptian desert. Excavations began in the late nineteenth century, and the work of transcription and decipherment during the ensuing twelve decades has brought the corpus of *Oxyrhynchus Papyri* to at least seventy volumes. The story of the discoveries at the "city of the sharp-nosed (*oxyrhynchus*) fish" is more arresting than anything from the age of Petrarch and Poggio. In January 1905, at a depth of six feet, the remains of twelve papyrus rolls were found.

> Apart from the usual classics (Thucydides, Plato, Isocrates), there were a number of major works which had gone missing in the Middle Ages—the *Paeans* of Pindar, Euripides' *Hypsipyle*, an anonymous History of Greece. Three days later, the excavators had moved to a different mound, where they found a second hoard of literary items: eight feet down, Greek lyric poets; 25 feet down, a mass of other texts. There were songs by Sappho and Alcaeus and Ibycus, dithyrambs and paeans by Pindar, dithyrambs and drinking songs by his rival Bacchylides, learned elegies and learnedly offensive satires by Callimachus, sermons in verse by the Cynic Cercidas—all lost treasures of the Classical inheritance.[26]

And there were others besides. It bears repeating that these resurrected treasures were not found in a library, not even a neglected one of the sort Boccaccio described at Monte Cassino, but rather in a garbage dump. The discovery at Oxyrhynchus offered scholars a bonus of another kind. Alongside formal, literary texts there were more ephemeral, even trivial documents that had been discarded: letters, notes, bills of sale, every kind of private or bureaucratic record went into the dump along with the discarded classical manuscripts, providing a massive, detailed archive, over a period of centuries, concerning the activities of everyday people as well as scholars. Researchers were thus able to resurrect not only literature but everyday life from the spectrum of buried documents. Many papyri had been deliberately ripped up, making it evident that their owners had no use for them, either because they were letters and other ephemera, or because mutilated or worn-out literary manuscripts needed replacement. Such work combines the features of archaeology in the modern sense—reconstructing a vanished civilization from its artifacts—with the older literary, philological archaeology in place before 1800. As a result, the Oxyrhynchus papyri can receive somewhat the same kinds of attention that scholars dedicate to medieval or even modern archives, including the partial reconstruction of individual lives that would otherwise have remained in oblivion because lacking the illustrious accomplishments that attracted the attention of ancient poets and historians. After all, papyrus was a relatively fragile medium: not only could it mildew and rot, as parchment and paper can, but because it was relatively brittle, it could be easily broken if crushed or sharply folded.

This fact makes the resurrection of papyrus documents an even more wondrous accident. Buried deep underground, in an airless, arid tomb of sand, the Oxyrhynchus papyri were protected from all their natural enemies: water, fire, vermin, and vandals. Rather than disintegrating in a matter of years or decades, they survived for many centuries, just as the corpses of people sometimes remain intact and lifelike for improbable spans of time in Roman catacombs. And because the dump was used for so long a period, there were Christian as well as pagan writings. The first of the *Oxyrhynchus Papyri* to be published (1898) contained *Logia*, or sayings of Jesus. It was part of the apocryphal gnostic *Gospel of Thomas*, one of the thirteen Nag Hammadi papyrus codices that would be discovered in Egypt in 1945. As we saw in chapter six, these texts documented the beliefs of the cultic rival that "orthodox" early Christians all but destroyed.

At least as curious was the recovery of the longest extant text in the mysterious Etruscan language, which was also yielded by the sands of Egypt. Like the swaddling clothes of Heliodorus's Chariclea, the book was written on linen; it contained a ritual, liturgical text of about twelve hundred words, and had been repurposed in order to make wrappings for the mummy of a young girl; a Croatian traveler bought it in Egypt in 1848 and bequeathed it to the Zagreb museum in 1867.[27] The linen book had been recycled by Egyptian embalmers, who "needed enormous quantities of linen for swaddling their corpses and . . . were not particular about where it came from," so that what had been a long linen scroll was roughly torn into strips.

> The fact that the text was not written on papyrus, which was cheaper than linen in Egypt, indicates that the linen [book-]roll was brought to Egypt by Etruscan colonists, either at the time of their initial migration from Lydia in the ninth century BC or at the time of their final conquest by the Romans around the first century BC The practice of preserving books on linen was certainly characteristic of early Mediterranean peoples, for we

hear of the holy books of the Samnites being recorded in this way, as well as the registers of the Roman magistrates in early Republican times.[28]

A similar case is represented by the *genizot*, or "graveyards," of Bibles and other worn-out or mutilated Hebrew writings containing the name of God, as well as heretical or blasphemous texts. These "cemeteries" were intended as way stations on a journey to real burial grounds, where such problematic texts are reverently entombed as if they were cadavers. The best-known cache is at the Ben Ezra synagogue in a township of Old Cairo. It came to prominence in the last third of the nineteenth century, when European scholars began sifting through approximately a thousand years of accumulated texts it contained. Some of its documents became important sources for the history of the Hebrew Bible during the second millennium CE. Moreover, like the Oxyrhynchus garbage dump, the Cairo Geniza also contains important documentation relating to the lives and activities of merchants and other laypeople, as well as of rabbis and scholars.[29]

Yet another massive discovery happened in the early twentieth century in Gansu province of China, sealed in one of the Mogao Caves of Dunhuang. Not being Western, the hoard of documents, dating from 406 to 1002 CE, might seem extraneous here, except that it included the earliest known text printed with movable type, the Diamond Sutra of 868 CE. The Gansu manuscripts, numbering perhaps 50,000, were heaped in piles, and were said to have filled a space of 500 cubic feet. "A whole cave of unknown books! [Paul] Pelliot . . . must have been ecstatic," exclaims a recent writer, who adds, "this was akin to suddenly being given access to a secret room in the Library of Alexandria, in which everything had been preserved. How his heart must have raced. There's a photo of him sitting among piles of antique texts, reading them by candlelight. Feeling incredibly blessed, we can be sure."[30]

IS THE BIBLE DERIVATIVE?

Consequences of the archaeological revolution reached deep into perceptions of the religious heritage. Literate, inquisitive Jews and Christians gradually faced up to a choice: they could either declare allegiance to the mythological history of writing inherited from the Bible and a few ancient pagans, or they could take seriously the kinds of evidence that were rapidly creating a very different story of the subject. Previously, the history of writing had been overdetermined by interpretation of the Bible; but between 1822 and 1872 it became evident that the biblical stories were anything but factual, historical accounts. Henceforth, the archaeologically based history of writing would be fundamental to contextualizing the Bible, rather than the reverse. Awareness of the changes did not instantly overthrow belief in the Bible's account of history; indeed, it has still not done so. Through much of the nineteenth century, the archaeological evidence was often interpreted as supporting traditional interpretations of Genesis. Alternatively, one could go on objecting that any evidence contradicting the history of writing implied by the revealed word of God was necessarily false or misunderstood. Both approaches have continued to appeal to biblical literalists.

Nonetheless, in 1872 the idea of a separate, sacred, divine guarantee of the Bible's reliability as history was seriously compromised by George Smith's announcement of "the original" story of Noah's Ark. Smith was a bona fide genius: the son of laborers, he was plebeian and, lacking a benefactor, too poor to attend Oxford or Cambridge. As was common in Christian households until recently, the Bible was likely the only book his family owned. He was fascinated by it, especially by the Babylonian history it recounted. At fourteen, Smith's artistic talent and literacy gained him apprenticeship as a bank-note engraver in London, and this job, as David Damrosch observes, nurtured the lad's omnivorous intellectual curiosity. Through hard work and unusual talent, Smith eventually turned himself into one of the greatest pioneer Assyriologists. Frequenting the British Museum during his scarce free time, he trained his tireless enthusiasm on cuneiform script and the Akkadian language. Smith's remarkable talents convinced Henry Rawlinson and other professional scholars at the Museum to employ him at reassembling and deciphering fragmentary tablets brought back from Iraq after the spectacular discovery of the Library of Ashurbanipal in 1853.[31]

In his epoch-making lecture of December 1872, Smith announced that he had deciphered the "original" story of Noah's flood from a tablet of *Gilgamesh*; its detailed similarities to the biblical story of Noah figured prominently three years later in Smith's book *The Chaldaean Account of Genesis*.[32] The implications were profound: here was a precedent, if not an immediate source, for the story of Noah, right down to the roles of the raven and the dove (Genesis 8:6–12). But "Moses's predecessor" was not another biblical patriarch. Textual parallels might be evidence that the deluge was historical in some sense, but where did they leave the idea that authentic knowledge of the "Great Flood" was originally the exclusive possession of the Hebrews, or that the Bible's chronology was worthy of belief—sacred in every sense? From early Christians through the eighteenth century, pagan flood-myths, from Deucalion onward, had been dismissed as faint echoes of "Moses's original."

During the next few decades, discoveries by Smith and other Assyriologists and Egyptologists had a profound effect on ideas about the Bible's reliability as history. Indeed, two decades before Smith, a fascinating document of transition appeared in S. G. Goodrich's *History of All Nations, from the Earliest Periods to the Present Time* (1852). Chapter 35, "The Antediluvians," dates the biblical patriarchs to the period 3074–2348 BCE, but Goodrich concedes that Plato, the Hindus, and the Chaldeans gave immensely longer chronologies than the Bible. Like Eusebius and Scaliger, he acknowledges Berossos's story of antediluvian times, giving particular attention to Xisouthros, the Flood, and "a certain irrational [*sic*] animal called Oannes," who taught arts, letters, urban planning, agriculture, and so on to the Chaldeans. Goodrich also refers to "other antediluvian stories, preserved by Sanchoniathon, a Phoenician writer." Although "it is hardly necessary to say that there is no authentic foundation for these stories," Goodrich agreed with tradition that "many of the writers who relate them appear to have had some knowledge of the books of Moses," which were obviously "the most ancient writings extant."[33] In fact, there were "many accounts of the creation" and "many statements relative to the origin of human society" among ancient Gentiles. "But these accounts are various and contradictory," Goodrich declared, repeating that "it is hardly necessary to say that these accounts are supported by no evidence." Unsurprisingly, he declined to note that the Bible's chronologies also lack evidence.

Like Melanchthon's *Chronicon Carionis* three centuries earlier, Goodrich admitted that Moses "has left us almost entirely in the dark as to the particular events of antediluvian history," but observed that "the Jews, and other Eastern nations . . . have made ample amends for his silence, by the abundance of their traditions" and quoted Josephus's anecdote of the antediluvian stelae in detail. However, "in regard to the customs, policy, and other general circumstances of the antediluvians, we must rest satisfied with conjectures." He noted that based on Genesis's account of the antediluvians' extreme longevity, some writers estimated that humanity grew from Adam and Eve to a population of 400 billion (more than fifty times today's world population) "before the year of the deluge," that is, in 1656 years.[34] Goodrich shows no awareness that Voltaire had mocked such speculations a century earlier in one of his novellas, "estimating" that by 285 years after the Flood, "either" Shem or Japheth counted 623,612,358,000 descendants.[35]

Discoveries concerning the archaeology of writing were not the only kind of evidence that tended to disqualify Genesis as historical truth. Since the 1600s, advances in geology and paleontology had controverted ever more strongly the old traditions that fossil seashells found on high mountains were either relics of Noah's Flood or mere "jokes" of Nature, naturally occurring sculptures unrelated to living organisms.[36] Newly discovered fossils and their geological context implied that life on Earth was much older, and more strangely various, than previously suspected. The most spectacular evidence came from fossils of large animals that bore no comparison to existing species. In Greek antiquity, skeletons of mastodons and other ice-age fauna had been accepted as "giants' bones," and even reburied in "giants' tombs."[37] For Christians, these references proved that biblical descriptions of "giants in the earth in those days" (Genesis 6:4) were accurate. But in the same years as Champollion's announcements, Bible-based explanations were being menaced by descriptions of extinct reptiles such as the Iguanodon (1822) and the Megalosaurus (1824), unmentioned in Holy Writ or elsewhere.[38] Unlike more timid contemporaries,

Goodrich embraced innovations in paleontology, geology, and even cosmology. He provided a detailed engraving of iguanodons and plesiosaurs, describing the former as "that stupendous reptile, whose very existence had never been imagined until a recent period."

Goodrich expressed no particular pessimism that fossils and geology threatened the "history" recounted in Genesis. They moved him to wonder, not dismay: "the imagination, in turning back to this period, pictures to itself this mighty reptile rioting in the waves where the solid earth of the British islands now stands, and . . . discovers flying reptiles in the air, crocodiles and turtles sporting in the fens, and lizards and fishes, now blotted out of existence, making the waters boil with their gambols."[39] It was still possible for Goodrich to square scientific curiosity with biblical revelation: "with respect to the periods of time at which the Deity executed his several works of creation, mankind have received no particular information." Despite centuries of scholarly scrutiny, "the sacred scriptures do not fix the era of creation with perfect precision. They leave it in some measure undetermined whether we are to understand what they say as applicable to the whole contents of created space, or only to our earth and its inhabitants. Critics disagree as to the meaning of the word *day* in the Mosaic account of the creation; some understanding by it the time of twenty-four hours, and others a period of indefinite extent."[40] Geology and paleontology did not inspire these interpretations: Christians had invented them centuries earlier to refute or reinterpret the chronologies of Pliny, Plato, and Berossos.[41] By reminding themselves that "Moses" only measured years when detailing human chronology, nineteenth-century Christians could acknowledge fossil evidence of mass extinctions as readily as they refuted the revelations of Oannes. The creation of Adam and Eve on the "sixth day" might have occurred long after the first through fifth "days," which could have each lasted centuries or millennia, so there was no need to argue dogmatically that the entire universe began about 4000 BCE, following the biblical scholars Voltaire had mocked.[42] Despite his up-to-date coverage of cosmology, geology, paleontology, and archaeology, Goodrich could still reassure his audience, "History is far from being decisive as to the age of the world."[43]

Half a century later, Daniel Defoe's view of the Bible was even more foreign to Samuel Rolles Driver, Regius Professor of Hebrew at Oxford. In his commentary *The Book of Genesis,* first published in 1904, Driver argued that scientific evidence accumulated during the nineteenth century had rendered the traditional, Bible-based history of the world untenable. Referring to Noah and other patriarchs, Driver declared it "impossible that these figures—or at least the majority of them—can be historical." Archaeological and other evidence made him "certain that man existed upon the earth long before" the supposed creation of Adam and Eve, datable to 4157 BCE by the Hebrew Bible or 5328 BCE by the Septuagint. The multi-century life spans of antediluvian patriarchs "are incompatible with the constitution of the human body; and could only have been attained if that constitution had differed from what it now is, to an extent which we are entirely unwarranted in assuming to have been the case." If Noah's Flood actually happened, it was not universal—it did not cover the highest mountains to a depth of fifteen cubits, as Genesis 7:20 (and commentators down through Athanasius Kircher) maintained. It was confined to Babylonia, and "we possess no independent information" about when it occurred. If it was "the basis of both the Babylonian and the Biblical narratives of the Flood," then Moses could not be its original historiographer; nor would God have dictated ("revealed") the events of Genesis to him, though Driver did not belabor the latter point.[44]

Driver saw no harm in admitting that even after the Flood and the Tower of Babel Genesis is packed with impossibilities. "It remains to consider the historical character of . . . the narratives of the patriarchal period [Genesis chapters 12–50]. Here it must at the outset be admitted that these narratives do not satisfy the primary condition which every first-class historical authority must satisfy: they are *not contemporary* (or nearly so) with the events which they purport to relate: even if Moses were their author, he lived many centuries after Abraham."[45] Driver went on to address traditional Christian quibbles about the history of writing more explicitly. What if Moses was just editing records inherited from more ancient writers (an argument as old as Josephus, revived by Godfrey and Annius of Viterbo)? That would mean that Genesis was at least based on eyewitnesses. Driver could have enlisted nineteenth-century discoveries in Egypt and Mesopotamia to update this old speculation. Yet he was unmoved. "The supposition that the writer (or writers) of Genesis may have based his (or their) narratives upon written documents contemporary with the events described, does not alter the case. There is no evidence, direct or indirect, that such documents were actually used as the basis of the narrative; and upon a mere hypothesis, for the truth of which no positive grounds can be alleged, and which therefore may or may not be true, it must be apparent that no further conclusions of any value can be built."[46] Contemporaries of Defoe had conjectured that the Hebrews possessed writing before God gave the Decalogue to Moses because inscribed tablets would be useless to an illiterate society. But Driver slammed the door on this argument as well. He would not deny "that the patriarchs possessed the art of writing," but so what? The idea that Jewish writers existed before Moses "leads practically to no consequences; for we do not know what they wrote, and there is no evidence that they left any written materials whatever behind them."[47]

Until the nineteenth century, the history of writing had been overdetermined: ancient Greek and Roman texts recounted, and the Bible implied, specific and detailed stories about heroic inventors of writing, such as Cadmus and Moses, whom archaeological evidence would inexorably reveal as mythical. Faith in the Bible as God's revealed word had previously been sufficient to dismiss as wildly exaggerated such pagan chronologies as Pliny's assertion that Babylonian astrological records were "either" 480,000 or 720,000 years old (figures that premodern Christians regularly quoted as 480 and 720 years).[48] But biblically based chronologies were equally overdetermined, even though they were arrayed on a far less alarming scale.

By contrast, the textless evidence of rocks and fossils had been *underdetermined*. Not being written, geology and paleontology permitted a range of comforting interpretations involving Noah's Flood. Champollion and Goodrich's contemporary Georges Cuvier resolutely avoided biblically based arguments, but his catastrophism, a cyclical paleontology of mass extinctions followed by the advent of different flora and fauna, was predictably adopted as evidence for a series of divine creations and destructions, including, of course, Noah's Flood.

By itself, neither archaeological, paleontological, nor geological evidence necessarily threatened the credibility of the "history" recounted in Genesis. However, between 1822 and 1872 the emerging paleographic evidence brought back into the limelight the long-standing historical uncertainties provoked by ancient pagan historians. The newly translated Egyptian and Mesopotamian texts called for ways of arguing that differed from traditional attempts to defend Genesis against ancient Greek and Roman histories of writing such as Pliny's. Evidence in these recently discovered texts provided a disturbing new interpretive link to the unwritten evidence of geology and paleontology, making the biblical chronology look even less plausible as history. More than ever, faith in Scripture was revealing itself as, in fact, a matter of faith in *scriptura*—in writing itself.

ANOTHER SMITH AND A NEW SCRIPTURE

This is the context in which we should situate one of the strangest products of nineteenth-century biblical culture, the *Book of Mormon, An Account Written by the Hand of Mormon Upon Plates Taken from the Plates of Nephi.*[49] It purports to be a two-fold revelation, both divine and historical, resulting from encounters that began in 1823 between the founder of Mormonism, Joseph Smith, and an angel named Moroni, who presented him a set of golden tablets or "plates" on which ancient revelations were engraved. From the plates Smith learned that Native Americans belonged to the fabled lost tribes of Israel, who did not disappear, as previously believed, but emigrated by means of wondrous oceanic voyages in ark-like "barges" and, moreover, that Christ appeared to their descendants shortly after his resurrection. Inscribed in the "Reformed Egyptian alphabet," the plates were translated by Smith, who professed ignorance of their language but was able to read it because the angel loaned him a wonderful ophthalmic device, "two smooth three-cornered diamonds set in glass," resembling "old-fashioned spectacles." Strangely, the device was attached to a breastplate; Smith referred to the ensemble as "the

Urim and Thummim," the Hebrew name for a mysterious component of the priestly "breastplate of judgment" worn by Aaron, the brother of Moses (Exodus 28:30).

In terms of both its content and the way successive audiences have received it, the Book of Mormon raises most of the issues explored in my previous chapters, beginning with wonder. Early converts accepted Smith's book as "a marvelous work and a wonder," echoing Isaiah 4:14.[50] The events of its plot or "history," the identities of its supposed ancient authors, along with its modern discovery, translation, and dissemination, are suffused with a portentous rhetoric of revelation borrowed from the King James Bible. "Behold," "Yea," "Wherefore," and "It came to pass" are repeated with tedious frequency, sometimes in adjacent verses. Although the events of the Book are mainly situated in the future United States, they are thoroughly reminiscent of the Old Testament story of the Jews and have provoked accusations of plagiarism.

One ostensible purpose of the Book of Mormon is to corroborate the truthfulness of the Bible. This role was reenergized a century and a half after Joseph Smith, in 1982, when his book first received the official subtitle "Another Testament of Jesus Christ."[51] However, aside from its supposed history of pre-Columbian North America, the most salient feature of the Book of Mormon is not its biblicism but its autoreferentiality: as with Old Testament apocrypha and pseudepigrapha fifteen hundred years earlier, it is principally about itself, and its main preoccupation is to establish itself as scripture, as a divine revelation. Critics have noticed that, in terms of doctrine, the Book adds very little to American Evangelical biblical culture of the 1820s and 1830s. As one critic puts it, "In a very essential way . . . the 'message' of the Book of Mormon *was* its manner of origin."

> Looking at the Book of Mormon in terms of its early uses and reception, it becomes clear that this American scripture has exerted influence within the [Mormon] church and reaction outside the church not primarily by virtue of its substance, but rather its manner of appearing, not on the merits of what it says, but what it enacts. . . . [T]he history of the Book of Mormon's

place in Mormonism and American religion generally has always been more connected to . . . its role as a sacred sign rather than its function as persuasive theology. The Book of Mormon is preeminently a concrete manifestation of sacred utterance, and thus an evidence of divine presence, before it is a repository of theological claims.[52]

To put it crudely, the Book of Mormon has been an effective social instrument simply because it insistently proclaims itself Scripture rather than mere literature.

In case the title were not sufficient to advertise its pretension to scripturality, the Book is prefaced by three "testimonies" to its authenticity. The first, "The Testimony of Three Witnesses," proclaims that "we, through the grace of God the Father, and Our Lord Jesus Christ, have seen the plates which contain this record . . . of the people of Nephi." The Three Witnesses aver that "we also know that they have been translated by the gift and power of God, for his voice hath declared it unto us, wherefore we know of a surety that it is true." They testify further "that we have seen the engravings which are upon the plates, and they have been shown to us by the power of God, and not of man." This came about because "an Angel of God came down from heaven, and he brought and laid before our eyes, that we beheld and saw the plates, and the engravings thereon." The "Testimony of Eight Witnesses," which follows, says that Smith "has shown us the plates," and that they "have the appearance of gold." The Eight Witnesses "saw the engravings," which have "the appearance of ancient work, and of curious workmanship." They "have seen and hefted, and [we] know of a surety that the said Smith has got the plates." The third testimony, dated May 2, 1838, is that of Smith himself, who narrates how the angel revealed the location of the golden plates and attendant paraphernalia, along with instructions for translating them. Once Smith had finished translating, "according to arrangements, the messenger [i.e., angel] called for them, and I delivered them up to him, and he has them in his charge until this day," meaning, presumably, that humans can no longer consult them.[53] (The seeming contradiction with the "testimonies" of the Three and Eight Witnesses is striking.) The actual process of translation was apparently far more complicated and more than a bit suspect,[54] but we are left with twelve people's testimony (including Smith's) that the plates existed and could be "hefted," and that eleven witnesses are convinced that "the power of God, and not of man" enabled Smith to translate them. As one scholar remarks, "It was not believing the Book of Mormon's teachings . . . but believing the story of its origin that set Mormons apart."[55]

Aside from these paratexts, the Book of Mormon relies entirely on itself to establish its claim to scripturality. Like Old Testament apocrypha and pseudepigrapha, it obsessively dramatizes the circumstances of its own composition throughout its fifteen pseudonymous "books."

> Behold, I make an abridgment of the record of my father, upon plates which I have made with my own hands. . . . (1 Nephi 1:17)
>
> . . . many of which sayings are written upon mine other plates; for a more history [*sic*] part are written upon mine other plates. (2 Nephi 4:14)
>
> Nephi gave me, Jacob, a commandment concerning the small plates upon which these things are engraven. (Jacob 1:1)

The next four books are extremely short, comprising only one chapter each, so their autoreferentiality is even more accentuated.

> I did cry unto God that he would preserve the records; and he covenanted with me that he would bring them forth . . . in his own due time. (Enos 1:16)
>
> Now behold, I, Jarom, write a few words. . . . And as these plates are small. . . . For what could I write more than my fathers have written? (Jarom 1:1–2)
>
> Behold, it came to pass that I, Omni, being commanded by my father, Jarom, that I should write somewhat upon these plates. . . . And now I Amaron, write the things . . . which are few, in the book of my father. . . . Now I, Chemish . . . write in the same book with my brother. . . . Behold, I, Abinadom, am the son of Chemish. . . . And behold, the record of this people is engraven upon plates. . . . (Omni 1:1–11)

> And now I, Mormon, being about to deliver up the record which I have been making into the hands of my son Moroni... after I had made an abridgment from the plates of Nephi ... I searched among the records which had been delivered into my hands, and I found these plates. (Mormon 1:1, 3)

The Book of Mosiah returns to the long format of the initial three books, but dedicates much of its first chapter to various "plates," expatiating upon their religious and cultural utility: "And he also taught them concerning the records which were engraven on the plates of brass, saying: My sons, I would that ye should remember that were it not for these plates, which contain these records and these commandments, we must have suffered in ignorance, even at this present time, not knowing the mysteries of God" (Mosiah 1:3). The remaining seven books, some of which are quite long, all find room to mention the "plates," and all fifteen speak in the prophetic authorial first person: "Behold, I." Unfortunately, its characteristic autoreferentiality and its obsession with scripturality cause the Book of Mormon to resemble not only the Old Testament apocrypha but also—and unintentionally—the parabiblical forgeries of Annius and the fake forgeries produced by early modern satirists such as Rabelais.

Both the Book of Mormon and other early Mormon documents assert that it demonstrates not only that "the holy Scriptures are true" but "also, that God doth inspire men, and call them to his holy work, in these last days as well as in days of old."[56] However, the Book of Mormon, like Annius's Pseudo-Berosus, donates less authority to the Bible than it receives from it. Smith's book has traditionally been seen as a peculiarly American adaptation of the Bible. But overall, if it did not claim to be both historically true and a divine revelation, its "history" might be mistaken for the outline of a grand historical novel. In terms of its theme, it could have been a sequel to Lew Wallace's *Ben-Hur: A Tale of the Christ* (1880), another American classic that offered emotional reinforcement to the Bible's message.

The early nineteenth century was a time of religious turmoil in the United States, when competing Protestant sects raised bitter controversies over the possession of religious truth, and American millenarists, looking forward to a Second Coming, an apocalypse, and a new heaven and Earth, demonstrated intense desire for a historically plausible revelation. It is significant that, as "history," the Book of Mormon answers questions for an American audience that the Bible never addresses, beginning with the existence of the entire Western Hemisphere. Geography had posed problems for the universality of Christian revelation ever since Saint Augustine. He refuted the concept of the "Antipodes, that is to say, men on the opposite side of the earth," because "Scripture, which proves the truth of its historical statements by the accomplishment of its prophecies, gives no false information." Thus it was "too absurd to say that some men might have taken ship and traversed the whole wide ocean, and crossed from this side of the world to the other."[57] But when the old myth of the Antipodes as the Southern Hemisphere was replaced by the reality of the Western Hemisphere, the question of geography racked European brains throughout the sixteenth and seventeenth centuries. If all humans were descended from Adam, were Native Americans included? If so, why had God allowed mere geography to deprive them of Christian revelation?[58] Joseph Smith had no need to be aware of polemics like Augustine's or the early modern biblical controversies to understand that the Bible did not address the problem of the Americas, and thence to remedy the omission explicitly in the Book of Mormon.

My point here is not to discuss the validity of historical claims in or about Joseph Smith's masterpiece but rather to emphasize its place in the emotional history of writing as "a marvelous work and a wonder." That place is largely a result of its claims to documentary truth. Yet despite its insistence on writing, documentation, and historicity, the Book's proofs of scripturality are ultimately emotional, as such proofs always are. In the closing chapter of its final book, the Book of Mormon issues a notorious challenge to the reader: "And when ye shall receive these things, I would exhort you that ye would ask God, the Eternal Father, in the name of Christ, if these things are not true; and if ye shall ask with a sincere heart, with real intent, having faith in Christ, he will manifest the truth of it unto you, by the power of the Holy Ghost" (Moroni 10:4). Here, three centuries after Martin Luther, Joseph Smith declared an American version of *sola scriptura*: the truth of the Book of Mormon cannot be found through scholarship or archaeology. The only necessary—or allowable—intermediary between the scripture and the reader is the Holy Spirit. A 1984 Mormon publication told of "a Yugoslavian woman who gained fervent conviction of the Book of Mormon's truthfulness, though her copy was printed in a language she could not read: she believed because she had 'looked through the book and studied the pictures and . . . prayed about the truthfulness of the book,'" as Moroni 10:4 recommends.[59] This emphasis on sheer mystique is reminiscent of Hermann Hugo's and Olaudah Equiano's descriptions of illiterate persons' first contact with writing. My 1986 printing of the Book of Mormon, acquired in a secondhand store, contains a pasted-in color photograph of a fresh-faced young Salt Lake City woman, testifying to the Book's importance in her life and admonishing the reader to "follow the promise given in Moroni Chapter 10 Verses 3–5." Nonetheless, despite this insistence on inner conviction as the criterion of truth, Mormons from Smith onward have been fascinated with ancient writing systems and other archaeological topics.[60]

The desire for the oracular, "scriptural" truth of a primordial document is displayed as frequently by Americans' veneration of the US Constitution as by their cult of the Judeo-Christian Bible. Nominally secular defenses of the "original intent" behind the language of the Constitution mirror the dynamics of religious reformations, with their clarion calls of "back to the sources," their denunciation of explanatory commentary, and their ironic inability to defend "literal" readings without it. In the United States, tendentious literalist interpretations of the Second Amendment's "right to keep and bear arms" now jostle in right-wing propaganda with the First Amendment's guarantee of "religious freedom" (invoked as a justification to legalize discrimination against "foreign" religions, ethnic groups, and sexualities). Outside the confines of explicit religious worship, Americans' desire for provable scripture lives on, undiminished—the yearning for a guarantee inherited from the "original patriarchs," the country's "Founding Fathers," that would be both rationally explainable and reliably documentable. (Events occurring since 2015, especially the Black Lives Matter movement, are now challenging such interpretations as expressions of white supremacy, much as Frederick Douglass and other abolitionists protested them in the 1850s. The deadly assault on the US Capitol by a mob of enraged supporters of defeated candidate Donald Trump on January 6, 2021, confirms the compatibility of white supremacy and appeals to the "Original" Constitution.)

The evolution that Bible-centered cultures underwent in the nineteenth century is an important test case for the cultural implications of the history of writing. Just as knowledge of geological time engulfed and discredited the minuscule chronology of Genesis, so texts recovered through archaeology revealed the old bookish history of writing as little more than fairy-tales. Joseph Smith's descriptions, in and after the Book of Mormon, of his experiences with "ancient documents" gesture excitedly toward his contemporaries' real discoveries in Egyptology. The transition from a history of writing founded on literary evidence to one grounded in archaeological excavation and decipherment relegated documents like the Book of Mormon to the sphere of literary fantasy. Yet Mormon archaeologists still search for proof of their scripture's accuracy in the landscape of the Western United States.[61]

But archaeological research inspired the literary imagination even in isolation from scriptural controversies. Like *Asclepius* many centuries previously, Percy Shelley's "Ozymandias" romanticizes the history of writing through its apocalyptic atmosphere. In *Asclepius*, imaginary, fragmentary archaic inscriptions hint at, but never actually describe, a vanished Egyptian past, far more magnificent than the dreary present. "Ozymandias" supposedly debunks and disenchants this vision; such vanished "greatness" was the temporary product of tyranny and megalomaniacal delusion. To idealize it is folly. Nonetheless, before "Ozymandias" can denounce the folly, it must necessarily evoke the grandeur, thus, ironically, perpetuating the threat of madness.

A THIRD SMITH

The romance of Egyptology was intense in the years between Napoleon's invasion of Egypt and Champollion's announcement of deciphering the hieroglyphics. We know from Mary Shelley and other sources that her husband's poem was written quickly, and in friendly competition with another poet, Horace Smith (1779–1849).[62] They agreed to compose sonnets on the same subject and chose the phrase "Ozymandias, King of Kings" as their inspiration. The poet-entrepreneur Leigh Hunt published both poems in his magazine *The Examiner*, Shelley's on January 11, 1818, and Smith's two weeks later. Reading Horace Smith's sonnet alongside Shelley's throws the mystique of writing into high relief, revealing wonder as the sentiment that inspired both poems. Smith's sonnet is poetically inferior to Shelley's, but it provides a valuable commentary by flat-footedly explaining the wonder that Shelley subtly evokes through typically Romantic suggestions of "the sublime."

Smith's garrulous title is almost comically prosaic: "On a Stupendous Leg of Granite, Discovered Standing by Itself in the Deserts of Egypt, with the Inscription Inserted Below." The title—really more of a caption—solicits our wonder (the *stupor* behind "stupendous") before we ever experience the poem, which, disappointingly, is naively didactic and, with its speaking leg of stone, faintly ridiculous.

> In Egypt's sandy silence, all alone,
> Stands a gigantic leg, which far off throws
> The only shadow that the desert knows.
> "I am great Ozymandias," saith the stone.
> "The King of Kings; this mighty city shows
> The wonders of my hand." The city's gone!
> Naught but the leg remaining to disclose
> The sight of that forgotten Babylon.

In the verses following, the wonder enjoined by the inscription gives way to a cruel dramatic irony much like Shelley's: "Babylon in Egypt" was an old name for Cairo, but Smith's generic evocation of "Babylon" insinuates the divinely punished hubris of Sennacherib and Belshazzar in the Bible. The poet clumsily steers the irony toward his own culture, threatening its civilized pride with a similar fate.

> We *wonder*, and some hunter may express
> *Wonder* like ours, when through the wilderness
> Where London stood, holding the wolf in chase,

He meets some fragment huge, and stops to guess
What *wonderful*, but unrecorded race
Once dwelt in that annihilated place.

The idea that our civilization could someday be as forgotten as Ozymandias's, which Shelley merely intimated, is explicit here. But hubris and its punishment are not the only issue in Smith's poem. Despite his desire to moralize, his repetition of *wonder* functions as a kind of Freudian slip that undermines his attempt to reduce his poem's message to pieties.

Strangely, both poets make it easy to overlook the essential role that writing plays in their poems. Both poets' Ozymandian ruins contain an inscription, but both poems emphasize the oral over the written: Shelley's narrator quotes the words of Ozymandias as *spoken* by a "a traveler from an antique land." As with Smith's "saith the stone," the words on Shelley's pedestal seem almost to be spoken by Ozymandias rather than written. Crucial to the effect of Smith's poem is the future absence of written evidence about London's "wonderful, but unrecorded race," which he mentions as abruptly as the punchline to a joke. Unlike Shelley's "traveler from an antique land," Smith's conjectural hunter encountering an "annihilated" London apparently sees no writing on the "fragment" he encounters, since none is mentioned.

NOVELISTS AND WRITING LORE

Percy Shelley's wife, Mary, explored the mystique of writing in *Frankenstein* (1818; revised 1831) by dwelling on the learned Doctor's love of discredited pseudo-science such as the spurious fourth book of Heinrich Cornelius Agrippa's *Occult Philosophy*. Conversely, she dramatized the abandoned, illiterate Monster's improbably easy self-education: after serendipitously discovering a parcel of books, including Milton's *Paradise Lost*, the Monster invokes that poem repeatedly as the major subtext of *Frankenstein*. But in a later, less famous novel, Mary asked one of the most interesting questions of the nineteenth century, one that surpasses *Frankenstein* in evoking a theological horror of solitude: What if, rather than a single book surviving a universal cataclysm, a single *person* survived, while books remained safe? By the age of twenty-two, Mary had survived her husband, three children, and several other loved ones; in 1826, five years before her definitive revision of *Frankenstein,* she imagined a man whose loss was absolute except for books. In *The Last Man,* her protagonist, Verney, is the sole human survivor of a universal plague, a kind of worldwide Robinson Crusoe. After traveling the globe and ascertaining that he is truly alone, he stands "on the height of St. Peter's" and allows his "wild dreams" to "rule [his] imagination." Collecting Homer, Shakespeare, and a few other masterpieces, he sets off in a boat "without hope or joy" to visit the libraries of the world and live entirely through books. At the end of his tale, he imagines his future in this tiny, book-centered ark: "Thus around the shores of deserted earth, while the sun is high, and the moon waxes or wanes, angels, the spirits of the dead, and the ever-open eye of the Supreme, will behold the tiny bark, freighted with Verney—the LAST MAN."[63] And, we must add, with his books, which symbolize any and all books.

Like the solitary, mad doctor at the end of *Frankenstein,* Verney imagines himself surrounded and surveilled by invisible presences, including the "ever-open" eye of God, rather than by emptiness. The last man also stares fixedly, but only at books, substituting perpetual reading for direct interaction with living human beings. And when Mary Shelley declares "THE END" of *The Last Man,* it coincides, albeit at the distance of a few days or years, with the end of humanity. While the texts we have seen thus far imply that writing created humanity, Shelley imagined it might actually outlive us. Many praisers of writing have imagined it as the gift of a god, but in Mary Shelley's apocalypse, the Christian God is the only reader left.

Given the multitudes of their readers, nineteenth-century writers of fiction and poetry were appropriately fascinated by the mystique of writing, books, and libraries. Sometimes the fascination had sinister repercussions. Uncanny intimations were particularly unsettling in the fantastic, a mode of narration that mixed everyday realism with the kinds of supernatural experiences recounted in folklore and demonology.[64] Jan Potocki's *Manuscript Found in Saragossa,* a bizarre, dreamlike novel composed of one hundred-plus concatenated fantastic tales, was supposedly discovered in a manuscript during the

Napoleonic wars. In a textbook case of life imitating art, the novel was indeed discovered in a French manuscript after its Polish author committed suicide with a homemade silver bullet in 1815, and its complicated publication history is appropriately weird.[65]

The title of Honoré de Balzac's novel *La Peau de chagrin* (*The Wild Ass's Skin*, 1831), puns on *chagrin* as both anxiety or embarrassment, and *shagreen*, a tough equine rawhide sometimes used for bookbinding. A dissipated young man in despair foolishly accepts a wild ass's skin (*peau de chagrin*) inscribed with mysterious "Sanskrit" or "Arabic" characters, which promise to grant his every wish. The inscription comes true, but the skin shrinks, along with its owner's health, each time it grants him a wish. Unable to rid himself of the skin or its curse, he dies after consummating his belated love for the woman he had wrongly neglected. Could this be an allegorical indictment of compulsive reading?

Mr. Lockwood, the frame-narrator of Emily Brontë's *Wuthering Heights* (1847), undergoes a kind of spirit-haunting when he reads the obsessive signatures covering the walls of the sleeping-cabinet that encloses his bed: "Catherine Earnshaw," "Catherine Heathcliff," "Catherine Linton." Finding several Protestant devotional books, he sees that the young Catherine used blank spaces of their pages to record a diary of her unhappy relations with three men: her brother, her soulmate Heathcliff, and her mismatched husband. Falling asleep, Lockwood dreams—or does he observe?—the despondent attempt of the child Catherine's ghost to enter the house that witnessed her experiences. It is as if by inhabiting the inscribed cabinet, Lockwood had been emotionally transported into her story. Subsequently, Lockwood learns the fuller tale of Catherine's life from Nelly Dean, the housekeeper in the lodgings he is renting.[66]

In George Eliot's *Middlemarch* (1872), the Reverend Mr. Casaubon, who ironically shares his surname with the debunker of the *Corpus Hermeticum*, squanders his life, and the hopes of his devoted Dorothea Brooke, as he pursues the chimera of his scholarly masterpiece, the *Key to All Mythologies*. His *Key* should have performed the same kind of syncretic, synoptic vindication of universal truth that Pico della Mirandola was preparing to unveil in 1486 and that

Giambattista Vico debunked in the three editions (1725, 1730, 1744) of his *New Science*.

The nineteenth century's fascination with the mystique of writing also manifested itself in what we could call popular bibliography. Before 1800, imaginary books and libraries had occupied a small but significant place in the pages of scholarly works in Latin, ranging from Sixtus of Siena's *Bibliotheca sancta* to the great works of Fabricius, and even in the vernacular essays of the *Encyclopédie*. But in the nineteenth century, interest in the topic jumped the bounds of decorum previously respected by scholars. The triumph of Romanticism and the vogue of folkloric study epitomized by the Grimm brothers brought new attention to bear on the works of François Rabelais. His comic romances had been regaining attention and a measure of respectability since the mid-eighteenth century, when Pierre Le Duchat's massive annotated edition treated *Pantagruel* and *Gargantua* to the kind of serious scholarly apparatus formerly reserved for the Bible and the classics.[67] A few decades previously, English publishers, stimulated by a nonclassical tradition stretching from Chaucer to Shakespeare, had anticipated French interest in Rabelais. The Scottish Royalist Thomas Urquhart had returned Rabelais to scholarly attention—including Le Duchat's—with his learned English translation of the first two books (1653), while the Huguenot expatriate Peter Anthony Motteux had translated the third though fifth (1693–1694).

Continued interest in Rabelais's sprawling, encyclopedic, and facetiously erudite romances brought them to the attention of a different scholarly demographic during the nineteenth century. Attentive bibliographers realized that the "ghosts" their profession pursued were not always the result of biblioclasms, flawed transcription of titles, or other accidents, nor even of the eighteenth century's optimistic projects for useful books as yet unwritten. In 1851, Gustave Brunet published his *Essay on Imaginary Libraries*, centered on Rabelais; eleven years later, his friend Paul Lacroix, alias "le Bibliophile Jacob," reprinted the *Essay*, adding his own *Catalogue of the Library of the Abbey of Saint-Victor in the Sixteenth Century, Compiled by François Rabelais*.[68] Because nineteenth-century France was rediscovering Rabelais as a pre-Enlightenment national treasure, the satirical bent of his excursions into imaginary bibliography convinced Brunet that the "genre" was essentially sardonic. "Bibliographic science . . . sometimes concerns itself with books that never existed, but which some writers, always in obedience to satirical instincts, have mentioned as published, an assertion which, clearly, no one took seriously. We owe the invention of this vein of sarcasm to Maître François Rabelais; no trace of it can be found, we believe, in Antiquity or the Middle Ages." For his part, Lacroix/Jacob concurred that Rabelais had invented the genre but conceded that imitators, such as Thomas Browne and John Donne in the seventeenth century, had often composed catalogs that were equally "spirited and naughty." However, Rabelais's imitators had reduced the genre to "a sort of epigram, a satirical barb," with no real bibliographic content, whereas the master had always taken aim at a particular printed book or manuscript.[69] The reductive notion that Rabelais "invented" imaginary bibliography has appealed to the twentieth and twenty-first centuries. Despite their positivistic limitations, however, Brunet and Lacroix and, more broadly, cultists of Rabelais in France and elsewhere, expanded the awareness of bibliographic imagination by investigating the capacity of fiction to square itself, to create metafictions, mimeses of unreality and inexistent mimeses of reality, imagining books as paradoxes, or even mystical symbols.

In 1914, Anatole France published *La Révolte des anges*,[70] which redefined the imaginary library in still another way, describing an ideal building filled with books whose historical existence was either attested or eminently plausible but whose effects were devastating. The novel revolves around a private library that is inconceivably rich in rare historical works of natural philosophy and theology. At a certain point the librarian begins noticing valuable books drawn from their shelves and spread in disorder about the rooms. Eventually books go missing altogether: an extremely ancient manuscript of Flavius Josephus in sixty volumes, a Lucretius with annotations by Voltaire, a manuscript of Richard Simon (d. 1712), whose *Critical Histories* of the Old and New Testaments declared, among other things, that Moses could not have written the Pentateuch, earning Simon the enmity of both Catholics and Protestants.[71] Ultimately, we learn that the thefts were perpetrated by a guardian angel curious to know more about humans than his job description permits. Owing to his readings, the angel loses his faith and "falls," going on to involve heaven and Earth in disasters on an apocalyptic scale. Echoes of Genesis, the Books of Enoch, and *Paradise Lost* suggest ironically that books are the poisoned apple of the Tree of Knowledge, and that ignorance was as essential to angels' well-being as was Adam and Eve's to humans.

BIBLIOMANIA

Like knowledge itself, the books that contained it could be baneful. "The mania for acquiring books" as prestigious physical objects rather than for their intellectual content had been a serious concern of the eighteenth century, and an article on *bibliomanie* already figured prominently in the *Encyclopédie* by 1751. "Bibliomania" entered the English lexicon in 1809–1811 with Thomas Frognall Dibdin's *Bibliomania, or Book-Madness: A Bibliographical Romance*, a series of mock-heroic dialogues on the malady.[72] Daniel Desormeaux has ably documented how the 1800s were in many ways the century par excellence of bibliomaniacs; he traces the history of the figure from intimations in Richard of Bury, in Sebastian Brant's *Ship of Fools* (1494), and in Rabelais's Library of Saint Victor, through the formal, philosophical naming of the malady in the Age of Grand Collections, culminating in the *Encyclopédie*, and onward to the fictionalization of bibliomaniacs by great French authors of the nineteenth century, including Anatole France. Desormeaux concludes that, in the twentieth century, a kind of poetics invaded the arena of bibliomania, in such authors as Jorge Luis Borges and Umberto Eco.[73] They will figure more generally as meditators on writing in the next chapter.

CHAPTER FIFTEEN

The Age of Media, 1950–2020

> Let us imagine that in Toledo a paper is discovered containing a text in Arabic which the paleographers declare to be in the handwriting of the Cide Hamete Benengeli from whom Cervantes derived the *Quixote*.
>
> — JORGE LUIS BORGES, "A Problem"

In previous chapters I have had to limit—more than may be apparent—historical evidence of the notion that "It is written." I have mentioned several traditions (Islamic, postbiblical Jewish, Chinese, Mesoamerican) but have not dared survey them for lack of both expertise and space. Even so, given the panorama sketched over these pages, the number and variety of opinions and hypotheses about the relationship between humanity and the art of writing escapes any easy historical synthesis.

Likewise, for this final chapter, a true survey of the past seventy-plus years would be difficult to achieve. Instead, I will concentrate on a few suggestive examples that demonstrate how themes identified and pursued so far have been transformed in what, following Marshall McLuhan, we can call the Age of Media.[1]

INSCRIPTION

In his article on papyrus, Pliny the Elder intimated that, as transmitter and guardian of thought, writing is the essential preserver of humanity. Two millennia later, Umberto Eco remarked on the fundamental, physiological humanity of writing. "We can think of writing as an extension of the hand, and therefore as almost biological. It is the communication tool most closely linked to the body. Once invented, it could never be given up. As I said about the book, it was like the invention of the wheel. Today's wheels are the same as wheels in prehistoric times. Our modern inventions—cinema, radio, Internet—are not biological."[2] Wonder and mystery have drained away from modern people's awareness of writing. Media technology has inherited the emotions once reserved for writing by hand. What Eco called "not biological" forms of writing break the charmed physiological connection between the eye, the hand, and the pen. Keyboards, screens, and voice-recognition software are now ubiquitous, and these technologies have made writing with pencils, pens, and chalk largely unnecessary.

To the average educated person, "technology" still implies machinery. But since the 1980s, the prostheses employed for uniform writing are no longer mechanical in the cumbrous, ponderous way that typewriters and typesetting were. Writing by hand, the simplest technology of all, made possible all the others.

BOOKS

For most of history, to discuss writing as the guardian of memory and humanity was to eulogize the book. The book was the original archive, both portable and reproduceable. Libraries were in effect superarchives, archives of archives. But since the 1980s, electronic archives often seem to challenge the hegemony of paper-based books, which have steadily migrated from library shelves to remote storage, or worse, to recyclers or rubbish dumps. In 1996, Eco contributed the afterword to a volume titled *The Future of the Book*; ten years later, another anthology bore the ominous title *The Future of the Book in the Digital Age*, and the question has persisted, despite recently commanding less attention from scholars and "the media." Eco's own meditations continued in 2011, in a coauthored book polemically titled *This Is Not the End of the Book*. Eco agreed with Plato's Critias, Isidore of Seville, and their many legatees before the 1980s that to preserve cultural memory, writing in physical scrolls and codices is indispensable. As Eco phrased it, "modern media formats quickly become obsolete. Why run the risk of choosing an object that may become mute and indecipherable? It is proven that books are superior to every other object that our cultural industries have put on the market in recent years. So, wanting to choose something easily transportable and that has shown itself equal to the ravages of time, I choose the book."[3] Museums of information technology, particularly those showcasing the ubiquitous personal computer, are vividly dramatizing the paradox of obtaining uniform written artifacts through endlessly varied and mutually incompatible electronic devices. A number of institutions are soliciting donations of obsolete computers, software, and paraphernalia.[4] In some cases, these museum pieces have enabled the rescue of data that had remained dormant and inaccessible for years.

As a medium for preserving knowledge and cultural memory, the book offers distinct advantages over electronic media. While print media have always had a relatively restricted readership, they do not "crash" or "black out" when electricity is interrupted, whereas the electrical grid is now considered a primary target of terrorist attack. The loss of a country's or a continent's electrical power—or a sophisticated hacker at any time—could almost instantaneously destabilize every cultural field, from accounting and banking to agriculture and war. Free of batteries and power cords, "hard copy" on paper and other material supports requires orders of magnitude longer than electronic media to degrade or obliterate. In the global twenty-first century economy, the primary disadvantage of paper-based media, their physical dispersion, becomes an advantage when nature, economics, or warfare imperil electricity. As Polydore Vergil recognized in the Gutenberg age, the multiplication of physical copies and their wide distribution guarantee safety from erasure far better than unique or scarce exemplars in even the sturdiest library buildings. Electronic copies fail to provide even the guarantees of the manuscript age: in a national or global emergency, dependence on electrically powered servers could propel us into a new dark age.

OBLITERATION

Clearly, this is not to claim that books and other archives on paper are the infallible savior of writing and cultural memory. Modern warnings of epochal biblioclasms have been accumulating for decades. Ray Bradbury's nightmarishly prophetic *Fahrenheit 451*, published in 1953, imagined a strangely familiar future American society—deprived of even the will to read, kept docile by addiction to tranquilizers and electronic media. Wireless "earbuds" transmit soothing music from some central location, without any individual's need or right to own a private music source. Flat-screen televisions take up entire walls in every home; domestic life consists of an individual's virtual interactions with generic but mysterious and compelling television characters called "family members." Although television was still primitive and a relatively scarce commodity in 1953, Bradbury's novella was prescient: it foreshadowed our global society's enthrallment by personalized information technology, ubiquitous video screens, and nonstop audio. The technology in *Fahrenheit 451* lacked only the portability of "smart" phones and the feedback loop that harvests an individual's clicks and postings on social media to create individualized marketing campaigns that target and exploit their consumer preferences, or channel "fake news" and other disinformation to them.

Bradbury's dystopia did not foresee our current level of willing technological thralldom; instead, its rulers adopted the primordial expedient of outlawing books altogether. Firemen have been redefined as book burners, and are on duty at all times. When they receive an emergency call—that is, an anonymous accusation—they speed to the locale and torch all the books. Yet despite the iron-fisted enforced illiteracy he describes, Bradbury's apocalyptic vision offers a paradoxical glimmer of hope at its end. The protagonist, a deserter from the fire brigade, is outlawed for surreptitiously collecting and reading books rather than incinerating them. Escaping, he discovers a resistance movement of unrepentant fugitive readers, who have found a solution to the deliberate, near-total biblioclasm. Each outlaw "becomes" a book by memorizing its entire contents and reciting its text to other fugitives who incarnate other books. In this curious, paradoxical throwback to Socrates's disdain for writing as mere reminder, only the books still contained verbatim in individual human memories offer hope for a rebirth of civilization. Bradbury inverted the constants in the emotional history of writing: rather than preserving human memory more or less durably, inscription (at some future date) must now be safeguarded by it. Yet we have glimpsed this paradox before. *Fahrenheit 451* radically updated the Fourth Book of Ezra and Renaissance scholars like Pico della Mirandola and Leo Hebraeus, who imagined that an oral Kabbalah sustained the Old Testament or the memoirs of the earliest patriarchs until they could be more securely preserved in writing.

Fahrenheit 451 somewhat revised the archetype of obliteration by suggesting that human biblioclasts finally possessed the technology and social organization to supplant the superhuman forces of time and nature that readers had dreaded for thousands of years. Electronic audiovisual media could make reading for knowledge or entertainment obsolete, and render writing superfluous for all but the most ephemeral practical tasks.[5] Presumably, only bureaucrats would need to perform reading or writing of any sort, and then only to extend the reach of obliterative technology into the lives of everyone else. In Bradbury's terms, the ultimate wonder of writing would be survival of the book as something more than a pastime for people deprived of or opposed to electronic media. The major difference between his dystopia and universal biblioclasms that humans had previously imagined lies in the narrowly belletristic nature of many endangered books he mentions—that is, "great" Western literature. But from Berossos to the *Encyclopédie*, the value of writing was its function, to quote Pliny, as preserver of "the immortality of human beings [*immortalitas hominum*]" and "the humanity of life or at all events our memory [*humanitas vitae . . . certe memoria*]." For many centuries, the mystique of writing had been based on its potential to safeguard human knowledge from obliteration, despite its own notorious vulnerability.

The specter Bradbury raised of a deliberate print apocalypse and technological enslavement to "media" was an ironic development in the history of writing and a prophetic one.

EXTINCTION

The obverse of total obliteration—a single surviving reader—also caught the popular imagination in the mid-twentieth century. "Time

Enough at Last," a 1959 episode of the television series *The Twilight Zone,* portrayed a compulsive reader, Henry Bemis, frustrated by the insistent demands of marriage and work, who became a modern version of Mary Shelley's Verney, as the only human lucky enough to survive the nuclear apocalypse.[6] But Bemis never received the minimal pleasure his tragic predecessor enjoyed. Having just gathered a vast pile of books from his local public library, the poor man watched in horror as his thick spectacles dropped from his face and shattered on the library steps. Sixty years after "Time Enough at Last," the question arises: Would it be easier for the last survivor to find a usable pair of glasses or to reactivate the technology necessary for reading an electronic text? Verney's final pilgrimage to the world's great libraries is at least thinkable, but the updated Last Man would be unable to read for long on an electronic screen—or even on all of them.

Quite suddenly, the COVID-19 pandemic that began in early 2020 made Mary Shelley's two-century-old meditations on writing, plague, and the obliteration of humanity seem even more timely. In *The New Yorker* magazine, a historian included *The Last Man*—with Boccaccio's prologue to the *Decameron* (1352), Defoe's *Journal of the Plague Year* (1722), and Albert Camus's *The Plague* (1947)—in her reflection on books as "a salve and a consolation" in times of enforced isolation and quarantine, and on reading itself as "an infection, a burrowing into the brain." Jill Lepore recalls that "wandering amid the ruins of Rome, [Shelley's Verney] enters the home of a writer and finds a manuscript on his writing table. 'It contained a learned disquisition on the Italian language.' This final book is a study of language, humanity's first adornment. And what does our narrator do, alone in the world? 'I also will write a book, I cried—for whom to read?' . . . It will have no readers. Except, of course, the readers of Shelley's book."[7] The final book—or rather the penultimate, since Verney obviously kept his promise to write one—would, like Verney's own manuscript, be an enduring material presence with no surviving human senses to perceive it. Having no readers, both manuscripts would be found books with no finder, revelations with no initiates, literary oxymorons. And so would every other book.

In one sense, the opposite of writing is not obliteration but the as-yet unwritten: not the missing bibliographic ghost, or even the erased book, but the wholly imaginary one. Once the concept of the imaginary book debuted on electronic media, it underwent a rapid proliferation, inspiring both semi-scholarly research into existing references in literary works, and do-it-yourself projects for the imagination, such as the contest sponsored by a museum near Newcastle, England.[8] On the internet, imaginary books preoccupied many readers of the 1990s and early 2000s. From 1999 until 2002, the internet hosted a site called The Invisible Library, which presented itself as "a collection of books that only appear in other books."[9] As with any cultural fad, the vogue of imaginary bibliography generated a good deal of banality. And in fact, the basic concept remains difficult for many people, even the most avid readers, to define and designate. A discussion among contributors to Wikipedia in 2006 betrayed considerable uncertainty over the differences between "books of fiction," "fictional books," "fictitious books," "fiction books," and "imaginary books." On August first, one contributor voiced opposition to the designation "imaginary books" by observing that "the name implies that the books don't exist." A colleague replied: "Entirely correct: that is exactly what the category is 'meant' to contain. The comment precisely demonstrates the problem the use of this term produces. 'Imaginary books' is clearer [than 'fictional books'], unless someone can think of a better term." Strangely, however, it appears that the term "fictional books" prevailed on Wikipedia.[10]

It is perhaps significant that imaginary books gained such prominence on an electronic, paperless medium during the same years when "the future of the book" was so widely discussed. Thanks to the internet, even imaginary books were freed from their historic dependence on paper and could transit seamlessly from waves in the ether to those in the human brain. In other ironic moves, a French author published a book describing how to discuss books one has never read, while an American writer published another tome detailing how to invent books and "review" them at cocktail parties with the aim of entrapping pretentious know-it-alls into discussing the books as though they had read them, thereby exposing themselves as phonies.[11]

We have seen how imaginary books are much older than the internet, even as an explicit narrative gambit, but the Wikipedia discussants and others were inspired by preelectronic developments within the lifetime of their parents. In 1947, the science fiction writer L. Sprague Decamp published an article titled "The Unwritten Classics" in *The Saturday Review of Literature*, and coined the term *pseudobiblia*, which has been adopted by numerous writers, many of them predominantly interested in science fiction and other forms of fantasy literature.[12] In 1972, two Italian authors published a rather substantial essay called "Books that Don't Exist (And Those That Should Not)" as an appendix to the Italian translation of a French mass-market work, *Les livres maudits* (*The Damned Books*).[13] Jacques Bergier's book participated in a vein of writings in the 1960s and 1970s that he and others referred to as "fantastic realism" (not to be confused with "magical realism" in fiction, popularized by Gabriel García Márquez and other postmodern Latin American writers). The most notorious examples of fantastic realism were Bergier's coauthored *Morning of the Magicians* (1960, English 1963)[14] and *Chariots of the Gods?* (1968, English 1969) by the professional huckster Erich von Däniken.[15] These books and their offshoots updated the Atlanteans and Egyptians of Plato's *Critias* through science-fiction explanations of real-world artifacts, arguing, for instance, that "ancient cosmonauts" (mistaken for gods by earthlings) had visited Earth and, thanks to their superior technology, created artifacts such as the Egyptian and Central American pyramids. On its surface, "fantastic realism" also differed from "the fantastic" in literature (e.g., *The Manuscript Found at Saragossa*) by purporting to explain well-known concrete phenomena in the real world rather than fictional encounters with ghostly or uncanny beings. For writings such as Bergier's, Italian has a clearer designation than "fantastic realism": *fantastoria*, or fantasy-history, a genre-label modeled on *fantascienza*, or science fiction; both terms imply somewhat ironically that fantasies cannot ultimately be confirmed as instances of realism. Imaginary books of all sorts have blossomed in the science-fiction hothouse, inspiring another useful Italian term, *fantabibliografia*, or bibliographic fantasy. Examples could be multiplied, but the ones given here suffice to illustrate that even in the Age of Media, myths of writing metamorphose rather than die.

THREE RELATED METAMORPHOSES

To illustrate thematic transformations since 1950, I will concentrate on Jorge Luis Borges, and his legacy to two like-minded novelists who composed works featuring *fantabibliografia*. Although he had been writing since the 1920s, Borges's international renown dates from the early 1960s, when his short stories and essays were translated from Spanish into French, English, and other languages, and he began receiving international literary prizes. Fittingly for an author who became director of the Argentine National Library in 1955, Borges's fictions and essays reflect a deep immersion in the themes traced throughout this book.

Borges's signature gambit, inspired by decades of reading in philosophy and religions, was the vertiginous blurring of boundaries between everyday reality, historically attested reality, and a purely literary fantastic reality. He was particularly drawn to the hallucinatory claims in apocryphal and pseudepigraphic texts produced by major Christian heresies, especially gnosticism. Referring to imaginary books by the suggestively named fictitious author Nils Runeberg, Borges's "Three Versions of Judas" begins with the abrupt statement that "in Asia Minor or in Alexandria, in the second century of our faith, when Basilides disseminated the idea that the cosmos was the reckless or evil improvisation of deficient angels, Nils Runeberg would have directed, with singular intellectual passion, one of the Gnostic conventicles." Borges's narrator also opines that "Dante would have assigned [Runeberg], perhaps, a fiery grave" among the Epicureans of *Inferno*, canto 10, who denied the immortality of the human soul and symbolized heresy in general. "His name would extend the list of lesser heresiarchs, along with Satornilus [or Saturninus] and Carpocrates; some fragment of his preachings, embellished with invective, would survive in the apocryphal *Liber adversus omnes haereses* or would have perished when the burning of a monastery library devoured the last copy of the *Syntagma*. Instead, God afforded Runeberg the twentieth century and the university town of Lund."[16] In the fictive reality of this paragraph, hypothetical or imaginary personages and works—Runeberg and his treatise that exalts Judas rather than Jesus as the true God-Become-Man—rub elbows with historically attested authors—Dante and several infamous gnostic heresiarchs—to create a literary "reality" that straddles the lines between fiction and forgery, as between well-known theological propositions and conclusions drawn from their logic that are only superficially aberrant. In a final twist, Runeberg's imaginary heresy closely resembles the extant pseudepigraphic gospels of Judas and Thomas, which were, respectively, unknown and newly discovered when Borges wrote.[17]

In "Tlön, Uqbar, Orbis Tertius," Borges's narrator reveals that "the metaphysicians of Tlön," an imaginary planet described by generations of assiduous literary forgers, "do not seek for the truth, or even for verisimilitude, but rather for the astounding. They judge that metaphysics is a branch of fantastic literature." As Borges himself did. The narrator tells us that one of the several schools of philosophy on Tlön averred that "the history of the universe—and in it our lives and the most tenuous detail of our lives—is the scripture produced by a subordinate god to communicate with a demon."[18] *A First Encyclopedia of Tlön* is the ultimate forgery, "a vast methodical fragment of an unknown planet's entire history, with its architecture and its playing cards, with the dread of its mythologies and the murmur of its languages, with its emperors and its seas, with its minerals and its birds and its fish, with its algebra and its fire, with its theological and metaphysical controversy. And all of it articulated, coherent, with no visible doctrinal intent or tone of parody."[19] The fantastic realities of the forged encyclopedia eventually overwhelm reality as it was understood previously. "The contact and the habit of Tlön have disintegrated this world. . . . Already the schools have been invaded by the (conjectural) "primitive language" of Tlön; already the teaching of its harmonious history (filled with moving episodes) has wiped out the one which governed in my childhood; already a fictitious past occupies in our memories the place of another, a past of which we know nothing with certainty—not even that it is false." Faced with this usurpation, the narrator sees little hope for reality as he remembers it and concludes gloomily that soon "the world will be Tlön."[20] The book has displaced the world, substituted its own fictions for ordinary "consensus reality."

Borges's most famous creation was the short story "The Library of Babel," which daringly reproposed the hoary commonplace of the world as a book. Borges imagined the universe as a literal bibliocosm: "The universe (which others call the Library) is composed of an indefinite and perhaps infinite number of hexagonal galleries, with vast air shafts between. From any of the hexagons one can see, interminably, the upper and lower floors."[21] In one of the axioms proposed by the narrator, "the Library exists *ab aeterno*," and "Man, the imperfect librarian," the only species inhabiting the bibliocosm, "may be the product of chance or of malevolent demiurgi."[22] This maxim recalls not only the malevolent impostor-god of the Old Testament that gnostic Christians imagined but also the eternal, uncreated world of Aristotle and the spiritless materialism of Epicurean philosophers. In an Epicurean-seeming pattern noticed by the librarians, the indefinite or infinite numbers of books, although uniform in format and number of pages, are "formless and chaotic" in their contents, which are randomly composed of the atom-like "twenty-two letters of the alphabet" (a number that coincides with the Hebrew alphabet) along with the comma, the period, and the blank space separating words.[23] Five hundred years previous to the narrator's time, "the chief of an upper hexagon came upon a book as confusing as the others, but which had nearly two pages of homogeneous lines. He showed his find to a wandering decoder who told him the lines were written in Portuguese; others said they were Yiddish. Within a century, the language was established: a Samoyedic Lithuanian dialect of Guarani, with Classical Arabian inflections. The content was also deciphered: some notions of combinative analysis, illustrated with examples of variation with unlimited repetition."[24] In other words, the book was a microcosm of the library itself, its self-contradictory linguistics belied by the simplicity of its constitutive principle.

Having discerned the pattern, we read, this long-ago genius librarian brilliantly "deduced that the Library is total and that its shelves register all the possible combinations of the twenty-odd orthographical symbols (a number which, though extremely vast, is not infinite)": "in other words, all that it is given to express, in all languages. Everything: the minutely detailed history of the future, the archangels' autobiographies, the faithful catalogue of the Library, thousands and thousands of false catalogues, the demonstration of the fallacy of those catalogues, the demonstration of the fallacy of the true catalogue, the Gnostic gospel of Basilides, the commentary on that gospel, the true story of your death, the translation of every book in all languages, the interpolations of every book in all books."[25] Two propositions undergird the story: Borges combines the orthodox idea that the world is a book, a *liber mundi,* with Basilides's terrifying and ultimately innavigable cosmology of 365 heavens, thus creating a bibliocosm which is as incomprehensible as our own universe. At the same time, Borges's story provides the definitive gloss on the ancient and medieval idea of the alphabet's potential for both precision and infinitude.[26]

The most famous, and in many ways the ultimate, homage to Borges's scriptive and scriptural fantasies is Umberto Eco's *Name of the Rose,* first published in 1980 and soon translated into multiple languages as well as a visually suggestive film and a recent television series. Borges's bibliocosm inspired the blueprint of Eco's *Aedificium,* the vast, mysterious, and labyrinthine library housed in a fictitious northern Italian monastery, the stacks of which are closed to everyone but its prefect. The novel is set in the year 1327, six years after the death of Dante, and is saturated with allusions to Borges's favorite themes of libraries, obsolete scholarship, and the history of Christian heresies. In the *Aedificium,* canonical works of philosophy, literature, and medieval theology are shelved alongside mutilated or ghostly treatises and tomes bearing completely invented or impossible titles.

Appropriately, the story commences with an exaggerated wink to the reader. A preface signed by "Umberto Eco" and titled "Naturally, a Manuscript," begins: "On August 16, 1968, I was handed a book written by a certain Abbé Vallet."[27] The book's title claims that it was written by a Benedictine monk, Adso of Melk, edited centuries later by Jean Mabillon (whose very real work on paleography was mentioned above in chapter twelve), and printed for Abbé Vallet in 1842. "Eco" hastily translated the book out of sheer enthusiasm, but a traveling companion absconded with the original volume. Since no other copy has ever been found, the reader must be content with "Eco's" Italian translation of Vallet's nineteenth-century translation into archaizing "neo-Gothic" French of an eighteenth-century scholar's edition of a fourteenth-century Latin memoir by a German monk. Given the absence of all the prior documents, why should "Eco" publish his translation? "Out of pure love of writing," he answers, and quotes a medieval mystic that "I have sought tranquillity everywhere, but never found it except when sitting in a corner with a book."[28] *The Name of the Rose* contained so much playful, plausible misinformation that it inspired reams of scholarly commentary, including Eco's own *Postscript to the Name of the Rose* (1983; English 1984).

Whereas Borges suggested chaos, infinity, or heresy in a few lines or pages, *The Name of the Rose,* like each of Eco's other novels, takes about five hundred pages to unfold meticulous encyclopedias of literature, theology, philosophy, and natural history. Still, if Borges had written a novel, it would have probably resembled *The Name of the Rose,* a complex mishmash of themes and characters drawn from all the disciplines just mentioned, and supplemented by Eco's extensive researches into narratology, which he elsewhere decanted into sophisticated theoretical treatises such as *The Role of the Reader* (1979) and *The Limits of Interpretation* (1990).

The fact that the "villain" of *The Name of the Rose* is an ancient, blind, humorless, and censorious librarian named Jorge da Burgos has puzzled some readers and led others to surmise that Eco nourished some sort of animus against Borges. Instead, the novel is in great measure an homage of gratitude toward the sightless librarian of Buenos Aires, whose fictions and essays have inspired theorists of literature and signification as well as novelists and other storytellers. Eco was a theorist of signification (semiotics) and literature who turned to writing fictions, in part because Borges was a writer of fictions that implied theories of signification and literature. Told through the aged monk Adso's reminiscences, *The Name of the Rose* recounts his attempts, as a naive and confused young Benedictine novice, to absorb the lessons of "a learned Franciscan, Brother William of Baskerville," a proto–Sherlock Holmes charged with investigating a series of murders centered on the library, which led inexorably to its fiery obliteration. In a nod to Borges's bibliocosmic "Library of Babel," Eco refers to that fire as the *ecpyrosis,* the fire that will end the cosmos forever.

In one of the most inspired scenes in the narrative, the aged Adso tells of returning after many years to the ruins of the monastery. "Poking about in the rubble, I found at times scraps of parchment that had drifted down from the scriptorium and the library and had survived like treasures buried in the earth; I began to collect them, as if I were going to piece together the torn pages of a book. . . . At times I found pages where whole sentences were legible; more often, intact bindings, protected by what had once been metal studs." Adso refers to these empty covers as "ghosts of books," using modern bibliographers' term for titles of unlocatable books.[29] "I spent many hours trying to decipher those remains," he confides. "Often from a word or a surviving image I could recognize what the work had been. When I found, in time, other copies of those books, I studied them with love, as if destiny had left me this bequest." In a poignant evocation of inscription, obliteration, and rediscovery, Adso concludes that "at the end of my patient reconstruction, I had before me a kind of lesser library, a symbol of the greater, vanished one: a library made up of fragments, quotations, unfinished sentences, amputated stumps of books."[30] In other words, an object that resembles the mutilated *Fragments of Greek Historians* where Berossos resides, or the scanty, oracular remains of pre-Socratic philosophers.

At the end of the novel, as Adso mourns the intractability and opacity of his "lesser library," the voice of Eco, the empirical or real-world author of the fictional old monk's memoir, winks ironically—again—at the reader, who has long ago understood that Eco's novel is both a book about specific books and a writing about writing in general. "I have almost had the impression that what I have written on these pages, which you will now read, unknown reader, is only a cento, a figured hymn, an acrostic that says and repeats nothing but what those fragments have suggested to me, nor do I know whether thus far I have been speaking of them or they have spoken through my mouth."[31] Indeed: a cento is a poem or other composition built entirely of quotations from one or more previous works; what Eco calls "figured hymns" were devotional poems printed in shapes appropriate to their contents, such as a cross or an altar (figure 15.1). In an acrostic, such as the one "Francesco Colonna" used to claim

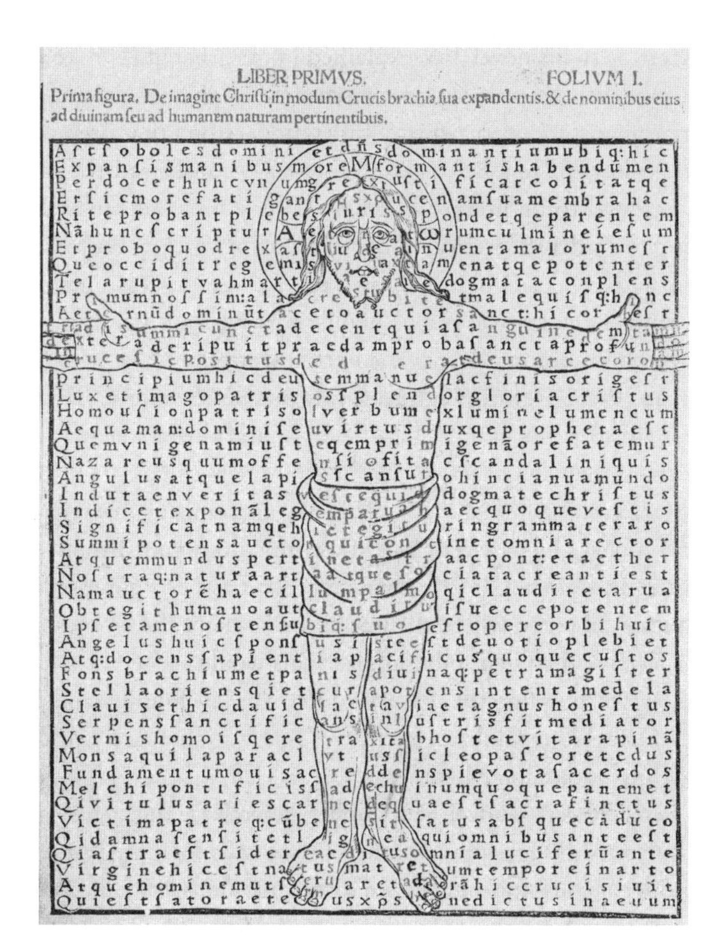

FIGURE 15.1. An extreme example of acrostic signification. "First Figure: On the Image of Christ spreading his arms in the manner of a cross, and on the names pertaining to his divine or to his human nature." Hrabanus Maurus, *De laudibus sanctae crucis* (ca. 814), ed. Jakob Wimpheling. Pforzheim: Thomas Anshelm, 1503. Sheridan Libraries Special Collections, Johns Hopkins University, PA8420.R11 A65 1503 QUARTO c. 1.

authorship of the *Hypnerotomachia Poliphili,* the reader does not only proceed straightforwardly but also obliquely, against the grain of normal syntax. Each of these techniques arranges letters, words, and texts to build scaffoldings of sense that are not detectable in ordinary speech and everyday writing.

Referring to his novel, Eco explained in 1980 that "if [I] have written a novel, it is because [I] have discovered . . . that what we cannot theorize about, we must narrate."[32] Eco understood that premodern writers—learned scholars as well as poets—considered the myths they repeated to be literary history, yet they were not unaware of groping toward a theoretical understanding of writing. Postmodern fiction writers like Eco and Borges reverse the process, inheriting their fantastic narratives about books, writing, and literature from this age-old tradition of quasi-theoretical stories.

Arturo Pérez-Reverte's extravagantly metaliterary novel *The Club Dumas* (1993) is the grandchild of Borges and the offspring of Eco. After more than 300 pages, as the novel nears its climax, its reader is apt to feel a sense of relief on finally seeing an unmistakable reference to Eco: "Look who's arrived. You know him, don't you? Professor of semiotics in Bologna."[33] The character Eco has been glimpsed at a lavish cocktail party, where the guests are some of the world's most enthusiastic and affluent bibliophiles. This would indeed be a fitting milieu in which to find the real-world Eco, the late founder and president of the Aldus Club, an exclusive "International Association of Bibliophily" located in Milan.

In Pérez-Reverte's novel, a bibliophile detective is engaged by a sinister bookseller (suggestively named Borja, thus recalling Borges via the proper Spanish pronunciation of the infamous surname Borgia) to search for the autograph manuscript of Alexandre Dumas's *Three Musketeers*. Suddenly the detective finds himself embroiled in passionate, deadly struggles to possess a mysterious book of necromancy. *The Nine Doors of the Kingdom of Shadows* is supposedly based on a treatise authored by the devil himself, and printed by a publisher whom the Inquisition burned in 1666. Quite helpfully, the bookseller shows the detective (and the reader) the edition's title page and its nine enigmatic woodcuts illustrating the "nine doors." The edition may be a forgery, the detective learns, but he is informed that "these illustrations are really satanic hieroglyphs. Interpreted with the aid of the text and the appropriate knowledge, they can be used to summon the prince of darkness." The original, Satan-authored treatise was titled *Delomelanicon*, "from the Greek: *delo*, meaning to summon. And *melas*: black, dark." The bookseller explains that the devil's treatise was, in point of fact, the aboriginal book, composed long before humans discovered writing, and making Satan, not God, the first author of a book.

> The prophet Daniel, Hippocrates, Flavius Josephus, Albertus Magnus, and [Pope] Leo III all mention this wonderful book. People have been writing only for the last six thousand years, but the *Delomelanicon* is reputed to be three times that old. The first direct mention of it is in the Turis papyrus, written thirty-three centuries ago.... [I]t is quoted several times in the *Corpus Hermeticum*.... And in an incomplete inventory of the Library at Alexandria, before it was destroyed for the third and last time in the year 646, there is a specific reference to the nine magic enigmas [the book] contains.... We don't know if there was one copy or several, or if any copies survived the burning of the library. Since then, its trail has disappeared and reappeared throughout history, through fires, wars, and disasters.[34]

Like *The Name of the Rose*, *The Club Dumas* is a postmodern metafiction, a novel about novels, books, and writing. Although the description of Satan's wondrous book is purest malarky, it, like the metafictions of Borges and Eco, mimics the kinds of narratives that poets, novelists, and serious scholars—classicists, theologians, and even scientists—told each other in all earnestness, from ancient times until after 1800. Its closest analogue is Gregory IX's imaginary *Book of the Three Great Imposters*, whose hide-and-seek trajectory and impious message it mimics.

PLAGIARISMS AND FORGERIES OF PARODY

The blurred lines between recreational fantasy and cynical literary forgery are a well-established feature of postmodern narrative, as Eco again illustrated in his second novel, *Foucault's Pendulum* (1988).[35] There he staged a deadly feedback loop connecting lunatic self-financed authors of occult fantastic realism to the contemptuous vanity-press editors who enticed them and published their unreadable books for profit—but also to enjoy their foolishness. Ironically,

Eco thereby became the unwitting father of a monster in 2003, when the reality of publishing copycatted his parodic novel with a vengeance. To many readers, numerous resemblances made Dan Brown's *Da Vinci Code* look like a solemn plagiarism of *Foucault's Pendulum*.[36] The affair appears to have amused Eco; he had remarked years previously that "this is parody's mission: it must never be afraid of going too far. If its aim is true, it simply heralds what others will later produce, unblushing, with impassive and assertive gravity."[37] *Foucault's Pendulum* was a prophetic parody of the as-yet unwritten *Da Vinci Code*. True to Eco's prediction, Brown's potboiler proposes Leonardo's *Last Supper* as a vehicle of historical "truth" about Jesus, his "wife" Mary Magdalene, and their descendants among French royalty.

Three years after *The Da Vinci Code*, Eco facetiously repaid Dan Brown in his own coin. Eco's short story "The Temesvar Codex" describes an even crazier interpretation of the *Last Supper* created by an imaginary author, a Borgesian figure whom Eco had mentioned in the preface to *The Name of the Rose*.[38] Eco now claimed, with impeccable, sarcastic literary logic, that Dan Brown "is a character from *Foucault's Pendulum*! I invented him. He shares my characters' fascinations—the world conspiracy of Rosicrucians, Masons, and Jesuits. The role of the Knights Templar. The hermetic secret. The principle that everything is connected. I suspect Dan Brown might not even exist."[39] This is the ultimate twist of fantastic realism: the real author Eco demotes his real copycat to a mere literary character that Eco himself invented; he implies that, like the imaginary works of other literary characters, Brown's text is also nonexistent or a cento.

In *The Prague Cemetery* (2010, English 2011), Eco went on to expound a conspiracy theory about conspiracy theories, in line with the opening essay of *Costruire il nemico* (*Constructing the Enemy*).[40] *Baudolino* (2000, English 2001) stars a medieval mythomaniac from Eco's hometown of Alessandria in Piemonte, whose escapades reenact—in both his adventurous experiences and the consummate lies he tells about them—the tall tales of ancient and medieval paradoxology and travel narratives that Eco so enjoyed.

The shadow of Scripture haunts the fictions of Borges and Eco, not to mention Dan Brown. In a final, postscriptive irony (or else an improbable coincidence), demonstrable literary forgery caught up to and imitated Dan Brown in 2012, when a small papyrus fragment surfaced at a conference of Coptic scholars. The "rediscovered" scrap was described as originating in a hypothetical *Gospel of Jesus's Wife*, since it bore the words "Jesus said to them, 'My wife . . .'" Within three years, the insubstantial scrap was exposed as a ponderous forgery utilizing genuine ancient papyrus, counterfeit approximations of ancient ink, and snippets of text traceable to the Nag Hamadi *Gospel of Thomas*.[41]

APOCALYPSE AND SCRIPTURE

Two narratives from the first decade of the twenty-first century typify the hold that the concept of Scripture continues to exert over the emotional history of writing. A novel by Will Self updates and intensifies the scenario evoked two centuries ago by Horace Smith's Ozymandias poem, while echoing later biblioclastic fictions such as *Fahrenheit 451*. *The Book of Dave* narrates how the diary of a despondent, divorced London cabbie becomes enshrined as the holy book of a postapocalyptic society. In this futuristic milieu, writing receives its due as a source of wonder, though in the bleakest possible fashion. Torn from their historical context and inflated to cosmic significance, the misunderstood banalities of the rediscovered journal provide an armature of wonder and devotion that supports the tenuous social and spiritual organization of a *Mad Max*–like society. Self's protagonist in the future dystopia is Symun, whose culture has, like Homeric Greece, reverted to orality while retaining a naive reverence for what they think of as the Gospel of Dave. Symun is inspired by the "old wive's tales" of his mother, a "knee woman and a rapper," and therefore "a power to be reckoned with on the island of Ham. Effi told little Symun the old legends of Ham, from before the Breakup and the Book that had ordained it, legends that, she maintained, went back to the MadeinChina, when the world had been created."[42]

Writing of any sort preserves its wondrous aura in this hardscrabble, hardscribble society. Effi's tales drive her son's archaeological curiosity, leading to a dangerous epiphany. "After another hundred paces Symun sat down on a mound and lowered his head between his legs, inhaling the atmosphere of the place, its brooding silence redolent of ancient abandonment. Muttering to himself he scrabbled in the mud: Vare ass 2 B sum, vare awlways iz, awl U gotta do iz dig. Sure enough, he soon exposed a corner of brickwork." As if channeling the voice of Horace Smith's wondering hunter, the narrator continues: "London bricks: the very stuff of Dave, created by Him, the material that Old London had been built from. . . . When the Hamstermen dug up courses of these sacred artefacts from the undergrowth . . . there were almost always one or two that retained their vivid redness, their sharp edges and their incised legend: LONDON BRICK."[43] When discovered, long after the cataclysm, the physical *Book of Dave* revealed itself as no ordinary tome: like the Book of Mormon, it had been engraved on metal plates. Although Dave claimed no otherworldly origin for it, and no angel transported it back to heaven, it became a scripture: its burial and rediscovery inspired a religious awe, a wonder stronger than any described by Horace Smith's or Joseph Smith's fictions, and rivaling George Smith's celebration of the Chaldean Noah. *The Book of Dave* chronicles the human drive to divinize writing—almost any writing—by assigning it a primordial origin and an ultimate, cosmic, or world-historical meaning. In a sense, "Jesus said 'my wife'" forms the perfect epitome of this scripturalizing urge.

The same drive to divinize writing animates *The Book of Eli* (2010). This film, directed by brothers Albert and Allen Hughes, brings our researches back to *The* Book, the Bible itself. Denzel Washington portrays a postapocalyptic loner in the mold of traditional Western movie heroes, who confronts a corrupt town, its tyrannical ringleader, and his violent henchmen. After decimating the henchmen in a spectacular showdown, Eli reluctantly joins forces with Solara, a young woman abused by the ringleader. After each of them has rescued the other from the remnants of the ringleader's gang, they escape to California together.

The Book of Eli has been criticized for too closely resembling other postapocalyptic adventure films in its desolate setting, grim visual texture, and violent characters. But insufficient attention has been paid to Denzel Washington's principal costar—not The Girl (Mila Kunis) but The Book. It is gradually revealed to be the last remaining copy of the Bible (New King James Version, first edition 1982). All other copies of the Scriptures were deliberately destroyed thirty years earlier in the despondent backlash among survivors of the nuclear holocaust. Captured and wounded by the ringleader, Eli is compelled to surrender The Book. This appropriation climaxes the criminal's long and violent search for any remaining copy of The Book, which he remembers and reveres only as a powerful instrument of social control. To the ringleader's dismay, however, he discovers that Eli's Book is printed in Braille. Eli and Solara eventually reach Eli's destination, a safe haven on Alcatraz Island occupied by a group dedicated to collecting and preserving the remnants of literary and musical culture. Eli himself is revealed to be blind (hence the Braille Bible), though, curiously, his handicap has never affected his marksmanship nor his skill at hand-to-hand combat. Likewise, Eli's loss of his book has not erased his Bible; having read it every day for thirty years, he dictates the text word-for-word to an amanuensis over an extended period, and the Alcatraz conventicle eventually prints it on their reconstructed press.

The biblical Eli was a high priest and the mentor of Samuel, the last judge of the Israelites. Eli's two sons were also priests, but they were corrupt, and Eli did not discipline them, earning a disastrous punishment from God. After the Philistines defeated the Israelite army in battle, Eli's sons carried the Ark of the Covenant into a second battle, expecting it to guarantee victory. Instead, the Israelites were routed again, the sons of Eli were killed, and the Philistines captured the Ark. When he heard the news, the blind and aged Eli fell from his chair and broke his neck. The Philistines retained the Ark for seven months, until a plague convinced them to return it to the Israelites.[44]

In the film we recognize the biblical motifs of Eli's blindness, his imperative to protect the holy text, its capture by his enemies, and its ultimate triumph after his death. The film employs these echoes pointedly to critique the social role of the Bible in the history and societal character of the United States. Eli tells Solara that postapocalyptic Americans tried to obliterate the Scriptures because they held them responsible for the war that ended civilization. Ironically, the ringleader envisions his enclave as an ideal revival of bibliolatrous preapocalyptic American society, with himself as its Messiah. Yet his quest to exploit the Bible's charisma destroys his miniature state, and himself, reminding viewers that survivors of the nuclear war blamed the Bible for its devastation. Nonetheless, the film's ending is optimistic in a humanistic way, since it imagines a future for the Bible as literature, and only literature, shelved alongside the *Tanakh* and the Qur'an in a bunker dedicated to human culture rather than religion.

Still, scriptural echoes and parallels complicate and potentially contradict the film's secular humanist message. Eli could as aptly be named Ezra, since he restores the lost text of the Bible to writing. Fittingly, he says the task was assigned to him by an inner voice that told him where to disinter the sole surviving copy of the Bible and then protected him supernaturally as he bore it to its destined ark-like refuge on Alcatraz. In its somewhat schizophrenic message, this apocalyptic Western surveys "the good, the bad, and the ugly" of how writing made—and makes—us human or rather, to quote Nietzsche, makes us human, all too human.

For more than half a century, writers have wondered about "the future of the book." But the postapocalyptic stories that close out this chapter pose the more fundamental question whether humanity is destined to outlive writing, or vice-versa. Will all writing go the way of cuneiform, will it be revived in some way, or will all human society follow Sumeria into oblivion? The emotional history of writing continues to trace *Homo scribens*'s process of continual, multidimensional metamorphosis, with no guarantees of survival or happiness ever after.

Acknowledgments.

This book has accumulated debts over many years. I dedicate it to the high school where I was first immersed in the life of the mind, in honor of the two teachers who most encouraged and inspired me during the scant two years I attended. I trace my earliest interest in the emotional history of writing to Dr. Armstrong's senior elective course in philosophy and my enthusiasm for languages to Monsieur Draper's love of French.

William J. Kennedy and Carol V. Kaske nurtured these interests in their comparative literature courses at Cornell. At the Scuola Normale Superiore di Pisa, my dissertation adviser, Eugenio Garin, first showed me what vast horizons in the history of scholarship my interests entailed and immediately on my arrival pointed me to the works of the arch-forger Annius of Viterbo, which led eventually to this book. Garin's ongoing generosity to advisees lasted, in my case, nearly thirty years. The example of Umberto Eco taught me how to marry Philologia with Mercurius, historical with hermeneutic scholarship, preserving (I hope) a measure of humor.

Gerald Gross of Baltimore, the agent's agent, encouraged me to have faith in the project over several years. Alas, he, like several others mentioned here, did not live long enough to receive my thanks in person.

Earle Havens, the Nancy H. Hall Curator of Rare Books and Manuscripts at Johns Hopkins University, has been my coconspirator on many teaching and research projects since 2008. Thanks to the generosity of Johns Hopkins donors and the inspired direction of Winston Tabb, Sheridan Dean of University Libraries, Archives and Museums, Earle has been able to add numbers of medieval and early modern manuscripts and imprints to Johns Hopkins's already formidable holdings in my fields of interest, foremost among them the Bibliotheca Fictiva, the Arthur and Janet Freeman Collection of Literary and Historical Forgery. Many of these works figure prominently in the present volume. Mackenzie Zalin, Lael Ensor-Bennett, and Kelsey Champagne have provided invaluable assistance with images and texts.

Anthony Grafton, the Henry Putnam University Professor of History at Princeton, has been a generous and inspiring friend and occasional collaborator for many years, as has Ingrid D. Rowland at the University of Notre Dame School of Architecture.

Numbers of other colleagues favorably influenced this project: at Yale and Cornell, my adviser Giuseppe Mazzotta; at Johns Hopkins, my close friend and biking companion Gérard Defaux; my colleagues Douglas Collins at the University of Washington and Daniel Desormeaux at Dartmouth and Johns Hopkins. Lately, Richard Jasnow and Paul Delnero in the Department of Near Eastern Studies at Johns Hopkins have discussed parts of chapter one with me and recommended readings that improved my project. Colleagues in the Charles Singleton Center for the Study of Premodern Europe have frequently inspired me, particularly Lawrence Principe and Christopher Celenza. Conversations with Thomas Hendrickson and Frederic Clark, and their scholarship, were invaluable.

All mistakes and omissions are obviously mine.

Publication was facilitated by a generous grant from the Virginia Fox Stern Center for the History of the Book in the Renaissance.

This book might never have been written without the generous opportunities afforded me by a number of fellowships and visiting professorships. From 1975 to 1977, I enjoyed liberal fellowship support from the Scuola Normale Superiore di Pisa, where I first encountered many of the books discussed here, often in company with Patrizia Castelli, now at the University of Ferrara. In 1987–1988 I was privileged to learn from Salvatore Camporeale, OP, at Villa I Tatti, the Harvard University Center for Italian Renaissance Studies in

Florence. "Campo" mentored dozens of scholars at I Tatti, on his frequent visiting professorships at Johns Hopkins, and at universities in several countries, until his death in 2002.

During winter term of 1997, I had occasion to serve as principal investigator of the Eighth Annual Dartmouth College Humanities Research Institute, "Books and the Imaginary," funded by a grant from the Mellon Foundation and the National Endowment for the Humanities.

Oxford University has granted me a number of precious opportunities for research. As a visiting fellow in All Souls, during 2004–2005, I first experienced prolonged access to the riches of the Bodleian and the many abundant collections of other Oxford libraries, especially the Codrington, Balliol, Queens, and the Taylorian. Ian Maclean was an especially generous host, and Richard Yeo of Griffith University, Australia, an inspiring "fellow fellow" through his interest in book history and encyclopedism. In spring and summer of 2009, I was the Fowler Hamilton Visiting Fellow, Christ Church, for which I am grateful to the dean and fellows. At the Oxford Centre for Hebrew and Jewish Studies, Oriental Institute, I was granted a Dorset Foundation Fellowship in Hilary and Trinity terms of 2014, to participate in the seminar The Reception of Josephus in the Early Modern Period. I am especially indebted to Joanna Weinberg and Martin Goodman for this opportunity to interact with scholars of Josephus from several countries and disciplines.

In spring of 2012, I spent a delightful Epiphany term at Durham University as International Senior Research Fellow in the Institute of Medieval and Renaissance Studies and as Slater Fellow of University College. The book collections at Durham are a precious resource, and Stefano Cracolici of the Italian department was an attentive host, as were many of his colleagues throughout the university. I am particularly indebted to Johannes Haubold for the chance to participate in "The World of Berossos," the Fourth International Colloquium on the Ancient Near East, at Hatfield College in July 2010.

In January 2008, I was Professeur Invité at the Centre d'Études Supérieures de la Renaissance in the Université François-Rabelais of Tours. I am grateful to Marie-Luce Demonet for this opportunity and for the invitation to lecture at the Louvre in the conference "Babel et la diversité des langues" on the occasion of the international exposition *Babylone* (March–June 2008).

At the Folger Institute in Washington, DC, I was privileged throughout the spring term of 2008 to direct a weekly seminar on Writing and Wonder: Books, Memory and Imagination in Early Modern Europe in the Folger Shakespeare Library. Several of the participants, notably Ellie R. Truitt of the University of Pennsylvania, made valuable contributions to my outlook.

At Johns Hopkins University Press, Matthew McAdam and Adriahna Conway have guided the manuscript toward publication with a light touch.

As always, my wife, Janet Stephens, inspires me daily with her love and her own scholarship. Our daughter, Catherine, following in both our bookish footsteps, is a source of constant joy.

Notes.

PREFACE.
Homo scribens: Humanity and Writing

1. See "Complement" in this volume.

2. Goldberg 1990: 174.

3. Aristotle, *Parts of Animals*, III.x.673; Appius Claudius Caecus, *Sententiae* ("Every man is the maker of his own destiny"); Huizinga 1949; Carl Linnaeus, *Systema naturae per regna tria secundum classes, ordines, genera*, etc., 10th ed. Stockholm: Laurentius Salvius, 1758, 1:20–22.

4. Technically, the difference is now defined as separating *Homo sapiens neanderthalensis* from *Homo sapiens sapiens*.

5. Clarke 1999: 2.

6. Podany 2019: 60; see also Jiménez 2019: 152: "Around the year 2500 BC, the first literary texts written in Akkadian and Sumerian appeared. . . . A wisdom text known as *The Instructions of Shuppurak*" was "faithfully transmitted" by Mesopotamian scribes for over two thousand years.

INTRODUCTION.
The Mystique of Writing

1. Damrosch 2006: 12.

2. Netz and Noel 2007.

3. Freeman 2014: 22 on W. H. Ireland.

4. Minois 2012.

5. *Oxford English Dictionary* online, June 2022, s.v. "literature."

COMPLEMENT.
Writing as Technology, from Myth to History

Epigraph: Vanstiphout 2004: 15; but see Woods 2015: 16.

1. The clearest and most comprehensive discussions are in Woods 2015 and Hooker 1990. Fischer 2001: 11–33, gives the best summary of the emergence of complete writing.

2. Adapted from Fischer 2001: 12. Fischer includes computer programming as complete writing.

3. Woods 2015: 20, quoting Cooper 2004: 83.

4. Forms of partial writing existed elsewhere, perhaps even earlier than at Uruk, but "in practice any meaningful discussion has to start with the tablets found at Uruk . . . ca. 3300–2900 BC[E]" (C. B. F. Walker, "Cuneiform," in Hooker 1990: 18–19).

5. The strongest case for a direct relationship is in Schmandt-Besserat 1992: 7–12 and Schmandt-Besserat 2007: 1–12. Her theory is summarized by Fischer 2001: 26–27, and contested by Woods 2015: 46–49.

6. Woods 2015: 42–46, 49–50, gives an admirably clear, illustrated synopsis of this process; a longer and more detailed synopsis is C. B. F. Walker, "Cuneiform," in Hooker 1990: 17–31.

7. Woods 2015: 43.

8. Woods 2015: 43.

9. Robinson 2007: 118.

10. Hooker 1990: 210–13, 221–34. On the Egyptian "consonantal alphabet," Fischer 2001: 83–86.

11. Hebrew and Arabic eventually developed supplementary systems of "vowel pointing," small marks representing vowel sounds without actually being full-fledged letters.

12. Hooker 1990: 229–35 (Etruscan alphabet, derived from Greek, and Latin, derived from Etruscan, on 234–35). Fischer 2001: 82–98 summarizes well the development from Proto-Sinaitic through Phoenician to Greek and Hebrew.

13. The classic account of the early twentieth-century fieldwork and interpretation of oral formulaic poetry of Yugoslav bards by Milman Parry and Albert Lord is in Lord 1960.

14. Godzich and Kittay 1987: xi–xii. As to why this is so, "[Northrop] Frye and others have held that meter is in fact closer to speech than prose, in being a less complex form of stylization, and have used this claim to explain why prose does not develop in some cultures, whereas verse has been developed in every known culture, and why, even when prose does develop, it is 'normally a late and sophisticated development in the history of a literature'" (Brogan 1993: 1349). Herodotus (d. 425 BCE) was the first Western author of a long work in prose, though there were shorter ones before (Herodotus 2007: ix). In the early eighteenth century, Giambattista Vico maintained that primitive peoples thought metaphorically rather than analytically and thus that poetry predated prose (Vico 2020: ¶¶409, 472, esp. 460. Note that although I have consulted Vico 2020 throughout, for the sake of uniformity I have preferred to cite the paragraph numbers of the 1744 edition as established by Fausto Nicolini in 1953 rather than the page numbers of any particular recent edition).

ONE. An Age of Wonder and Discovery, 2500–600 BCE

1. Vanstiphout 2004: 173: "Unknown, probably mythical city . . . known only from literary references."

2. Vanstiphout 2004: 85.

3. Vanstiphout 2004: 87; cf. Woods 2015: 34–35.

4. Vanstiphout 2004: 54.

5. Berossos 1996: 44; Berossos is being quoted by a ninth-century Byzantine writer, George Syncellus.

6. John Dillery, "Greek Historians of the Near East," in Marincola 2011: 223.

7. Cluzan 2005: 241–42. My translation.

8. Cluzan 2005: 242. See also Bottéro and Kramer 1989: 527–28; 198–202; 598–601.

9. See Cluzan 2005: 129–43 for comparisons.

10. Berossos 1996: 49.

11. Berossos 1996: 50.

12. John Dillery, "Greek Historians of the Near East," in Marincola 2011: 223; more parallels in Berossos 1996. But see Bottéro and Kramer 1989: 564–67.

13. Xisouthros, reflecting Ziusudra or Zi-ud-sura, attested in "The Instructions of Shuruppak" according to Kramer 1967: 12–18. Kramer concluded that "Ziusudra had become a venerable figure in literary tradition by the middle of the third millennium BC." Aside from Uta-napishtim, the other Akkadian Noah-name is Atrahasis.

14. Berossos 1996: 20; George 1999: xxxv. On the name Ziusudra's meaning, Bottéro and Kramer 1989: 564.

15. Bottéro and Kramer 1989: 540–45; 555–59.

16. Frahm 2019: 157; cf. 162.

17. Berossos 1996: 20; on the sources of *Gilgamesh*, see Foster 1995: 52–77.

18. Chronology in George 2003: lxc–lxi.

19. George 2003: 1–2.

20. "Legend of Naram-Sin," in Foster 1995: 171, 176–77 (*stela* regularized here to *stele*). "Son" here has the sense of descendant not offspring. An alternative translation in George 2003: xxxvi, refers to "tablet-box" and "stone tablet" rather than "foundation box" and "stele."

21. Winter 1985: 24.

22. Nor in the poem is invention at issue: "The story may be interpreted as an etiology, though one necessarily based on a much older oral tradition, for Uruk's ascendancy" (Woods 2015: 35, sidebar).

23. Civil 1983: 61 (verses 160–68), emphasis added.

24. "The Temple Hymns," The Electronic Text Corpus of Sumerian Literature, Faculty of Oriental Studies, University of Oxford, https://etcsl.orinst.ox.ac.uk/cgi-bin/etcsl.cgi?text=t.4.80.1#, vv. 543–44.

25. Escobar 2019: 186; see also Frahm 2019: 164–65.

26. "Nidaba (goddess)," Ancient Mesopotamian Gods and Goddesses, The Open Richly Annotated Cuneiform Corpus, http://oracc.museum.upenn.edu/amgg/listofdeities/nidaba/; "Nabu (god)," Ancient Mesopotamian Gods and Goddesses, The Open Richly Annotated Cuneiform Corpus, http://oracc.museum.upenn.edu/amgg/listofdeities/nabu/index.html.

27. Evidence suggests Egyptian writing came about for ceremonial rather than bureaucratic reasons (Woods 2015: 142–43).

28. Lichtheim 1973: 16. The *ka* was the life-force common to all humans; the *ba*, which remained with the dead body, was the set of an individual's distinctive spiritual characteristics. The purpose of funeral rituals was to release the *ba* so that it could rejoin the *ka*, which was nourished by the spiritual essence of the funeral-offering foods: Allen 2000: 79–80, 94–95.

29. Lichtheim 1973: 60; cf. 6–7, 59.

30. Lichtheim 1973: 52.

31. Lichtheim 1973: 51.

32. Lichtheim 1973: 169–84 at 181–82.

33. Lichtheim 1976: 176, emphasis in original.

34. Lichtheim 1976: 176.

35. Lichtheim 1976: 177.

36. Jasnow and Zauzich: 47.

37. Jasnow 2011: 300–301.

38. Jasnow 2011: 310.

39. Jasnow 2011: 313.

40. Homer 1951: 157–58 (*Iliad* 6.156–90).

41. Phemios (*Odyssey* 1.154; 22.330); Demodokos (*Odyssey* 8.44); Agamemnon's unnamed bard (*Odyssey* 3.267).

42. Knox 1996: 21.

TWO. An Age of Philosophy, 600 BCE–400 CE

1. Hesiod 1983: 2.

2. Hesiod 1983: 13–14.

3. Kirk and Raven 1969: 48.

4. Watts 2017.

5. Turner 2014: 8. Reynolds and Wilson 2013: 1–19; König et al. 2013.

6. Turner 2014: 7–14 on early Greek work. Homer's texts were stable in the 2nd c. BCE (Zenodotus 275); the tragedians a century earlier (Lycurgus 330).

7. Martin 1994: 60–61.

8. Drogin 1983; Drogin 1989.

9. Herodotus 2007: 391–92 (i.e., 5.58–59); Carpenter 1935.

10. Herodotus 2007: 391 (5.58); cited by Knox 1996: 21.

11. Herodotus 2007: 134 (2.36–37).

12. Herodotus 2007: 141 (2.50).

13. Herodotus 2007: 138 (2.45).

14. Herodotus 2007: 137 (2.43).

15. Herodotus 2007: 118 (2.2–3).

16. Herodotus 2007: 185 (2.145).

17. Herodotus 2007: 160–61 (2.102, 106).

18. Herodotus 2007: 318 (4.87).

19. *Iliad* books 3–6, *Odyssey* book 4.

20. Herodotus 2007: 168 (2.19).

21. Herodotus 2007: 168 (2.18).

22. Herodotus 2007: 168–69 (2.120).

23. Herodotus 2007: 169 (2.120).

24. Knox 1996: 20–21; Thompson 1912: 11–20.

25. Letter 2, 312d, 314b (Plato 1961: 1566–67).

26. Knox 1996: 21.

27. Plato 1961: 1720, 1729. Papyrus and styli are not mentioned, but tablets ("blocks of wax") and slates are.

28. Plato 2000: 7–8 (*Timaeus*, 22a–b). Also trans. Benjamin Jowett, in Plato 1961: 1157.

29. There is an apparent exception in 22d: "whenever the gods send floods of water upon the earth to purge it." This is probably a reminiscence of the Deucalion story; see Ovid, *Metamorphoses* 1.230ff.; Graves 1960: 138–43 (¶38). But the priest has already declared that Solon's other story, of Phaeton setting the Earth on fire, "is told as a myth, but the truth behind it" is astrological and meteorological (Plato 2000; 22c–d).

30. Plato 1961: 677–84; quotation from *Laws*, 678b, 677d, 679d.

31. Plato 2000: 8 (*Timaeus*, 23a).

32. Heliodorus in Reardon 1989: 420–21; Lucian in Reardon 1989: 640.

33. Robson 2013: 41–45.

34. Plato 1995: 79–80 (*Phaedrus*, 275A).

35. Michael Trapp, "Socrates," 895–97, in Grafton et al. 2010.

36. Derrida 1981: 67, 75, 91, 97; see also 103: "In the *Philebus* and the *Protagoras*, the *pharmakon*, because it is painful, seems bad whereas it is beneficial"; whereas the opposite happens in the *Phaedrus*. See further references to writing in the probably spurious second epistle, 170–71, and Plato 1961: 1566–67.

37. Derrida 1981: 97; 72–73. While Derrida briefly mentions the Mesopotamian god Nabu in relation to Hermes/Thoth, he does not pursue the connections in enough depth to be useful here (note on p. 85, note on p. 94–95).

38. Havelock 1963: 198.

39. Havelock 1963: vii.
40. Havelock 1963: 13.
41. Havelock 1963: 31.
42. Havelock 1963: 30–31, emphasis added.
43. Havelock 1963: 31.
44. Havelock 1963; Ong 2012; McLuhan 1962.
45. Diodorus 12.11.4 (Loeb 4:397).
46. Diodorus 12.12.4 (Loeb 4:399).
47. Diodorus 12.13.1–2 (Loeb 4:399, 401).
48. Diodorus 12.13.3 (Loeb 4:401), emphasis added.
49. Diodorus 12.13.4 (Loeb 4:401, 403).
50. Nehamas 1998, esp. 157–88 (on Michel Foucault and Socrates).
51. Diodorus 1.1.3–4 (Loeb 1:7).
52. Diodorus 1.9.3 (Loeb 1:33).
53. Diodorus 1.9.2 (Loeb 1:33).
54. Diodorus 1.1.1 (Loeb 1:5).
55. Thucydides 1.22 (Loeb 41).
56. Diodorus 1.2.3 (Loeb 1:9, 11).
57. Diodorus 1.3.6–7 (Loeb 1:17), emphasis added.
58. Diodorus 1.3.8 (Loeb 1:17).
59. Diodorus 1.9 (Loeb 33, 35).
60. Diodorus 1.81 (Loeb 1.279).
61. Diodorus 1.69 (Loeb 1.239); cf. 1.96 (Loeb 1.327, 329, 331).
62. Diodorus 1.24.1 (Loeb 1:77).
63. Diodorus 1.24.4 (Loeb 1:77).
64. Diodorus 1.27.3–6 (Loeb 1:87, 89).
65. Diodorus 1.96.1–6 (Loeb 1:327, 329); cf. *Odyssey* 24.1–2, 11–14.
66. Diodorus 3.67.1–4 (Loeb 2:305–9).
67. Diodorus 5.74.1 (Loeb 3:297–99).
68. Diodorus 5.57.1 (Loeb 3.251).
69. Diodorus 5.57.3–4 (Loeb 3:251).
70. Diodorus 5.57.5 (Loeb 3:251).
71. Diodorus 5.57.1–4 (Loeb 3: 251, 253).

THREE. Collections, Histories, and Forgeries, 300 BCE–400 CE

1. Diodorus 1.46–47 (Loeb 1: 167, 169).
2. Diodorus 1.49 (Loeb 1: 173).
3. Lutz 1979: 21.
4. "One wonders whether Diodorus, as he wrote these words, was recalling the inscription 'Healing-place of the soul,' which, he told us, stood on the library

of the Egyptian pharoah Osymandias (Book 1.49.3)," writes C. H. Oldfather, Loeb 1:400.

5. Müller 1848–1874; Jacoby et al. 1923–.

6. In addition to the works mentioned previously, see West 2003; Kirk and Raven 1969.

7. See Beyerlinck 1678: 219–27; *Encyclopaedia Britannica* 1797: 10.25–26; *Encyclopaedia Britannica* 1911: 16.545–48.

8. *Enciclopedia Zanichelli* 1992: 58, s.v. "Alessandria." See also Casson 2001: 31–47; Canfora 1990: 123–25.

9. Canfora 1990: 98.

10. Canfora 1990: 192.

11. Canfora 1990: 77–79; see also Casson 2001: 45–47; Grafton et al. 2010: 32, 531–36.

12. Canfora 1990; Damrosch 2006; more generally König et al. 2013.

13. See Grafton et al. 2010: 31–32; Bibliotheca Alexandrina, a new library and cultural center in Egypt, http://www.bibalex.org/English/index.aspx.

14. Iamblichus 2004: 5 (i.e., I.1–2).

15. Iamblichus 2004: 5 (I.1–2).

16. Iamblichus 2004: 9 (I.2.5–6).

17. Iamblichus 2004: 305–7 (VIII.1.260–61).

18. Iamblichus 2004: 309 (VIII.2.262).

19. Berossos 1996: 174; cf. 102. Copenhaver 1992: xv, reads *stelae* rather than *monuments*.

20. See chapters nine through eleven.

21. Copenhaver 1992: xxxii, quoting Garth Fowden.

22. Copenhaver 1992: 81.

23. Copenhaver 1992: 81.

24. Ovid 1916: 2:423 (i.e., 15.810–14).

25. The question of writing's role in magic is too complex for inclusion in this book, but to the extent that magic evokes or invokes superhuman or nonhuman forces, its role is not to "make us human." In fact, *theurgy*, as described by Iamblichus and other Neoplatonic philosophers, employed magical or magical-seeming rituals to seek union with divine forces.

26. Horace 2004: 244–47 (i.e., *Odes* 4.9.25–34).

27. Horace 2004: 216–17.

28. Horace 1929: 388–89 (i.e., *Epistles* 1.20.10–13, 17–18).

29. Yates 1966; Carruthers 1990; Rossi 2000; Bolzoni 2001; Carruthers and Ziolkowski 2002.

30. Quintilian 1920: 144–45 (i.e., 1.7.30–31).

31. Pliny the Younger 1969: 1:425–35 (i.e., VI.16).

32. Pliny 1942: 2:634–37 (i.e., *Natural History* 7.56.192–93).

33. Pliny 1945: 138–39 (i.e., *Natural History* 13.21.69–70).

34. *Natural History* 13.27.89.

35. *Natural History* 13.21.68, 70.

36. *Natural History* 13.27.88.

37. Pliny 1945: 148–51 (i.e., *Natural History* 13.27.84–85).

38. It is possible that Pliny is here referring to cedar-oil (*cedrus*) rather than oil of citron (*citrus*), as the former was a well-known preservative from decay and insects.

39. *Natural History* 13.27.86.

40. All quotations from Livy *History of Rome* 12:471 (book 40.29.3).

41. My emphasis for both.

42. E. T. Sage and A. C. Schlesinger, "Titus Livius (Livy), *The History of Rome, Book 40*," Perseus Digital Library, http://www.perseus.tufts.edu/hopper/text?doc=Perseus%3Atext%3A1999.02.0167%3Abook%3D40%3Achapter%3D29, note 3.

43. Livy *History of Rome* 1:64–65 (Book 1.18.4–5).

44. In addition to the primary texts mentioned, see Speyer, 1971: 51–55.

45. Livy *History of Rome* 1:62–65 (Book 1.19–20).

46. Livy *History of Rome*, 12:471 (book 40.29.3–14); Plutarch, *Numa* 22; Augustine, *City of God* 7.34; Valerius Maximus, *Memorable Deeds and Sayings* 1.1.12 depended on Livy's account; and Lactantius, *Divine Institutes* 1.22.5 depended on Valerius's account.

47. Dares (Pseudo-) and (Pseudo-) Dictys 2019: 20.

48. Dares (Pseudo-) and (Pseudo-) Dictys 2019: 20.

49. Dares (Pseudo-) and (Pseudo-) Dictys 2019: 20–21.

50. Dares (Pseudo-) and (Pseudo-) Dictys 2019: 19.

51. Dares (Pseudo-) and (Pseudo-) Dictys 2019: 19.

52. But see Dares (Pseudo-) and (Pseudo-) Dictys 2019: 9–10.

53. All translations from Frazer 2019: 133.

54. In addition to Speyer 1970: 55–59; see Speyer 1971, passim.

55. See Havens 2018: 60.

56. Stephens 1993: 71–80. Quotation from Heliodorus 1989: 434.

57. Athenaeus 1930: 4:543; Gulick translates *hupò kólpois* as "beneath its bosom."

58. Gulick (Athenaeus 1930: 4:545) translates as "epistle" and notes, "Of the feminine gender in Greek." Hermann Hugo (1617: 15) also translates as *epistola* (see chapter twelve).

59. Athenaeus 2009: 10.450e–451b (Loeb 4:157–59). Note by Olsen: "Important evidence for silent reading already in the 4th century BCE."

60. Fischer 2003: 90–92, 160–62.

61. Cooper 2004: 83; quoted in Woods 2015: 20.

FOUR. Writing and Scripture, 600 BCE–650 CE

1. Peters 2007 is a thorough and readable treatment of the questions examined here.

2. *Tanakh: The Holy Scriptures* notes (see Jewish Bible 1985: 88) that the Hebrew meaning is "uncertain, variously translated: 'I Am That I Am'; 'I Am Who I Am'; 'I Will Be What I Will Be': etc." *Tanakh* glosses the continuation of the verse as God referring to himself as "'I Am' or 'I Will Be.'" It notes that God's reference to himself in v. 15 as "YHWH" "is here associated with the root *hayah* 'to be.'" Christians transliterate the *Tetragrammaton* ("YHWH") as Yahweh or Jehovah, but devout Jews do not pronounce it and substitute epithets such as *Adonai* (the Lord) or *HaShem* (the Name). Presumably only God may pronounce his real name.

3. Capitalization indicates the conception of a very particular god or set of scriptures.

4. Peters 2007: 90.

5. Peters 2007: 244.

6. Peters 2007:140, emphasis added.

7. Peters 2007: 172–73. For the Well-Guarded Tablet see, in addition to Qur'an 28:20, Qur'an 3:7; 85:21–22; for the "Mother of the Book," see 13:39; 43:4.

8. Peters 2007: 98; other examples 98–100.

FIVE. The Jewish Scriptures

1. Defoe 2007. See chapter twelve.

2. All quotations from *Tanakh: The Holy Scriptures* (see Jewish Bible 1985).

3. Exodus 25:16, 21–22; cf. Numbers 7:89.

4. Numbers 12:2–8 asserts that God appears visibly and talks to Moses, whereas to others he appears only in dreams and speaks in riddles. In Numbers 23:4–5, "God appeared to Balaam, who said to Him. . . ." Yet Balaam is a Gentile not an Israelite, and God inspires him to prophesy the Israelites' success on seven occasions. Numbers 23:16; 23:4 and 16 refer to Balaam as "the man whose eye is true . . . him who hears God's speech, Who obtains knowledge from the Most High, / And beholds visions from the Almighty, / Prostrate, but with eyes unveiled."

5. Examples in Stephens and Havens 2018, and in Speyer 1970.

6. Sixtus of Siena 1742: vol. 1, p. 72 (bk 2, s.v. "Bellorum Domini Liber"). See chapter ten.

7. 4 Ezra 14:19–22. All quotations from 4 Ezra are taken from B. M. Metzger's trans. in Charlesworth 1983: 517–59.

8. 4 Ezra 14:39–41; cf. Ezekiel 3:1–3, Revelation 10:9–10.

9. 4 Ezra 14:38–42.

10. 4 Ezra 14:45–47.

11. Charlesworth 1985: xxi. He lists "at least" thirteen.

12. Charlesworth 1983: xxi–xxvii for definitions.

13. Charlesworth 1983: 5–89; trans. E. Isaac.

14. Augustine 1950: 511 (*City of God* 15.23).

15. Lawlor 1897; Kaske 1971.

16. All quotations from the "books of Enoch" (1–3 Enoch) from Charlesworth 1983.

17. Papyrus? The note on this passage in Charlesworth 1983: 1.48 does not explain.

18. 1 Enoch 82:1–2.

19. Charlesworth 1983: 91–221; trans. F. I. Andersen.

20. Charlesworth 1983: 156 (version J); 157 (version A).

21. Charlesworth 1983: 190, n. 64c; 2 Enoch 64:5 (J); 64:5 (A) refers only to the one who carries away our sins.

22. 67:3 (J); Andersen's note to this passage suggests that there might be confusion between the Slavonic words for "invisible" and "unknown," so that there would be an allusion to Paul's "unknown God" of Acts 23:17, but this seems far-fetched. Besides, Enoch "did not appear" after he walked with God (Genesis 5:24), which is a good way of saying he "became invisible."

23. Charlesworth 1983: 223–315; trans. P. Alexander.

24. Charlesworth 1985: 714; text on 35–142, trans. O. S. Wintermute.

25. See Grafton 1993: 583 on the significance of this disparity.

26. Charlesworth 1985: 249–95, trans. M. D. Johnson.

27. Charlesworth 1985: 259.

28. Charlesworth 1985: 292.

29. Charlesworth 1985: 250.

30. For the story of Seth and Solomon, see Charlesworth 1985: 294 (*Life* [*vita*] *of Adam and Eve* 50.4–9).

31. Charlesworth 1985: 310.

32. Charlesworth 1985: 320.

33. Charlesworth 1985: 338.

34. Charlesworth 1985: 388.

SIX. The Christian Scriptures

1. Kasser et al. 2006: 84, cf. 7, emphasis in original.

2. Pagels 1979: 101.

3. Kasser et al. 2006: 84.

4. Pagels 1979: 140.

5. Pagels 1979: 14–15.

6. Pagels 1979: 14–15; Mark 4:11; Matthew 13:11.

7. Pagels 1979: 145, quoting the *Gospel of Thomas*.

8. Layton 1987: xxi.
9. Layton 1987: 37, 38. See also Kasser et al. 2006: 100.
10. Layton 1987: xxi, xix.
11. Layton 1987: 12.
12. Borges 1999: 66; see "Basilides' Myth," in Layton 1987: 420–25 at 425.
13. Layton 1987: 425.
14. Layton 1987: xxii, emphasis added.
15. Layton 1987: 49–50.
16. Pagels 2005: 131–33.
17. Pagels 2005: 104, emphasis in original, unless noted.
18. Pagels 2005: 34.
19. Pagels 2005: 152.
20. Pagels 2005: 58.
21. Pagels 2005: 39.
22. Pagels 2005: 57.
23. Borges 1998: 163.
24. Kasser et al. 2006: 4.
25. Kasser et al. 2006: 101.
26. Kasser et al. 2006: 98–99.
27. Kasser et al. 2006: 38.
28. Layton 1987: 55, 64.
29. Layton 1987: 152.
30. Layton 1987: 189, quoting Epiphanius of Salamis (d. 403).
31. Layton 1987: 17.
32. Pagels 2005: 97.
33. Speyer 1970: 23–42.
34. See Havens 2016: viii–41 at 21–23 and fig. 5.
35. Freeman 2014: 104, no. 69. Now in Johns Hopkins University Libraries' *Bibliotheca Fictiva* Collection. Modern edition in Elliott 1993: 541–42.
36. Elliott 1993: 76 (*Infancy Gospel of Thomas*, Greek Version A, chap. 4); 81 (*Infancy Gospel of Thomas*, Greek Version B, chap. 4); 89–90 (*Gospel of Pseudo-Matthew*, chap. 29).
37. Elliott 1993: 164–204, at 197; cf. 203–4.
38. Elliott 1993: 197.
39. Josephus, *Jewish Antiquities* 18.64; (Loeb 10:48–51).
40. See chapter fifteen.

SEVEN. Cultural Clashes and the Defense of Uniqueness

1. Charlesworth 1985: 831–42 (Aristobulus, 2nd c. BCE), trans. A. Yarbro Collins. Texts on 839–42 (fragments preserved by Eusebius of Caesarea).
2. Charlesworth 1985: 898–99.

3. See Grafton 1993: 569–613. The first-century CE writer Alexander Polyhistor spread a myth that Moses was a woman named Mosò (Grafton 1993: 596–97).

4. Estimates vary from the third century BCE to the first century CE (Charlesworth 1985: 8).

5. Charlesworth 1985: 12.

6. Charlesworth 1985: 15.

7. Charlesworth 1985: 33.

8. Augustine 1872: 651 (i.e., 18.42); also Augustine 1997: 42–43.

9. Augustine 1872: 652 (i.e., 18.43).

10. James L. Kugel, foreword to Ginzburg 1998: 1.ix; Ginzburg 1998: 1.xxiii; Joseph Heinemann, "The Nature of the Aggadah," in Hartman and Budick 1986: 41–55, at 41. The only reference to Josephus in this large book occurs in another essay, p. 214.

11. Allen 1963: 73.

12. See Berossos 1996: 27–31; quotation from 27; ancient testimony and fragments, 35–67.

13. Josephus 1930: 31 (i.e., *Jewish Antiquities* 1.64); cf. Genesis 4:17–22.

14. Lutz 1956: 41–46.

15. In fact, Josephus never quoted the Chaldean directly but used extracts and paraphrases of Berossos that he found in second- and thirdhand compilations by intervening historians. Like other historians of his era, Josephus habitually presented such mediated quotations as if they were the original sources. Berossos's was a name to conjure with, and his authority was too prestigious to dilute. A similar dynamic existed between an earlier defender of the Jews, Artapanus of Alexandria (after 250 BCE), and the Egyptian chronicler Manetho (3rd c. BCE) on the cultural precedence of Jews over Egyptians.

16. Josephus 1926: 165–67 (1.7–9) echoing *Timaeus* 22B–C.

17. Josephus 1926: 185–87 (1.58).

18. Josephus 1926: 163–65, 169–71 (*Against Apion* 1.1–7, 16; hereafter AA).

19. Josephus 1926: 167 (AA 1.10–11). Homer's mention of writing in the *Iliad* as "baneful tokens" was such a vague reference that "it is a highly controversial and disputed question whether even those who took part in the Trojan campaign made use of [actual] letters." In fact, the Greeks can boast no texts earlier than Homer's poems. "[Homer's] date, however, is clearly later than the Trojan war; and even he, they say, did not leave his poems in writing" but transmitted them by memory, to be gathered, edited, and written down much later; Josephus 1926: 167 (AA 1.12). Josephus's claim that Homer was an oral poet would be revived in the eighteenth century. See chapter thirteen.

20. Josephus 1926: 215 (AA 1.128–31).

21. Josephus 1926: 192 (AA 1.71).

22. Josephus 1930: 45 (JA 1.93).

23. Josephus 1930: 45–47 (JA 1.94–95).

24. On Josephus's relation to previous Jewish and Gentile writers, see Berossos 1996: 28–30, and Sterling 2011: 231–43.

25. For this and other reasons, the story of Oannes was destined to remain buried until Joseph Scaliger retrieved it from Syncellus in the early 1600s. See Grafton 1993: 681–728. Sculptural representations of the *Apkallu,* and of priests whose robes mimicked Oannes's body, would not be discovered by archaeologists until the nineteenth and twentieth centuries.

26. Josephus 1926: 175 (AA 1.29).

27. Josephus 1930: 51 (JA 1.105–6), emphasis added.

28. Josephus 1930: 51–53 (JA 1.107–8).

29. Augustine 1997: 55–56.

30. Romans 1:19–20 in St. Paul 1972: 70.

31. Augustine 1997: 32.

32. Augustine 1997: 48.

33. Augustine 1997: 76.

34. Augustine 1997: 106.

35. Augustine 1997: 112.

36. Jerome 1933: 125–29 (Letter 22.30).

37. Augustine 1997: 76–77, 88.

CHAPTER EIGHT. An Age of Paradoxical Optimism, 650–1350

1. Reynolds and Wilson 2013: 86.

2. Glaber 1989a: 3.

3. Guglielmo Cavallo, "Introduzione," in Glaber 1989b: xxviii–xxxiii.

4. Augustine 1997: 47.

5. Isidore 2006: 11. In general, see Bloch 1983: 34–63.

6. Quoted in Isidore 2006: 24, "Introduction."

7. Isidore 2006: 39 (i.e., I.3), emphasis added.

8. Augustine, in chapter seven of this book.

9. Isidore 2006: 39.

10. Isidore 2006: 142; Vergil, *Eclogues* 10.67. In context, "bookman" is a closer translation of *librarius* than "copyist."

11. Isidore 2006: 341 (i.e., XVII.6).

12. Isidore 2006: 67 (i.e., I.42).

13. Isidore 2006: 142, 140 (i.e., VI.13–14; 12).

14. Lewis and Short, *A Latin Dictionary*, s.v. "penna," gives this as the sole example, qualifying it as "late Latin."

15. Isidore 2006: 142 (i.e., VI.14).

16. Isidore 2006: 40 (i.e., I.3).

17. Isidore 2006: 39 (i.e., I.3). See chapter ten on Petrus Crinitus and chapter eleven on the Salone Sistino.

18. Isidore 2006: 67 (i.e., I.42). For Isidore's *apud nos*, Isidore 2004: 348.

19. Ovid, *Metamorphoses* 1.649.

20. Isidore 2006: 39 (i.e., I.4).

21. Lucan 1928: 130 (i.e., *Civil War* 3.220–21).

22. Pliny, *Natural History* 7.56 says the primitive alphabet had sixteen letters, and quotes Aristotle as maintaining there were eighteen.

23. Isidore 2006: 39 (i.e., 1.4).

24. Netz and Noel 2007: 4.

25. Netz and Noel 2007: 84–85.

26. Netz and Noel 2007: 280.

27. Netz and Noel 2007: 182.

28. Bede 1930: 1:21.

29. Drogin 1983; Drogin 1989; Bischoff 1990; Reynolds and Wilson 2013, etc.

30. *Se pareba boves / alba pratalia araba / albo versorio teneba / negro semen seminaba*. Curtius 1953: 313–14.

31. Isidore 2006: 141 (i.e., VI.9).

32. Isidore 2006: 142 (i.e., VI.14).

33. Curtius 1953: 317.

34. Curtius 1953: 319.

35. Cervantes 2003: 67 (i.e., pt. 1, chap. 9).

36. Curtius 1953: 326.

37. Dante 2013: 3.

38. Curtius 1953: 328.

39. Dante 2011: 665 (i.e., *Paradiso* 33.85–87).

40. Curtius 1953: 326–27.

41. Schreckenberg 1972; Sterling 2011.

42. Hrabanus Maurus, *De universo* II, 1. (*Patrologia Latina* 111: 32). Sixtus of Siena (d. 1569) records (1742, 1: 209) a Talmudic legend that Abel's soul transmigrated into Seth and later into Moses.

43. Isidore 2006: 162, 175 (i.e., 7.6.9 and 8.5.15).

44. Cassiodorus 2004: 149 (*Institutions* 1.17); Schreckenberg 1972: 56–171.

45. Cassian 1997: 307.

46. Cassian 1997: 306.

47. Cassian 1997: 308, 310.

48. Schreckenberg 1972: 105, 192.

49. Isidore 2003: 16–19.

50. Lutz 1956: 43.

51. Schreckenberg 1972: 56–171.

52. Bang 1657: 16–21 at 19. Tertullian *De cultu foeminarum*, chap. 3 suggests that the Holy Spirit could have easily inspired Noah to rewrite the Book of Enoch in the same way that Esdras restored the books of Moses.

53. Lutz 1956: 43–44.

54. Lutz 1956: 47–49. My translation.

55. Although Jubal was a descendant of Cain rather than Seth, both Jubal and Noah had fathers named Lamech, and the two Lamechs were at times confused by commentators.

56. Freculf 2002: 2:17–20; Lutz 1956: 49.

57. Freculf 2002: 2:40–41 (Lutz 1956: 49; *Patrologia Latina* 126. 926). My translation, emphasis added.

58. *Patrologia Graeca* 139. 224–25.

59. Clark 2016.

60. Comestor 2005: 77 (*Patrologia Latina* 198. 1079)

61. Comestor 2005: 54.

62. Curtius 1953: 36–42.

63. Vincent 1624: 37 (i.e., I.101); Kircher 1675: 206–7.

64. Godfrey 1559: col. 585; Godfrey 1872: 105.

65. Godfrey 1872: 95–96, emphasis added. Cf. Josephus 1930: 50–53 (*Jewish Antiquities* 1.107): "My words are attested by all historians of antiquity . . . Manetho the annalist of the Egyptians, Berossos the compiler of the Chaldaean traditions; Mochus, Hestiaeus, along with the Egyptian Hieronymus, the authors of Phoenician histories."

66. Godfrey 1559: 36; Godfrey 1872: 95, 134.

67. Godfrey 1559: 6.

68. In general, see Festugière 1950: 1:308–54: *Les fictions littéraires du* topos *de révélation*.

69. Pingree 1968: viii–18 at 3–4. Plessner 1954: 51–57.

70. Plessner 1954: 52–53 mentions Latin alchemical treatises recalling antediluvian books mentioned by Abū Ma'shar; see also Ruska 1926: 61–64, and Festugière 1949–1954: 1:323 n. 1 for the Arabic motif of antediluvian codices in watertight citadels. My belated thanks to the late Professor Eugenio Garin for these references.

71. Nallino 1944: 228–231.

72. Pseudo-Thomas Aquinas 1488: fol. 11r; see Secret 1959: 9.

NINE. Pessimism in the Age of Rediscovery, 1350–1500

1. Celenza 2004.

2. Boccaccio 2013: 857.

3. Richard of Bury 1960: 70–79.

4. Aulus Gellius, *Noctes Atticae* (*Attic Nights*), VII, 17.

5. Richard of Bury 1960: 75.

6. See chapter seven.

7. Richard of Bury 1960: 75.

8. Richard of Bury 1960: 77.

9. Dante 2011: 664–65 (*Paradiso* 33.95–96: *la 'mpresa / che fé Nettuno ammirar l'ombra d'Argo*).

10. Sandys 1967: 2:36–37.

11. Sabbadini 1967: 2:6. Petrarch 1975: 116 (*Familiares* 3.1).

12. Petrarch 1975: 290–95, at 294: "In our travels through the remains of a broken city, there too, as we sat, the remnants of the ruins lay before our eyes."

13. Celenza 2017.

14. Petrarch 1975: 3:320–21 (*Familiares* XXIV, 4).

15. Petrarch 1975: 3:327–28 (*Familiares* XXIV, 6).

16. Petrarch 1975: 3:321 (*Familiares* XXIV, 4).

17. Petrarch 1975: 3:329 (*Familiares* XXIV, 7).

18. Petrarch 1975: 3:332 (*Familiares* XXIV, 8).

19. Petrarch 1975: 3:332 (*Familiares* XXIV, 8).

20. Petrarch 1975: 3:332 (*Familiares* XXIV, 8).

21. Petrarch 1953: 1247.

22. Freccero 1975.

23. Petrarch 1975: 3:329 (*Familiares* XXIV, 7).

24. Petrarch 1975: 3:341 (*Familiares* XXIV, 11).

25. Petrarch 1953: 1249.

26. Witt 2000: 117–73, 230–91.

27. Witt 2000: 1–116.

28. Conte 1994: 375; Reynolds and Wilson 2013: 131–32.

29. Quoted from Gaisser 2008: 97. Gaisser translates *brevia* as "breviaries." Conceivably Boccaccio could refer to tiny books of hours, but the term was regularly used to refer to little scrolls containing charms against bodily or spiritual harm, a likely use for narrow strips of paper and parchment, and a sense that fits better here.

30. Gaisser 2008: 96.

31. Quoted from Boccaccio 1930: 8, in preference to Boccaccio 2011: 13 for clarity and concision.

32. Boccaccio 1930: 9.

33. Boccaccio 1930: 9.

34. Ovid 1916: 381 (i.e., *Metamorphoses* 15.234–36).

35. Boccaccio 1930: 11.

36. Boccaccio 1930: 13, 11.

37. Poliziano, *Oration on Quintilian and the* Silvae *of Statius*, in Garin 1952: 878–79. My translation.

38. Vespasiano 1951: 442–43; George and Waters (Vespasiano 1926: 402) translates *così antico come era* as "old as he was," but the context is Niccoli's collection of *vasi antichi bellissimi*; Thompson and Nagel 1972: 44.

39. Thompson and Nagel 1972: 15–52.

40. Gordan 1974: 188.

41. Gordan 1974: 188–90.

42. Walker 1972: 1–21 at 19.

43. On these figures, see Hibbert 1975 (Lorenzo); Grafton et al. 2010: 360–61 (Ficino), 761–63 (Poliziano).

44. See Yates 1964: 62–83 (Pico), 84–116 (Ficino); Borghesi et al. 2012: 10–108.

45. Yates 1964: 42–43. See Giovanni di Stefano's famous 1488 inlaid pavement in the Duomo di Siena.

46. Pico, *Oration*; Borghesi et al. 2012: 261.

47. The topic of Kabbalah is immense, and its influence on Christian philosophical and theological literature was extensive. A minimum bibliography would include Scholem 1978; and Idel 1988, esp. 250–71, "From Jewish Esotericism to European Philosophy: An Intellectual Profile of Kabbalah as a Cultural Factor." Drucker 1994: 129–58 is illustrated and informative.

48. Borghesi et al. 2012: 233; I disagree with the note, which depends on Pico knowing Herodotus, whom he doesn't mention in the *Oration*. Whatever they thought of Ham/Zoroaster, few scholars would have been unaware of the figure.

49. See G. W. Bowersock's introduction to Valla 2007: vii–xvi.

50. Valla 2007: 31; I Maccabees 8:22 (Septuagint and Vulgate).

51. Valla 2007: 31–32.

52. Valla 2007: 32.

53. Valla 2008: 4.

54. Valla 2008: 50.

55. Valla 2008: 56.

56. Stewart 1991.

57. Valla 2008: 45

58. Lorenzo Valla, *Elegantiae linguae latinae*, general preface (Garin 1952: 596–97). My translation throughout.

59. Garin 1952: 598–99.

60. Valla, *Elegantiae*, book 3, preface (Garin 1952: 610–11).

61. Valla, *Elegantiae*, general preface (Garin, 1952: 599).

62. Grafton 1990.

63. Stephens 1989: 98–138.

64. Stephens 2016: 68–70.

65. Stephens 2013: 284–86.

66. Stephens 2011: 702–5.

67. Curran 2007: 123–31.

68. Stephens 2018.

69. Bizzocchi 1995.

70. Bibliography in Stephens 2011: 695n.

71. Grafton 2018.

72. Colonna 1999.

73. Mitchell 1960.

74. Bruun and Edmondson 2015.

75. Colonna 1999: 41.

76. Iversen 1961: 67–70.

77. Colonna 1999: 47.

TEN. Alternating Currents, 1450–1550

Epigraph: Curtius 1953: 315.

1. Steuco 1540: 5.

2. Sixtus of Siena 1742: 57 (Bk. II, preface). In reality this was probably a succession of temple libraries, more like the shelves of Ozymandias than the bibliopolis that the Library of Alexandria became in the scholarly imagination.

3. Neander 1565: 39–40.

4. Neander 1565: 45.

5. Xerxes, in Herodotus 2007: 518 (7.46).

6. Barreiros 1565: 2.

7. Barreiros 1565: 1–2.

8. Barreiros 1565: 8.

9. Stephens 2018.

10. Polydore Vergil 2002: 245.

11. Given the complexity and time-consumption of printing, copies of a book from the same press run could differ somewhat for causes ranging from oversight to correction "on the fly" during the process.

12. Polydore Vergil 2002: 247.

13. Blair 2003: 11–28; Blair 2010.

14. See chapter eleven.

15. The latest round of schoolbook-bannings in Texas and elsewhere in the United States (2020 to 2022) stems from conservatives' wrath over what they inaccurately term "critical race theory," meaning any substantive discussion of the *history* of racism in the US.

16. For the early colophons discussed in this section, see Josephson 1917: 1–14, except where noted.

17. "Punches and matrices" is translated from *patroonarum* [*sic*] *formarumque*.

18. Nider 1479: colophon: *Non per pennis ut pristi quidem, sed litteris sculptis artificiali certe conatu ex ere remota nempe indagine*.

19. Polydore Vergil 2002: 247.

20. Clarke 1999: xxx.

21. Trithemius 1974: 35.

22. Erasmus 2001.

23. *Phaedrus* 276c; Plato 1961: 522; Erasmus 2001: 81–82.

24. Erasmus 2001: 142, 137.

25. Erasmus 2001: 143.

26. Erasmus 2001: 144–45.

27. Erasmus 2001: 145–46.

28. But see Bibliander 2011.

29. Bibliander 2011: 22n.

30. Bibliander 2011: 23.

31. Bibliander 2011: 121.

32. Bibliander 2011: 125.

33. Bibliander 2011: 33; Postel 1538a, 1538b.

34. Bibliander 2011: 77.

35. Bibliander 2011: 129, modified.

36. Bibliander 2011: 137. Ref. to Eusebius *Praeparatio evangelica* bk. 1, chs. 9–10.

37. Brown 1946; Baumgarten 1981.

38. Bibliander 2011: 141

39. Bibliander 2011: 138. My translation, which is closer to the original than Amirav and Kirn's on p. 139.

40. Bibliander 2011: 139, 141. My translation.

41. Bibliander 2011: 141. My translation. See Crinitus 1955: 337 (i.e., bk. 17 chap. 1).

42. Isidore of Seville 2006: 39–42 (bk 1, chaps. 3–4).

43. Bibliander 2011: 128, 130.

44. Bibliander 2011: 234.

45. Bibliander 2011: 234.

46. Bibliander 2011: 236. My translation.

47. Bibliander 2011: 236, 238. My translation.

48. The double vision strategy is usefully described in Coleman 2018: 130–32.

49. More 1964: 12.

50. More 1964: 106.

51. Stephens 1989: 185–289.

52. Rabelais 2006: 11.

53. Rabelais 2006: 208.

54. Rabelais 2006: 20–21.

55. Sider 2005.

56. Rabelais 2006: 121.

57. Rabelais 2006: 210–15.

58. Rabelais 2006: 374–79.

59. Zatti 2006.

60. Stephens 1989: 210.

61. See Freeman 2014: 141; and Clark 2018.

62. Folengo 1927.

63. Rabelais 2006: 993–95.

64. Rabelais 2006: 37–44.

65. Cervantes 2003: 65–69.

ELEVEN. A Second Age of Scripture, 1500–1600

1. The story may be apocryphal, since Melanchthon only arrived in Wittenberg in August 1518; Marius 1999: 137–40 at 138–39; Roper 2016: 1–6.

2. On the history of *sola scriptura*, see Pelikan 1984: 4:118–26, 262–66, 276–77, 322–31.

3. Bielo 2018. Photographs of the "Ark" at: "Explore the Life-size Replica of Noah's Ark and the Creation Museum in Kentucky," *Washington Post* May 23, 2017, https://www.washingtonpost.com/lifestyle/explore-the-life-size-replica-of-noahs-ark-and-a-creation-museum-in-kentucky/2017/05/23/5bba236e-2c39-11e7-a616-d7c8a68c1a66_gallery.html.

4. Pelikan 1984: 187–203, 290–302.

5. Biller and Hudson 1994.

6. Pelikan 1996: 49–50.

7. Pelikan 1996: 130–31; Marius 1999: 359–60.

8. Lambert 1992: 381.

9. Alberigo et al. 1973: 663. My translation, emphasis added.

10. Sixtus of Siena 1742: 1:56–220.

11. Curtius 1953; Garin 1961.

12. Steuco 1540: 4, emphasis added.

13. E.g., Schmitt 1966: 527; Walker 1972: 40–41.

14. Leone Ebreo 2009: 238–39.

15. Stephens 1989: 116 and n. 45; Grafton 2018.

16. Annius 1498: sig. o3r; Stephens 1984: 318; Stephens 1989: 374n53. The passage is the linchpin of Annius's program of rewriting world history. Although it was mangled in all subsequent editions containing Annius's commentaries, his program remained clear to his imitators even without it.

17. Stephens 1989: 117.

18. Melanchthon 1532: sig. B4v. My translation.

19. Herodotus 2007: 514 (i.e., 5.58); Josephus 1926: 174, 190, 204 (i.e., *AA* 1.28–29, 1.70–71, 1.106–7).

20. Carion 1538: fol. 10v. My translation.

21. Melanchthon 1532: fol. B1v–C1r; Carion 1543: fol. 10v. My translation.

22. See the explanations of Melanchthon and Peucer 1580: sig. a1r, sig. a7v.

23. Melanchthon and Peucer 1580: 17. My translation.

24. See chapter five.

25. Melanchthon and Peucer 1580: 17. My translation.

26. Melanchthon and Peucer 1580: 16. My translation.

27. Melanchthon and Peucer 1580: 10. My translation.

28. Melanchthon and Peucer 1580: following c1r. My translation.

29. Melanchthon and Peucer 1580: sig. d6v.

30. *Chronica durch Magistrum Johan Carion*, sig. C2v. See Johannes Carion,

Chronica, 1532, Google Books, https://www.google.com/books/edition/Chronica/N-QsKZVWyCsC?hl=en&gbpv=1.

31. Melanchthon and Peucer 1580: sig. (c1v). My translation.

32. This refers to the promises made to Abraham in Genesis 12–17, although Genesis 11:10–26 makes Abraham/Abram the ninth generation from Shem.

33. Melanchthon and Peucer 1580: 25. My translation.

34. See Lotito 2019, esp. 112–206; Stephens 2005: 70–83. Duret 1611 shows many affinities and debts to Bibliander for extensive commentary on legendary books; Secret 1959.

35. Camozzi 1575. I owe this reference to Dr. Maude Vanhaelen of the University of Warwick.

36. Rocca 1591: 78–174.

37. The Vulgate says that Enosh (*iste*, "he") began to call on the name of God; the *Tanakh* says, "It was then that men began to invoke the Lord by name," a reading that agrees with King James Version. I have not encountered Enosh elsewhere as founder of devotional images.

38. Rocca 1591: 82–83.

39. Stoneman 2008: 73–77, 230–34, 238; Dronke et al. 1997: 272–75, 591–93, 694–95.

40. Rossi 2001: 331–32.

41. Fontana 1590.

42. Mercati 1981: 108–9. My translations from Mercati.

43. Mercati 1981: 109.

44. Mercati 1981: 112–13.

45. Mercati 1981: 113–14.

46. Tasso 2006: 511–14 (bk. 6, vv. 779–818). My translation. The English version by Joseph Tusiani (Tasso 1982) is more paraphrase than translation.

47. Rabelais 2006: 207, citing Plato, *Republic*, 376a–b.

48. Tasso 2006: 514 (6.819–29).

49. Iamblichus 1497: sig. a3r. My translation. Clark et al. (Iamblichus 2004: 5) translate literally, whereas Ficino condensed and paraphrased both Iamblichus and the other Neoplatonic works he edited. Tasso's annotated copy of Ficino's translation (Iamblichus 1497) survives in the Vatican Library.

50. Semiramis's inscription is mentioned by Diodorus Siculus II.xiii.2; on Jacob's, see Genesis 28:18–22 and Sixtus of Siena 1742: 110 (s.v. "Jacobi Patriarchae Scala").

51. Tasso 1998: 2:1120–21.

TWELVE. The Age of Grand Collections, 1600–1800

1. Illustrations in Campbell and Pryce 2013: 90–207.

2. See, e.g., Kenseth 1991; Mauriès 2002.

3. Redford 2008.

4. See Getto 1969. I know of nothing similar in other languages.

5. Yeo 2001.

6. Calmet 1741–1750: 1:86–101.

7. Vincent of Beauvais 1624; Franklin-Brown 2012.

8. All quotations from Lipsius 2017: 66–69.

9. Mader 1702–1705.

10. Morhof 1747.

11. Morhof 1747: 1:9.

12. Grafton 1990: 99–103, 118–23; Grafton 1993: 514–632.

13. Hugo 1617.

14. Bang 1657.

15. Bruun and Edmondson 2014.

16. Stephens 2004; Stephens 2005.

17. Inghirami 1637. The rather hilarious story of Inghirami's forgeries and the attacks by Allacci and others is told in Rowland 2004.

18. Allacci 1642: 112–207, at 168–69. Rowland 2004: 135 erroneously states that "only in 1700, forty-five years after [Inghirami's] death, did someone notice that the paper on which the texts were written bore the watermark of the state paper factory in Colle di Val d'Elsa."

19. Rowland 2004: 24, 32, 43, 54, 62–63, 76–77.

20. Inghirami 1645.

21. Mabillon 1681: 31–39.

22. Schwarz 1705–1717; Sallbach [Saalbach] 1705.

23. Nicols 1711.

24. Lucretius 1995: 51.

25. Lutz 1979: 17–21; Lutz's frontispiece shows the Saint Gall inscription.

26. Grafton 1999: 111–21.

27. Nicols 1711: 1. My translation.

28. Peignot 1823: 1:130: *Corneille disoit qu'il donneroit deux de ses meilleures pièces pour avoir fait les quatre vers si connus où Brébeuf, dans sa traduction de la Pharsale, peint l'art de l'écriture, qu'on attribue à Cadmus: "C'est de lui que nous vient cet art ingénieux / de peindre la parole et de parler aux yeux, / et par des traits divers de figures tracées / Donner de la couleur et du corps aux pensées."*

29. Brébeuf 1682: 80 (3.18–21). My translation.

30. Nicols 1711: 1–2.

31. The exception was Bang's *Eastern Sky* (*Caelum orientis*), several pages of which Defoe (2007: 34–41) reproduced, untranslated, in their original Latin.

32. Defoe 2007: 15.

33. Massey 1763: 36; as we saw, Exodus 24:4 shows Moses writing God's commands before God does.

34. Defoe 2007: 11.

35. Massey 1763: sig. A2r; pp. 1–3.

36. Defoe 2007: 15.

37. Massey 1763: 1–3.

38. Robert More, "A Compendious Essay, on the First Invention of Writing," in Massey 1763: 106–7.

39. Massey 1763: 168–69.

40. Massey 1763: 104, echoing More, "Compendious Essay."

41. Astle 1784: i (Introduction).

42. Astle 1784: 10, referring to William Warburton, *Divine Legation of Moses*, 4th ed. (1765) IV.4, pp. 69–105, which is closer to the historically accurate view of hieroglyphics that triumphed after Champollion. Etienne Fourmont (d. 1745) wrote *Meditationes Sinicae* (1737), which considered Chinese "hieroglyphics." Antoine Court de Gebelin (d. 1784) asserted that the Tarot images were descended from the ancient Egyptian *Book of Thoth* (see chapter one of this book) in *Le Monde primitif*, vol. 8 (1781). See Iversen 1993: 100–108.

43. Fabricius 1790–1812: 1:1–316.

44. Fabricius 1722.

45. Hugo 1617: 41.

46. Hugo 1617: 41, 53.

47. Boulduc 1630: 21–29, at 25.

48. Bartholin 1711: 71.

49. Mader 1702–1705: 1:6.

50. Mader 1702–1705: 1:7.

51. Saint Augustine 1950: 647. Mader's transcription appears defective.

52. Mader 1702–1705: 1:8; Godfrey of Viterbo 1559: 34; Bodleian Library MS Lat. hist. c. 1. fol. 19v; my translation.

53. Mader 1702–1705: 1:7–8.

54. Mader 1702–1705: 1:6–7; Isidore of Seville 2006: 39.

55. Synopsis in Allen 1963: 182–93.

56. Ginzburg 1998: 1:90–93, 154–57; 5:117–18, 177.

57. Mader 1702–1705: 1:8–9, quoting Kircher 1650.

58. The work in question, however, the so-called *Gelasian Decree* (*Decretum Gelasianum*), is a later work falsely attributed to this pope.

59. Mader 1702–1705: 1:9.

60. Vockerodt 1704.

61. Reimmann 1728: 342.

62. Reimmann 1709: 342.

63. Reimmann 1709: 36–37.

64. Lomeier 1680: 9, 11–12.

65. Lomeier 1680: 22–23.

66. Lomeier 1680: 9–10.

THIRTEEN. Skepticism and Imagination, 1600–1800

1. Massey 1763: 22, 23.

2. Vico 2004: 64–65; Vico 2020: 37.

3. Montgomery 1962: 3–4.

4. Nigel Wilson in Photius 1994: 1–8.

5. Fabricius 1790–1812: 1:208–9. Fabricius cites codex no. 190 (Photius 1612: cols. 485–486 [not translated in Photius 1994; see cols. 480–481 for Musaeus's daughter Helen]); he also cites the 12th c. commentary by Eustathius on the *Odyssey* (Eustathius 1825–1826: 1:2).

6. Homer 1951: 103 (*Iliad* 3.125–28).

7. See Barber 1994 for the importance of textiles since Neolithic times, and 209–16 for Homer's Helen, Troy, and Linear B.

8. Homer 1967: 71–72; see Bergren 2009.

9. Ryck 1684: 431, emphasis added; Pliny 1942: 636–37 (*NH* 7.56.192). For Ryck's list of Homer's sources, see his *Dissertatio* in Ryck 1692. See also Allacci 1640.

10. Struve 1754: 179.

11. Struve 1754: 179; Lipsius 2017: 66–69; on Eustathius, see n. 5 above.

12. *Encyclopédie* 2 (Paris, 1751): 229–30, s.v. "Bibliothèque."

13. Lutz 1979: 17–24.

14. Bartholin 1670: 1–42.

15. Bartholin 1670: 3.

16. Bartholin 1711: 19. All translations mine.

17. Bartholin 1711: sig. *****2r (*sic*).

18. Bartholin 1711: 20.

19. Bartholin 1711: 10, 33.

20. Bartholin 1670: 12.

21. Bartholin 1670: 6. Dionysius of Halicarnassus, *Roman Antiquities* IV.62 (repeated by Aulus Gellius I.19); Varro, according to a remark in Lactantius I.6; Pliny's *Natural History* XIII.27; none of these historians claims that the old woman selling the books was the Cumaean Sibyl, a connection that grew up later.

22. Charlesworth 1983: 317–472.

23. Bartholin 1670: 6.

24. Bartholin 1670: 7. The story is also mentioned by Mabillon 1681: 32; Lipsius 2017: 81–82, relying on the same Byzantine historians, says the intestine was 120 feet long, not 20, and that there were 120,000 books in the library.

25. Bartholin 1670: 10; see chapter three of this book.

26. Bartholin 1711: 3.

27. *Encyclopédie* 1 (1751): xxxviii–xxxix. My translation differs somewhat from that of Richard Schwab in D'Alembert 1995: 121–22.

28. Yeo 2001: 2–3.

29. Blair 2010.

30. Petrarch, *Remedia* 1.43 on *librorum copia.*

31. *Encyclopédie* 6:608; 9:610; 2:240.

32. Relayed by Fabricius 1790–1812: 1:76–77, reporting on Christopher Kriegsmann in *Tabula smaragdina vindicata.*

33. On the *Liber Hermetis Trismegisti* see Gundel 1936; Garin 1976: ix–x.

34. Ursin 1661.

35. Allen 1963: 134.

36. Allen 1963: 136–37.

37. Minois 2012: 203.

38. Struve 1703.

39. Its newness was methodological: see Burke 1985; and Mazzotta 1999.

40. Vico 2020 (*La scienza nuova,* 1744: ¶¶45–59); Manuel 1959; Rossi 1999.

41. Grafton 1993: 681–728; see chapter one of this book.

42. Voltaire 1962: 1:62.

43. Translations of *Zadig* my own. Freer, more literary translation in Voltaire 1981: 102–72.

44. See *Le Taureau Blanc, La Princesse de Babylone,* etc. Regarding "Orientalism," see Said 2003.

45. Sterne 1965: 25–26 (end of vol. 1, chap. 12); p. 357 (vol. 6, chap. 38) actually is blank, while pp. 169–70 (vol. 3, chap. 36) imitate the marbled endpapers of eighteenth-century book bindings.

46. Roberts 1972: 33.

47. Anderson 1734.

48. Anderson 1734: 7.

49. Anderson 1734: 8.

50. Roberts 1972: 37.

51. Mackey 1996: 44–49. See chapter eight on Isidore's *Cronicon.*

52. Anderson 1734: 7–8.

53. 3 Kings 7:13–22; 2 Chronicles 3:15–17; 4 Kings 25:13–17.

54. Roberts 1972.

55. Chambers 1738: 1, fol. 6Q2v.

56. *Encyclopédie* 9:605.

57. Chambers 1738: 1, fol. 6Q2v.

58. Hugo 1617: 15–17.

59. Hugo 1617: 17.

60. Wilkins 1694: 5–7.

61. Defoe 2007: 3–4.

62. Defoe 2007: 21.

63. Gates 1988, esp. chap. 4.

64. Equiano 1789: 75.

65. Equiano 1789: 97–98, 118.

66. Douglass 2001: 31.

67. Douglass 2001: 31–32.

68. Bingham 1837.

69. Douglass 2001: 35.

70. Douglass 2001: 35–36.

71. Douglass 2001: 37–38.

72. Equiano 1789: 83–84.

73. Douglass 2001: 37.

74. Equiano 1789: 24–31 (Introduction).

FOURTEEN. The Age of Decipherment, 1800–1950

1. On needlework and writing in antiquity, see Barber 1994.

2. Bender 2002: 25–41.

3. Cassirer 1969: 33–35, at 34.

4. On my exclusion of Chinese, Mayan, Indus Valley, and other scripts, see the introduction.

5. Eddy 2011.

6. Josephus's title in Greek was *Ioudaikes Archaiologias*.

7. Rowley-Conwy 2007. Thomsen's system was based on the tripartite scheme of Lucretius 1995: 195 (book 5).

8. Goodrum 2004: 172–80, 224–33.

9. Stephens 1989: 1–57.

10. Champollion followed up in 1824 with a more extensive proof, the *Précis du système hiéroglyphique*.

11. Robinson 2012: 15–27, 80–91; V.-David 1965; Iversen 1993: 124–47.

12. V.-David 1965: 13–14. "Code-talkers" of Navajo and other Native American languages in the US military during World War II provided an oral equivalent to written ciphers, defeating Axis attempts to decode vital messages.

13. Robinson 2012: 18–21.

14. Kramer 1963: 7; Pallis 1954.

15. Robinson 2007: 72.

16. Defoe 2007: 15–16; Pliny, *Natural History* 7.56.192–93 (see chapter three of this book).

17. Schoep 2018: 11.

18. Schoep 2018: 5–32, at 5 (abstract).

19. Margalit 2020.

20. Schoep 2018: 10.

21. Schoep 2018: 9–10.

22. Schoep 2018: 10.

23. The film *Anonymous* (see Emmerich 2011) carries this elitist daydream to ridiculous lengths, dismissing Shakespeare as functionally illiterate, Ben Jonson as a hack, and Queen Elizabeth I as a serial unwed mother who commits incest with her son the Earl of Oxford; the film promotes Edward de Vere, the Earl of Oxford, as not only the "real" Shakespeare but the rightful heir to the British crown.

24. Butler 1897: xvii–13, 105–270.

25. Parsons 2007.

26. Parsons 2007: 16–17; cf. 23–24.

27. Cristofani 1979: 88–89.

28. Wellard 1973: 76–85.

29. Hoffman et al. 2011.

30. Carrière and Eco 2011: 153–55 (some details apparently wrong).

31. Damrosch 2006: 9–18, 29–31.

32. Damrosch 2006: 33–48.

33. Goodrich 1852: 66–67, 25.

34. Goodrich 1852: 66–67. Although the Hebrew calculation was 1656 years, the Septuagint and Josephus (who depended on its account rather than the Hebrew) calculated 2262 years from Creation to the Flood.

35. Voltaire 1960: 338. Voltaire attributes the chronology to "Father [Denis] Pétau," the Jesuit author of a universal chronology published in 1627, which was translated into both English and French, and abridged as late as 1849 (see Manuel 1959: 91, 99–100).

36. Gillispie 1951; Rossi 1984: 152–67; Allen 1963: 92–112.

37. Mayor 2011: 106–56.

38. In 1852, an exhibition of "life-size" dinosaur statues opened in the "Dinosaur Court" at the Crystal Palace Park in London; like Goodrich's engravings, they were quite incorrect by modern standards.

39. Goodrich 1852: 36.

40. Goodrich 1852: 22.

41. Allen 1963; Gillispie 1951; Wendt 1968; Rossi 1984.

42. The best-known attempt was Archbishop James Ussher's *Annales Veteris Testamenti a prima mundi origine deducti* (London, 1650), which proposed the world was created in 4004 BCE.

43. Goodrich 1852: 22.

44. Driver 1904: xxviii.

45. Driver 1904: xliii, emphasis in original.

46. Driver 1904: xliii.

47. Driver 1904: xliii. The latest writer I have found who took antediluvian libraries seriously is Richardson 1914: 25–51.

48. Pliny, *Natural History* 7.56.192–93.

49. Smith 1830.

50. Givens 2002: 7, 66.

51. Givens 2002: 5. My own copies of the Book of Mormon from 1977 and 1986 show the change.

52. Givens 2002: 84, 63–64.

53. Smith 1830: vi–ix (unnumbered).

54. See Givens 2002: 19–42.

55. Givens 2002: 187.

56. Givens 2002: 188, 189–90, quoting 1 Nephi 38–40 and "Articles and Covenants of the Church of Christ."

57. Augustine 1950: 532 (*City of God* 16.9).

58. Gliozzi 1976.

59. Quoted by Givens 2002: 307 n. 90.

60. See Givens 2002: 26–37, 89–184.

61. Givens 2002: esp. 89–184.

62. For Shelley, Smith, and Ozymandias, see Rodenbeck 2004.

63. Shelley 2021: final sentence.

64. Todorov 1973.

65. Potocki 1995: xii–xiv.

66. Brontë 1988: 63–75 (*Wuthering Heights*, chap. 3).

67. Rabelais 1741.

68. Lacroix and Brunet 1862.

69. Lacroix and Brunet 1862: 1 (Brunet), x (Lacroix). My translations.

70. France 1922.

71. On Simon, see Manuel 1959: 111–114.

72. Dibdin 1811.

73. Desormeaux 2001: 242–44.

FIFTEEN. The Age of Media, 1950–2020

1. See McLuhan 1964.

2. Carrière and Eco 2011: 7–8; Early Modern observations are in Goldberg 1990.

3. Carrière and Eco 2011: 31.

4. An example among many: Daniel P. Dern, "9 Museums That Want Your Legacy Tech," *Computerworld*, October 3, 2013, https://www.computerworld.com/article/2474004/121888-Museums-want-your-legacy-technology.html.

5. This, as of 2007, was approximately Bradbury's own interpretation of *Fahrenheit* 451.

6. Brahm 1959.

7. Lepore 2020: 23.

8. In January 2005, BALTIC, a museum of contemporary art in Gateshead, England (near Newcastle), posted this call for participation: "You are invited to contribute to an anthology of imaginary books. Submissions should include title, author and publisher and can be in any language. Any further description should not exceed 300 words. The anthology will be edited by Olaf Nicolai and published by BALTIC." Searches on the BALTIC website in 2008 and 2012 produced no trace of the project, prompting the question whether the catalog of imaginary books itself was perhaps a deliberate joke, an unlocatable unreadable book about unlocatable unreadable books.

9. The Invisible Library (webpage), http://web.archive.org/web/20041208232527/http://www.invisiblelibrary.com/index.html.

10. "Fictional book," Wikipedia, http://en.wikipedia.org/wiki/Fictional_books.

11. Bayard 2007; Tuleja 1989.

12. Decamp 1947: 7–8, 25–26.

13. De Turris and Fusco 1972.

14. Bergier and Pauwels 1964.

15. Von Däniken 1970.

16. Borges 1964: 95.

17. *Thomas* was discovered in 1945, *Judas* in 1983. "Three Versions of Judas" appeared in 1944.

18. Borges 1964: 10.

19. Borges 1964: 7.

20. Borges 1964: 18.

21. Borges 1964: 51.

22. Borges 1964: 52.

23. Borges 1964: 53, note by the imaginary editor.

24. Borges 1964: 54.

25. Borges 1964: 54.

26. Best exemplified in Kabbalah.

27. Eco 1994: 1.

28. Eco 1994: 5.

29. Eco 1994: 500.

30. Eco 1994: 500.

31. Eco 1994: 501.

32. Quote on dust jacket of Eco 1980.

33. Pérez-Reverte 1996: 323.

34. Pérez-Reverte 1996: 57–58.

35. Eco 1988.

36. Brown 2003. Although the subtitle of *The Da Vinci Code* is *A Novel*, it begins (prior to the prologue) with a page titled "Fact" and proceeds to repeat a debunked modern legend about the Templars.

37. Eco 1993: 5.

38. Eco 2005. On Milo Temesvar, see Eco 1994: 3; Carrière and Eco 2011: 282–84.

39. Lila Azam Zanganeh, "Umberto Eco, The Art of Fiction No. 197," *Paris Review* 185, Summer 2008, https://www.theparisreview.org/interviews/5856/the-art-of-fiction-no-197-umberto-eco; see https://en.wikipedia.org/wiki/The_Da_Vinci_Code.

40. Eco 2011.

41. Bernhard 2015. See chapter six of this book.

42. Self 2006: 58.

43. Self 2006: 65.

44. 1 Samuel 1–7.

Works Cited.

Acta sanctorum quotquot toto orbe coluntur. Ed. Johannes Bolland et al. 68 vols. Various Publishers. 1643–1940.

Alberigo, Giuseppe, et al., eds., 1973. *Conciliorum oecumenorum decreta*. Bologna, Italy: Istituto per le Scienze Religiose.

Allacci, Leo, 1640. *De Patria Homeri*. Lyon: Laurent Durand.

———, 1642. *Leonis Allatii in antiquitatum etruscarum fragmenta ab Inghiramio edita animadversiones*. Additur eiusdem *Animadversio in libros Alphonsi Ciccarelli et auctores ab eo confictos*. Rome: Mascardus.

Allen, Don Cameron, 1963. *The Legend of Noah: Renaissance Rationalism in Art, Science, and Letters*. Urbana: University of Illinois Press. First published 1949.

Allen, James P., 2000. *Middle Egyptian: An Introduction to the Language and Culture of Hieroglyphs*. New York: Cambridge University Press.

Anderson, James A. M., 1734. *The Constitutions of the Free-Masons (1734). An Online Electronic Edition*. Ed. Paul Royster. Faculty Publications, UNL Libraries. 25. https://digitalcommons.unl.edu/libraryscience/25. First published 1723 (London).

Annius, Joannes, of Viterbo, 1498. *Commentaria Fratris Ioannis Annii Viterbensis super Opera Diversorum Auctorum de Antiquitatibus Loquentium*. Rome: Eucharius Silber.

———, 1515. *Antiquitatum Variarum Volumina XVII a Venerando et Sacrae Theologiae et Praedicatorii Ordinis Professore Ioanne Annio*. Paris: Jean Petit and Josse Bade.

Aquinas, Thomas (Pseudo-), 1488. *Doctoris Aquinatis de esse et essentiis*. Venice: Ioannis Lucilii santritor de fonte salutis & Hieronymi de Sanctis.

Astle, Thomas, 1784. *The Origin and Progress of Writing: As Well Hieroglyphic as Elementary, Illustrated by Engravings Taken from Marbles, Manuscripts and Charters, Ancient and Modern: Also, Some Account of the Origin and Progress of Writing*. London: Printed for the Author; sold by T. Payne and Son, B. White, P. Elmsly, G. Nicol, and Leigh and Sotheby.

Athenaeus, 1930. *The Deipnosophists*. Trans. Charles Burton Gulick. 7 vols. Reprint. Cambridge, MA: Harvard University Press, 1969.

———, 2009. *The Learned Banqueters*. Vol. 4. Trans. S. Douglas Olson. Cambridge, MA: Harvard University Press.

Augustine, Saint, 1950. *The City of God*. Trans. Marcus Dods, et al. Reprint. New York: Modern Library. First published 1872.

———, 1997. *On Christian Teaching*. Trans. R. P. H. Green. Oxford: Oxford University Press.

Bang, Thomas, 1657. *Et ha-kedem. Caelum orientis et prisci mundi triade exercitationum literariarum repraesentatum*. Heidelberg: Typis P. Morsingii.

Barber, Elizabeth Wayland, 1994. *Women's Work: The First 20,000 Years: Women, Cloth, and Society in Early Times*. New York: W. W. Norton.

Barreiros, Gaspar, 1565. *Censura in quendam auctorem qui sub falsa inscriptione Berosi Chaldæi circunfertur*. Rome.

Bartholin, Thomas, 1670. *On The Burning of His Library and On Medical Travel*. Trans. Charles D. O'Malley. Lawrence: University of Kansas Libraries, 1961.

———, 1711. *Viri doctissimi Thomae Bartholini De libris legendis dissertationes*. 2nd ed. Ed. Johann Gerhard Meuschen. Hagae-Comitium: Nicolaus Wildt. First published 1676.

Baumgarten, Albert I., 1981. *The Phoenician History of Philo of Byblos: A Commentary*. Leiden: Brill.

Bayard, Pierre, 2007. *How to Talk about Books You Haven't Read*. Trans. Jeffrey Mehlman. New York: Bloomsbury.

Bede, 1930. *Bedae Opera Historica*. Trans. J. E. King. Reprint. 2 vols. Cambridge, MA: Harvard University Press, 1979.

Bender, Margaret, 2002. *Signs of Cherokee Culture: Sequoyah's Syllabary in Eastern Cherokee Life*. Chapel Hill: University of North Carolina Press.

Bergier, Jacques, and Louis Pauwels, 1964. *Morning of the Magicians*. Trans. Rollo Myers. New York: Stein and Day.

Bergren, Ann L. T., 2009. "Helen's 'Good Drug.'" 314–335 in *Oxford Readings in Classical Studies: Homer's* Odyssey. Ed. Lillian E. Doherty. Oxford: Oxford University Press.

Bernhard, Andrew, 2015. "The Gospel of Jesus' Wife: Textual Evidence of Modern Forgery." *New Testament Studies* 61: 335–55.

Berossos, 1996. In *Berossos and Manetho, Introduced and Translated. Native Traditions in Ancient Mesopotamia and Egypt*. Trans. Gerald P. Verbrugge and John M. Wickersham. Ann Arbor: University of Michigan Press.

Beyerlinck, Laurentius, 1678. *Magnum Theatrum Vitae Humanae, Hoc est rerum divinarum humanarumque syntagma catholicum, philosophicum, historicum, et dogmaticum, ad normam polyantheae universalis dispositum*. 8 vols. Lyon: Johannis Antonij Huguetan.

Bible, Jewish, 1985. *Tanakh: The Holy Scriptures. The New JPS Translation According to the Traditional Hebrew Text.* Philadelphia: Jewish Publication Society.

Bible, Roman Catholic (Douai-Reims Version), 1899. *The Holy Bible Translated from the Latin Vulgate.* Reprint. Rockford, IL: Tan Books, 1971

Bibliander, Theodore, 2011. *De ratione communi omnium linguarum et litterarum commentarius.* Trans. Hagit Amirav and Hans-Martin Kirn. No. 475. Geneva: Droz.

Bielo, James S., 2018. *Ark Encounter: The Making of a Creationist Theme Park.* New York: New York University Press.

Biller, Peter, and Anne Hudson, 1994. *Heresy and Literacy, 1000–1530.* Cambridge: Cambridge University Press.

Bingham, Caleb, 1837. *The Columbian Orator: Containing a Variety of Original and Selected Pieces; Together with Rules; Calculated to Improve Youth and Others in the Ornamental and Useful Art of Eloquence.* First published 1797, Boston: J. H. A. Frost.

Bischoff, Bernard, 1990. *Latin Palaeography: Antiquity and the Middle Ages.* Trans. Dáibhí Ó Cróinín and David Ganz. Reprint. Cambridge: Cambridge University Press, 2008.

Bizzocchi, Roberto, 1995. *Genealogie incredibili: scritti di storia nell'Europa moderna.* Reprint. Bologna, Italy: Società editrice il Mulino, 2009.

Blair, Ann, 2003. "Reading Strategies for Coping with Information Overload, 1550–1700." *Journal of the History of Ideas* 64: 1.

———, 2010. *Too Much to Know: Managing Scholarly Information before the Modern Age.* New Haven: Yale University Press.

Bloch, R. Howard, 1983. *Etymologies and Genealogies: A Literary Anthropology of the French Middle Ages.* Chicago: University of Chicago Press.

Boccaccio, Giovanni, 1930. *Boccaccio on Poetry: Being the Preface and the Fourteenth and Fifteenth Books of Boccaccio's* Genealogia Deorum Gentilium. Trans. Charles G. Osgood. Reprint. Indianapolis: Bobbs-Merrill, 1956.

———, 2011. *Genealogy of the Pagan Gods.* Vol. 1. Trans. Jon Solomon. Cambridge, MA: Harvard University Press.

———, 2013. *The Decameron.* Trans. Wayne A. Rebhorn. New York: W. W. Norton.

Bolzoni, Lina, 2001. *The Gallery of Memory: Literary and Iconographic Models in the Age of the Printing Press.* Trans. Jeremy Parzen. Toronto: University of Toronto Press.

Borges, Jorge Luis, 1964. *Labyrinths: Selected Stories and Other Writings.* Ed. Donald A. Yates and James E. Irby. New York: New Directions.

———, 1998. *Collected Fictions.* Trans. Andrew Hurley. Reprint. New York: Penguin, 1999.

——, 1999. "A Defense of Basilides the False." 65–68 in his *Selected Non-Fictions*. Ed. Eliot Weisberger. New York: Viking.

Borghesi, Francesco, Michael Paio, and Massimo Riva, eds., 2012. *Pico della Mirandola, Oration on the Dignity of Man.* A New Translation and Commentary. Cambridge: Cambridge University Press.

Bottéro, Jean, and Samuel Noah Kramer, 1989. *Lorsque les dieux faisaient l'homme: mythologie mésopotamienne*. Paris: Gallimard.

Boulduc, Jacques, 1630. *De ecclesia ante legem*. 2nd ed. Paris: Joseph Cottereau.

Brahm, John, dir., 1959. *The Twilight Zone*. Season 1, Episode 8, "Time Enough at Last." Aired November 20, 1959, on CBS. Culver City, CA: MGM.

Brébeuf, Georges de, 1682. *La Pharsale de Lucain, ou les guerres civiles de César et de Pompée, en vers françois*. Paris: Jean Cochart.

Brogan, T. V. F., 1993. "Verse and Prose." 1346–51 in Preminger and Brogan 1993.

Brontë, Emily, 1988. *Wuthering Heights*. Ed. Heather Glen. London: Routledge.

Brown, Dan, 2003. *The Da Vinci Code*. New York: Doubleday.

Bruun, Christer, and Jonathan Edmondson, eds., 2015. *The Oxford Handbook of Roman Epigraphy*. Oxford: Oxford University Press.

Burke, Peter, 1985. *Vico*. Oxford: Oxford University Press.

Butler, Samuel, 1897. *The Authoress of the Odyssey: Where and When She Wrote, Who She Was, The Use She Made of the Iliad, and How the Poem Grew under Her Hands*. Reprint. Chicago: University of Chicago Press, 1967.

Calmet, Antoine Augustin, 1741–1750. *Il tesoro delle antichità sacre e profane*. 6 vols. Trans. Lamberto Gaetano Ponsanpieri. Verona: Dionisio Ramanzini; Venice: Francesco Pitteri.

Camozzi, Giovanni Battista, 1575. *Ioannis Baptistae Camotii Oratio de antiquitate litterarum*. Rome: Antonio Blado.

Campbell, James W. P., and Will Pryce, 2013. *The Library: A World History*. Chicago: University of Chicago Press.

Canfora, Luciano, 1990. *The Vanished Library: A Wonder of the Ancient World*. Trans. Martin Ryle. Berkeley: University of California Press.

Carion, Johannes, 1532. *Chronica durch Magistrum Johan Carion vleissig zuraunnen gezogen, wenngleich nützlich zu lesen*. Wittenberg: Georg Rhaw.

——, 1538. *Chronicorum libellus, maximas quasque res gestas, ab initio mundi, apto ordine complectens*. Trans. Hermann Bonnus. Reprint. Frankfurt: Petri Brubachij, 1543.

Carpenter, Rhys, 1935. "Letters of Cadmus." *American Journal of Philology* 56.1: 5–13.

Carrière, Jean-Claude, and Umberto Eco, 2011. *This Is Not the End of the Book: A Conversation Curated by Jean-Phillipe de Tonnac*. London: Harvill Sacker.

Carruthers, Mary, 1990. *The Book of Memory: A Study of Memory in Medieval Culture*. Cambridge: Cambridge University Press.
Carruthers, Mary, and Jan M. Ziolkowski, 2002. *The Medieval Craft of Memory: An Anthology of Texts and Pictures*. Philadelphia: University of Pennsylvania Press.
Cassian, John, 1997. *The Conferences*. Trans. Boniface Ramsay, OP. New York: Paulist Press.
Cassiodorus, 2004. *Institutions of Divine and Secular Learning [and] On the Soul*. Trans. James W. Halporn, introd. Mark Vessey. Liverpool: Liverpool University Press.
Cassirer, Ernst, 1944. *An Essay on Man*. Reprint. New Haven: Yale University Press, 1969.
Casson, Lionel, 2001. *Libraries in the Ancient World*. New Haven: Yale University Press.
Celenza, Christopher, 2004. *The Lost Italian Renaissance: Humanists, Historians, and Latin's Legacy*. Baltimore: Johns Hopkins University Press.
———, 2017. *Petrarch: Everywhere a Wanderer*. London: Reaktion Books.
Cervantes, Miguel de, 2003. *Don Quixote*. Trans. Edith Grossman. New York: Harper Collins.
Chambers, Ephraim, 1738. *Cyclopædia, or, An Universal Dictionary of Arts and Sciences . . . by E. Chambers, F.R.S. with the Supplement and Modern Improvements incorporated in One Alphabet. By Abraham Rees, D.D., F.R.S.* 2 vols. London: D. Midwinter et al. First published 1728.
Charlesworth, James H., ed., 1983. *The Old Testament Pseudepigrapha*. Vol. 1. *Apocalyptic Literature and Testaments*. New York: Doubleday.
———ed., 1985. *The Old Testament Pseudepigrapha*. Vol. 2. *Expansions of the "Old Testament" and Legends, Wisdom and Philosophical Literature, Prayers, Psalms and Odes, Fragments of Lost Judeo-Hellenistic Works*. New York: Doubleday.
Civil, Miguel, 1983. "Enlil and Ninlil: The Marriage of Sud." *Journal of the American Oriental Society* 103: 43–64.
Clark, Frederic, 2018. "Forgery, Misattribution, and a Case of Secondary Pseudonymity: Aethicus Ister's *Cosmographia* and Its Early Modern Multiplications." 74–98 in Stephens and Havens 2018.
Clark, M. J., 2016. *The Making of the* Historia scholastica, *1150–1200*. Toronto: Pontifical Institute of Medieval Studies.
Clarke, Arthur C., 1999. *Profiles of the Future: An Enquiry into the Limits of the Possible*. First published 1962.
Cluzan, Sophie, 2005. *De Sumer à Canaan: L'orient ancien et la Bible*. Paris: Editions du Seuil / Musée du Louvre.
Coleman, James K. "Forging Relations between East and West: The Invented Letters of Sultan Mehmed II." 118–34 in Stephens and Havens 2018.

Colonna, Francesco, 1999. *Hypnerotomachia Poliphili: The Strife of Love in a Dream*. Trans. Joscelyn Godwin. Reprint. London: Thames and Hudson, 2005. First published 1499 (Venice: Aldus Manutius).

Comestor, Peter, 2005. *Petri Comestoris Scolastica Historia: Liber Genesis*. Ed. Agneta Sylwan. Corpus Christianorum Continuatio Mediaevalis, 191. Turnhout: Brepols.

Conte, Gian Biagio, 1994. *Latin Literature: A History*. Trans. Joseph P. Solodow, rev. Don Fowler and Glenn W. Most. Reprint. Baltimore: Johns Hopkins University Press, 1999.

Cooper, Jerrold S., 2004. "Babylonian Beginnings: The Origin of the Cuneiform Writing System in Comparative Perspective." 71–99 in Stephen D. Houston, ed., *The First Writing: Script Invention as History and Process*. Cambridge: Cambridge University Press.

Copenhaver, Brian P., 1992, trans. *Hermetica: The Greek* Corpus Hermeticum *and the Latin* Asclepius. Cambridge: Cambridge University Press.

Crinitus, Petrus, 1955. *De honesta disciplina*. Ed. Carlo Angeleri. Rome: Fratelli Bocca.

Cristofani, Mauro, 1979. *The Etruscans: A New Investigation*. Trans. Brian Phillips. London: Orbis Publishing.

Curran, Brian, 2007. *The Egyptian Renaissance: The Afterlife of Egypt in Early Modern Italy*. Chicago: University of Chicago Press.

Curtius, Ernst Robert, 1953. *European Literature and the Latin Middle Ages*. Trans. Willard R. Trask. Reprint. New York: Harper and Row, 1963.

D'Alembert, Jean Le Rond, 1995. *Preliminary Discourse to the Encyclopedia of Diderot*. Trans. Richard N. Schwab. Chicago: University of Chicago Press.

Damrosch, David, 2006. *The Buried Book: The Loss and Recovery of the Great Epic of Gilgamesh*. New York: Henry Holt.

Dante Alighieri, 2011. *Paradiso*. Trans. Robert Durling, introd. and notes Ronald Martinez. New York: Oxford University Press.

———, 2013. *Vita nuova*. Trans. Anthony Mortimer. Reprint. Richmond, UK: Alma Classics.

Dares (Pseudo-), and (Pseudo-) Dictys, 1966. *The Trojan War*. New Edition. *The Chronicles of Dictys of Crete and Dares the Phrygian*. Trans. R. M. Frazer (Jr.). Reprint. Bloomington: Indiana University Press, 2019.

De Turris, Gianfranco, and Sebastiano Fusco, 1972. "I Libri che non esistono (e quelli che non dovrebbero esistere)." 149–210 in Jacques Bergier, *I libri maledetti*. Trans. Roberta Rambelli. Rome: Edizioni Mediterranee.

Decamp, L. Sprague, 1947. "The Unwritten Classics." *Saturday Review of Literature* 30.13: 7–8.

Defoe, Daniel, 2007. *An Essay upon Literature: Or, an Enquiry into the Original of Letters*. Reprint. Baltimore: Archangel Foundation. First published 1726 (London: John Bowles).

Derrida, Jacques, 1981. *Dissemination*. Trans. Barbara Johnson. Chicago: University of Chicago Press.

Desormeaux, Daniel, 2001. *La Figure du bibliomane. Histoire du livre et stratégie littéraire au XIXe Siècle*. Paris: Nizet.

Dibdin, Thomas Frognall, 1811. *Bibliomania: Or, Book Madness: A Bibliographical Romance in Six Parts. Illustrated with Cuts*. London: Printed for the author, and sold by Messrs. Longman, Hurst, Rees, Orme, and Brown.

Diodorus of Sicily, 1933–1967. *The Library of History*. 12 vols. Trans. C. H. Oldfather. Cambridge, MA: Harvard University Press.

Douglass, Frederick, 2001. *Narrative of the Life of Frederick Douglass, An American Slave*. Ed. John W. Blassingame et al. New Haven: Yale University Press.

Driver, Samuel Rolles, 1904. *The Book of Genesis*. London: Methuen.

Drogin, Marc, 1983. *Anathema! Medieval Scribes and the History of Book Curses*. Montclair: Allanheld and Schram.

———, 1989. *Biblioclasm: The Mythical Origins, Magic Powers, and Perishability of the Written Word*. Savage, MD: Rowman and Littlefield.

Dronke, Peter, et al., eds., 1997. *Alessandro nel Medioevo occidentale*. Verona: Fondazione Lorenzo Valla / Arnaldo Mondadori.

Drucker, Johanna, 1995. *The Alphabetic Labyrinth: The Letters in History and Imagination*. London: Thames and Hudson.

Eco, Umberto, 1980. *Il nome della rosa* (hardback of first Italian edition). Milan: Bompiani.

———, 1988. *Il pendolo di Foucault*. Milan: Bompiani. English version: *Foucault's Pendulum*. Trans. William Weaver. Orlando, FL: Harcourt, 2007.

———, 1993. *Misreadings*. Trans. William Weaver. New York: Harcourt, Brace, and Co.

———, 1994. *The Name of the Rose. Including Postscript to* The Name of the Rose. Trans. William Weaver. Reprint. Orlando, FL: Harcourt Brace.

———, 2005. "Il codice Temesvar." *Almanacco del Bibliofilo* 15: 131–43.

———, 2011. *Costruire il nemico e altri scritti occasionali*. Milan: Bompiani. Trans. Richard Dixon, as *Inventing the Enemy and Other Occasional Writings*, Boston: Houghton Mifflin Harcourt, 2012.

Eddy, Matthew Daniel, 2011. "The Prehistoric Mind as a Historical Artifact." *Notes and Records of the Royal Society* 65: 1–8.

Eliot, George (Mary Ann Evans), 2000. *Middlemarch*. An Authoritative Text, Backgrounds, Criticism. 2nd ed. Ed. Bert G. Hornback. New York: Norton.

Elliott, J. K., trans., 1993. *The Apocryphal New Testament. A Collection of Apocryphal Christian Literature in an English Translation*. Oxford: Clarendon.

Emmerich, Roland, dir., 2011. *Anonymous*. Culver City, CA: Columbia Pictures.

Enciclopedia Zanichelli, 1992. *Dizionario enciclopedico di arti, scienze, tecniche, lettere, filosofia, storia, geografia, diritto, economia*. Bologna, Italy: Zanichelli.

Encyclopaedia Britannica, 1911. 11th ed.
———, 1797. 3rd ed.
Encyclopédie 1751–1772. *Encyclopédie, ou dictionnaire raisonné des sciences, des arts et des métiers, par une société de gens de lettres*. 28 vols. Ed. Denis Diderot and Jean Le Rond D'Alembert. Paris: Chez Briasson, 1754–1772.
Equiano, Olaudah, 1789. *The Interesting Narrative of the Life of Olaudah Equiano, Written by Himself. With Related Documents*. 2nd ed. Ed. Robert J. Allison. Boston: Bedford; New York: St. Martins, 2007.
Erasmus, Desiderius, 2001. *The Adages of Erasmus*. Selected by William Barker. Toronto: University of Toronto Press.
Escobar, Eduardo A., 2019. "Number Culture: Old Babylonian Mathematics at Yale." 286–91 in Lassen et al. 2019.
Eustathius, 1825–1826. *Eustathii . . . commentarii ad Homeri Odyssea*. 2 vols. Ed. J. G. Stallbaum. Leipzig: J.A.G. Weigel.
Fabricius, Johann Albert, 1722. *Codex Pseudepigraphus Veteris Testamenti, collectus, castigatus, testimoniisque, censuris et animadversionibus illustratus*. 2nd ed. 2 vols. Hamburg: Theodori Christoph Felginer.
———, 1790–1812. *Bibliotheca graeca, sive notitia scriptorum veterum graecorum quorumcumque monumenta integra aut fragmenta edita exstant tum plerumque e mss. ac deperditis*. 4th ed. 14 vols. Ed. Gottlieb Christophorus Harles. Hamburg: Bohn.
Festugière, A. M. J., 1949–1954. *La révélation d'Hermès Trismegiste*. 4 vols. Vol. 1 (1950), *L'astrologie et les sciences occultes. Avec un appendice sur l'Hermétisme arabe*. Paris: Lecoffre, J. Gabalda.
Fischer, Steven Roger, 2001. *A History of Writing*. London: Reaktion Books.
———, 2003. *A History of Reading*. London: Reaktion Books.
Folengo, Teofilo, 1927. "Le prefazioni." 2:274–84 in *Le maccheronee*. 2nd ed. 2 vols. Ed. Alessandro Luzio. Bari: Laterza.
Fontana, Domenico, 1590. *Della Trasportatione Dell'obelisco Vaticano et delle Fabriche Di Nostro Signore Papa Sisto V*. Rome: Domenico Basa.
Foster, Benjamin R., 1995. *From Distant Days: Myths, Tales, and Poetry of Ancient Mesopotamia*. Bethesda, MD: CDL Press.
Frahm, Eckart, 2019. "Scholars, Diviners, Learned Kings." 157–67 in Lassen et al. 2019.
France, Anatole, 1922. *The Revolt of the Angels*. Trans. Emilie [Mrs. Wilfred] Jackson. London: John Lane, The Bodley Head.
Franklin-Brown, Mary, 2012. *Reading the World: Encyclopedic Writing in the Scholastic Age*. Chicago: University of Chicago Press.
Freccero, John, 1975. "The Fig Tree and the Laurel." *Diacritics* 5.1: 34–40.
Freculf of Lisieux, 2002. *Frechulfi Lexoviensis Episcopi Opera Omnia*. Ed. Michael I. Allen. 2 vols. Turnhout: Brepols.

Freeman, Arthur, 2014. *Bibliotheca Fictiva: A Collection of Books & Manuscripts Relating to Forgery, 400 BC–AD 2000*. London: Bernard Quaritch.

Gaisser, Julia Haig, 2008. *The Fortunes of Apuleius and the Golden Ass: A Study in Transmission and Reception*. Princeton: Princeton UP.

Garin, Eugenio, 1961. "La nuova scienza e il simbolo del 'libro.'" 451–65 in Eugenio Garin, *La cultura filosofica del Rinascimento italiano*. Reprint. Florence: Sansoni, 1979.

———, 1976. *Lo zodiaco della vita: la polemica sull'astrologia del Trecento al Cinquecento*. Bari: Laterza.

———, ed., 1952. *Prosatori latini del Quattrocento*. Milan: Ricciardi.

Gates, Henry Louis, 1988. *The Signifying Monkey: A Theory of Afro-American Literary Criticism*. New York: Oxford University Press.

George, Andrew, 2003, trans. *The Epic of Gilgamesh: The Babylonian Epic Poem and Other Texts in Akkadian and Sumerian*. London: Penguin. First published 1999.

Getto, Giovanni, 1969. *Storia delle storie letterarie*. 2nd rev. ed. Florence: Sansoni.

Gillispie, Charles Coulston, 1951. *Genesis and Geology: A Study in the Relations of Scientific Thought, Natural Theology, and Social Opinion in Great Britain, 1790–1850*. Reprint. New York: Harper Torchbooks, 1959.

Ginzburg, Louis, 1998. *The Legends of the Jews*. 7 vols. Baltimore: Johns Hopkins University Press. First published 1909–1938.

Givens, Terry L., 2002. *By the Hand of Mormon: The American Scripture that Launched a New World Religion*. New York: Oxford University Press.

Glaber, Rodulfus, 1989a. *Historiarum libri quinque. The Five Books of the Histories*. Ed. John France. Oxford: Clarendon.

———, 1989b. *Cronache dell'anno mille (Storie)*. Ed. Guglielmo Cavallo and Giovanni Orlandi. Milan: Fondazione Lorenzo Valla / Arnaldo Mondadori Editore.

Gliozzi, Giuliano, 1976. *Adamo e il nuovo mondo. La nascita dell'antropologia come ideologia coloniale dalle genealogie bibliche alle teorie razziali*. Florence: La Nuova Italia.

Godfrey of Viterbo, 1559. *Pantheon, sive universitatis libri qui chronici appellantur*. Ed. B. J. Herold. Basel: ex Officina Iacobi Parci.

———, 1872. *Pantheon*. Ed. Georg Waitz. Monumenta Germaniae Historica . . . Scriptorum. Vol. 22. Hannover, Germany: Impensis Bibliopolii Aulici Hahniani.

———, ca. 1350–1400. *Pantheon*. Bodleian Library MS Lat. hist. c. 1.

Godzich, Wlad, and Jeffrey Kittay, 1987. *The Emergence of Prose: An Essay in Prosaics*. Minneapolis: University of Minnesota Press.

Goldberg, Jonathan, 1990. *Writing Matter: From the Hands of the Renaissance*. Stanford, CA: Stanford University Press.

Goodrich, S. G., 1852. *A History of All Nations, from the Earliest Periods to the Present Time, or Universal History*. Auburn, NY: H. W. Derby.

Goodrum, Matthew R., 2004. "Prolegomenon to a History of Paleoanthropology: The Study of Human Origins as a Scientific Enterprise, Part I: Antiquity to the Enlightenment"; "Part II: Enlightenment to the Twentieth Century." Evolutionary Anthropology 13.5: 172–80; 13.6: 224–33.

Gordan, Phyllis Walter Goodhart, 1974. *Two Renaissance Book Hunters: The Letters of Poggius Bracciolini to Nicolaus de Niccolis*. New York: Columbia University Press.

Grafton, Anthony, 1990. *Forgers and Critics: Creativity and Duplicity in Western Scholarship*. Princeton: Princeton University Press.

———, 1993. *Joseph Scaliger: A Study in the History of Classical Scholarship. II. Historical Chronology*. Reprint. Oxford: Clarendon, 2005.

———, 1999. *The Footnote: A Curious History*. Cambridge, MA: Harvard University Press.

———, 2018. "Annius of Viterbo as Student of the Jews: The Sources of His Information." 147–69 in Stephens and Havens 2018.

Grafton, Anthony, Glenn W. Most, and Salvatore Settis, eds., 2010. *The Classical Tradition*. Cambridge, MA: Belknap Press.

Graves, Robert, 1960. *The Greek Myths*. 2nd rev. ed. 2 vols. New York: Penguin.

Gundel, Wilhelm, 1936. *Neue astrologische Texte des Hermes Trismegistos. Funde und Forschungen auf dem Gebiet der antiken Astronomie und Astrologie*. Munich: Verlag der Bayerischen Akademie der Wissenschaften.

Hartman, Geoffrey H., and Sanford Budick, eds., 1986. *Midrash and Literature*. New Haven: Yale University Press.

Havelock, Eric A., 1963. *Preface to Plato*. Cambridge, MA: Belknap Press.

Havens, Earle A., 2016. "Catastrophe? Species and Genres of Literary and Historical Forgery." viii–41 in Havens 2016.

———, 2018. "Babelic Confusion." 33–73 in Stephens and Havens 2018.

———, ed. 2016. *Fakes, Lies, and Forgeries: Rare Books and Manuscripts from the Bibliotheca Fictiva Collection*. Baltimore: Sheridan Libraries, Johns Hopkins University.

Heliodorus of Emesa, 1989. *An Ethiopian Story* [*The Ethiopian Romance*]. Trans. J. R. Morgan. 349–588 in Reardon 1989.

Herodotus, 2007. *The Landmark Herodotus. The Histories*. Ed. Robert B. Strassler. Reprint. New York: Anchor Books, 2009.

Hesiod, 1983. *Theogony*. Trans. Apostolos N. Athanassakis. Baltimore: Johns Hopkins University Press.

Heumann, Christoph August, 1711. *De libris anonymis ac pseudonymis*. Jena, Germany: Apud Io. Felicem Bielckium.

Hibbert, Christopher, 1975. *The House of Medici: Its Rise and Fall*. New York: Morrow.
Hoffman, Adina, Peter Cole, and Solomon Schechter, 2011. *Sacred Trash: The Lost and Found World of the Cairo Geniza*. New York: Nextbook.
Homer, 1951. *The Iliad*. Trans. Richmond Lattimore. Reprint. Chicago: University of Chicago Press, 1970.
———, 1965. *The Odyssey*. Trans. Richmond Lattimore. Reprint. New York: Harper Torchbooks, 1967.
Hooker, J. T., 1990. *Reading the Past: Ancient Writing from Cuneiform to the Alphabet*. Introd. J. T. Hooker. Berkeley: University of California Press; London: British Museum.
Horace, 1929. *Satires, Epistles, and Ars poetica*. Trans. H. Rushton Fairclough. Reprint. Cambridge, MA: Harvard University Press, 1991.
———, 2004. *Odes and Epodes*. Trans. Niall Rudd. Cambridge, MA: Harvard University Press.
Hughes, Albert, and Allen Hughes, 2010. *The Book of Eli*. Culver City, CA: Sony Pictures.
Hugo, Hermann, 1617. *De prima scribendi origine et universa rei litterariae antiquitate*. Antwerp: Plantin.
Huizinga, Johan, 1949. *Homo Ludens: A Study of the Play-Element in Culture*. London: Routledge and K. Paul.
Iamblichus, 1497. *Jamblichus de mysteriis Aegyptiorum, Chaldaeorum, Assyriorum*. Trans. Marsilio Ficino. Facsimile reprint. Frankfurt am Main: Minerva, 1972.
———, 2004. *De mysteriis*. Trans. Emma C. Clarke, John M. Dillon, and Jackson P. Hershbell. Leiden: Brill.
Idel, Moshe, 1988. *Kabbalah: New Perspectives*. New Haven: Yale University Press.
Inghirami, Curzio, 1637. *Ethruscarvm antiquitatum fragmenta: Quibus urbis Romae aliarumque gentium primordia, mores, et res gestae indicantur a Curtio Inghiramio reperta Scornelli propè Vulterram*. Francofurti: Anno Salutis. Ethrusco verò MMMMCCCCXCV [4495].
———, 1645. *Discorso di Curzio Inghirami sopra l'opposizioni fatte all'antichità Toscane: diuiso in dodici trattati*. Florence: Amadore Massi e Lorenzo Landi.
Isidore of Seville, 2003. *Isidori Hispalensis Cronica*. Ed. José Carlos Martin. Turnhout, Belgium: Brepols.
———, 2004. *Etymologías* [Latin and Spanish]. Ed. José Oroz Reta, Manuel A. Marcos Casquero, introd. Manuel C. Diaz y Diaz. Reprint. Madrid: Biblioteca de Autores Cristianos, 2009.
———, 2006. *The Etymologies of Isidore of Seville*. Trans. Stephen A. Barney, W. J. Lewis, J. A. Beach, and Oliver Berghof. Cambridge: Cambridge University Press.

Iversen, Erik, 1961. *The Myth of Egypt and Its Hieroglyphs in European Tradition*. Reprint, with new preface by the author. Princeton: Princeton University Press, 1993.

Jacoby, Felix, et. al., eds., 1923–. *Die Fragmente der griechischen Historiker*. 4 vols. Berlin: Weidmann; Leiden: Brill.

Jasnow, Richard, 2011. "'Caught in the Web of Words': Remarks on the Imagery of Writing and Hieroglyphs in the Book of Thoth." *Journal of the American Research Center in Egypt* 47: 297–317.

Jasnow, Richard, and Karl-Theodor Zauzich, 2014. *Conversations in the House of Life: A New Translation of the Ancient Egyptian Book of Thoth*. Wiesbaden: Harrassowitz.

Jerome, Saint, 1933. *Select Letters of St. Jerome*. Trans. F. A. Wright. Reprint. Cambridge, MA: Harvard University Press, 1991.

Jiménez, Enrique, 2019. "Mesopotamian Literature." 149–55 in Lassen et al. 2019.

Josephson, Aksel G. S., 1917. "The Literature of the Invention of Printing: A Bibliographical Study." *Papers of the Bibliographical Society of America* 11.1.

Josephus, Flavius, 1926. *The Life* [and] *Against Apion*. Trans. H. St. J. Thackeray. Reprint. Cambridge, MA: Harvard University Press, 1997.

———, 1930. *Jewish Antiquities*, Books 1–3. Trans. H. St. J. Thackeray. Reprint. Cambridge, MA: Harvard University Press, 2001.

———, 1965. *Jewish Antiquities*, Books 18–19. Trans. L. H. Feldman. Reprint. Cambridge, MA: Harvard University Press, 2000.

Kaske, R. E., 1971. "*Beowulf* and the Book of Enoch." *Speculum* 46: 421–31.

Kasser, Rodolphe, Marvin Meyer, and Gregor Wurst, 2006. *The Gospel of Judas, from Codex Tchacos*. With additional Commentary by Bart D. Ehrman. Washington, DC: National Geographic Society.

Kenseth, Joy, ed., 1991. *The Age of the Marvelous*. Hanover, NH: Hood Museum of Art, Dartmouth College.

Kircher, Athanasius, 1650. *Obeliscus Pamphilius, hoc est, interpretatio nova*. Rome: Ludovico Grignani.

———, 1675. *Arca Noë*. Amsterdam: Apud Joannem Janssonium à Waesberge.

Kirk, G. S., and J. E. Raven, 1969. *The Presocratic Philosophers. A Critical History with a Selection of Texts*. Cambridge: Cambridge University Press.

Knox, Bernard, 1996. "Introduction." 3–64 in Homer, *The Odyssey*. Trans. Robert Fagles. Reprint. New York: Penguin, 1997.

König, Jason, Katerina Oikonomopoulou, and Greg Wolf, eds., 2013. *Ancient Libraries*. Cambridge: Cambridge University Press.

Kramer, Samuel Noah, 1963. *The Sumerians: Their History, Culture and Character*. Chicago: University of Chicago Press.

———, 1967. "Reflections on the Mesopotamian Flood." *Expedition* 9.4: 12–18.

Lacroix, Paul, and Gustave Brunet, 1862. *Catalogue de la bibliothèque de l'abbaye de Saint-Victor au XVIe siècle, rédigé par François Rabelais, commenté par le Bibliophile Jacob, et suivi d'un Essai sur les bibliothèques imaginaires par Gustave Brunet*. Paris: Techener.

Lambert, Malcolm, 1992. *Medieval Heresy: Popular Movements from the Gregorian Reform to the Reformation*. Oxford: Blackwell.

Lassen, Agnete W., Eckart Frahm, and Klaus Wagensonner, eds. 2019. *Ancient Mesopotamia Speaks: Highlights of the Yale Babylonian Collection*. New Haven: Yale University Press.

Lawlor, H. J., 1897. "Early Citations from the Book of Enoch." *Journal of Philology* 25: 164–225.

Layton, Bentley, ed., 1987. *The Gnostic Scriptures: A New Translation with Annotations and Introductions*. Anchor Bible Reference Library. Reprint. New York: Doubleday, 1995.

Leone Ebreo (Judah Leo Abravanel), 2009. *Dialogues of Love*. Trans. Cosmos Damian Bacich and Rossella Pescatori. Toronto: University of Toronto Press.

Lepore, Jill, 2020. "Don't Come Any Closer: What's at Stake in Our Fables of Contagion?" *New Yorker*, March 30: 22–25.

Lewis, Charlton D., and Charles Short, 1969. *A Latin Dictionary*. Oxford: Clarendon. First published 1879.

Lichtheim, Miriam, 1973. *Ancient Egyptian Literature*. Vol. 1, *Old and Middle Kingdoms*. Reprint. Foreword by Antonio Loprieno. Berkeley: University of California Press, 2006.

———, 1976. *Ancient Egyptian Literature*. Vol. 2, *The New Kingdom*. Reprint. Foreword by Hans-W. Fischer-Elfert. Berkeley: University of California Press, 2006.

Lipsius, Justus, 2017. *Ancient Libraries and Renaissance Humanism: The* De bibliothecis *of Justus Lipsius*. Ed. Thomas Hendrickson. Leiden: Brill.

Liverani, Mario, 2004. "Adapa, Guest of the Gods." 2–23 in his *Myth and Politics in Ancient Near Eastern Historiography*. Ed. Zainab Bahrani and Marc Van De Mieroop. Ithaca: Cornell University Press.

Livy, 1964–1967. *Livy, with an English Translation*. 14 vols. Trans. E. T. Sage and A. C. Schlesinger. Loeb. Cambridge, MA: Harvard University Press.

Lomeier, Johann, 1680. *De Bibliothecis liber singularis. Editio secunda, priori multo auctior, et addito rerum indice locupletior*. Utrecht: ex officina Johannis Ribbii. First published 1669.

Lord, Albert B., 1960. *The Singer of Tales*. Reprint. Cambridge, MA: Milman Parry Collection of Oral Literature; Washington, DC: Center for Hellenic Studies, 2019.

Lotito, Mark A., 2019. *The Reformation of Historical Thought*. Leiden: Brill.

Lucan, 1928. *Civil War (Pharsalia)*. Trans. J. D. Duff. Reprint. Cambridge, MA: Harvard University Press, 2014.
Lucian of Samosata, 1989. *A True Story*. Trans. B. P. Reardon. 619–49 in Reardon 1989.
Lucretius, 1995. *On the Nature of Things*. Trans. Anthony M. Esolen. Baltimore: Johns Hopkins University Press.
Lutz, Cora E., 1956. "Remigius' Ideas on the Origin of the Seven Liberal Arts." *Medievalia et Humanistica* 10: 32–49.
———, 1979. *The Oldest Library Motto and Other Library Essays*. Hamden, CT: Archon Books.
Mabillon, Jean, 1681. *De re diplomatica libri VI*. Paris: sumtibus viduae L. Billaine.
Mackey, Albert Gallatin, 1996. *The History of Freemasonry: Its Legendary Origins*. New York: Gramercy Books. First published 1906 [?].
Mader, Joachim Johann, 1702–1705. *De bibliothecis atque archivis*. 3 vols. Ed. Johann Andreas Schmidt. Helmstadt, Germany. First published 1666.
Manuel, Frank E., 1959. *The Eighteenth Century Confronts the Gods*. Reprint: New York: Athenaeum.
Margalit, Ruth, 2020. "Built on Sand: King David's Story Has Been Told for Millennia: Archaeologists Are Still Fighting Over Whether It's True." *New Yorker*, June 29, 2020, 42–51.
Marincola, John, ed., 2011. *A Companion to Greek and Roman Historiography*. Oxford: Blackwell.
Marius, Richard, 1999. *Martin Luther: The Christian between God and Death*. Cambridge, MA: Harvard University Press.
Martin, Henri-Jean, 1994. *The History and Power of Writing*. Trans. Lydia G. Cochrane. Chicago: University of Chicago Press.
Massey, William, 1763. *The Origin and Progress of Letters: An Essay*. London: J. Johnson.
Mauriès, Patrick, 2002. *Cabinets of Curiosities*. New York: Thames and Hudson.
Mayor, Adrienne, 2011. *The First Fossil Hunters: Dinosaurs, Mammoths, and Myth in Greek and Roman Times*. Princeton: Princeton University Press.
Mazzotta, Giuseppe, 1999. *The New Map of the World: The Poetic Philosophy of Giambattista Vico*. New Haven: Yale University Press.
McLuhan, Marshall, 1962. *The Gutenberg Galaxy: The Making of Typographic Man*. Toronto: University of Toronto Press.
———, 1964. *Understanding Media: The Extensions of Man*. New York: McGraw-Hill.
Melanchthon, Philipp, 1532. See Carion 1532.
Melanchthon, Philipp, and Caspar Peucer, 1580. *Chronicon Carionis, expositum et auctum . . . a Philippo Melanthone et Casparo Peucero*. Wittenberg: Haeredes Johannis Cratonis. First published 1572.

Mercati, Michele, 1981. *Gli obelischi di Roma*. Ed. Gianfranco Cantelli. Bologna, Italy: Cappelli. First published 1589.
Minois, Georges, 2012. *The Atheist's Bible: The Most Dangerous Book That Never Existed*. Trans. Lys Ann Weiss. Chicago: University of Chicago Press.
Mitchell, Charles, 1960. "Archaeology and Romance in Renaissance Italy." 455–83 in *Italian Studies: A Tribute to the Late Cecilia M. Ady*. Ed. E. F. Jacob. London: Fraser and Fraser.
Montfaucon, Bernard de, 1708. *Palaeographia Graeca*. Paris: Apud Ludovicum Guerin.
Montgomery, John Warwick, 1962. *A Seventeenth-Century View of European Libraries: Lomeier's* De bibliothecis, *Chapter X*. Berkeley: University of California Publications in Librarianship, vol. 3.
More, Sir Thomas, 1964. *Utopia*. Ed. Edward Surtz, SJ. Reprint. New Haven: Yale University Press, 1969.
Morhof, Daniel Georg, 1747. *Polyhistor, literarius, philosophicus et practicus*. 4th ed. 3 vols. Photographic reprint. Aalen: Scientia, 1970.
Müller, Karl, ed., 1848–1874. *Fragmenta Historicorum Graecorum*. 5 vols. Paris: Firmin Didot.
Nallino, C. A., 1944. *Raccolta di scritti editi e inediti*. Vol. 5. Ed. Maria Nallino. Rome: Istituto per l'Oriente.
Neander (Neumann), Michael, 1565. *Graecae linguae erotemata*. Basel: per Ioannem Oporinum.
Netz, Reviel, and William Noel, 2007. *The Archimedes Codex: How a Medieval Prayer Book Is Revealing the True Genius of Antiquity's Greatest Scientist*. Philadelphia: Da Capo Press.
Nicols, William, 1711. *De literis inventis, libri sex*. London: Henry Clement.
Nider, Johannes, 1479. *Preceptorium Divinae Legis*. Augsburg: Wiener.
Ong, Walter, 2012. *Orality and Literacy: The Technologizing of the Word*. Thirtieth Anniversary Edition. Abingdon, UK: Routledge.
Ovid, 1916. *The Metamorphoses*. Trans. Frank Justus Miller. Reprint: Cambridge, MA: Harvard University Press, 1977.
Pagels, Elaine, 1979. *The Gnostic Gospels*. Reprint. New York: Vintage Books, 1989.
———, 2005. *Beyond Belief: The Secret Gospel of Thomas*. New York: Random House.
Pallis, Svend Aage, 1954. "Early Exploration in Mesopotamia, with a List of the Assyro-Babylonian Cuneiform Texts Published before 1851." *Det Kongelige Danske Videnskabernes Selskab. Historisk-filologiske Meddelelser* 33.6: 3–58.
Parsons, Peter, 2007. *City of the Sharp-Nosed Fish: Greek Papyri beneath the Egyptian Sand Reveal a Long-Lost World*. London: Phoenix.
Patrologia Graeca (*Patrologia cursus completus series graeca*), 1857–1866. General editor Jacques-Paul Migne. 161 vols. Paris: Imprimerie Catholique.

Patrologia Latina (*Patrologia cursus completus series latina*), 1841–1855. General editor Jacques-Paul Migne. 221 vols. Paris: Imprimerie Catholique.
Paul, Saint, 1972. *The Writings of St. Paul*. Ed. Wayne A. Meeks. New York: W. W. Norton.
Peignot, Gabriel, 1823. *Manuel du bibliophile*. 2 vols. Paris: V. Lagier.
Pelikan, Jaroslav, 1984. *The Christian Tradition: A History of the Development of Doctrine*. Vol. 4, *Reformation of Church and Dogma (1300–1700)*. Chicago: University of Chicago Press.
———, 1996. *The Reformation of the Bible and the Bible of the Reformation*. With Valerie R. Hotchkiss and David Price. New Haven: Yale University Press.
Pérez-Reverte, Arturo, 1996. *The Club Dumas*. Trans. Sonia Soto. Reprint. Orlando, FL: Harcourt, 2006.
Peters, F. E., 2007. *The Voice, the Word, the Books: The Sacred Scripture of the Jews, Christians, and Muslims*. London: British Library.
Petrarch [Francesco Petrarca], 1953. "Petrarch's Coronation Oration." Ed. Ernest Hatch Wilkins. *PMLA* 68: 1245–49.
———, 1975. *Rerum Familiarium Libri (Letters on Familiar Matters)*. 3 vols. Vol. 1, Albany: State University of New York Press; vols. 2–3, Baltimore: Johns Hopkins University Press.
Photius, 1612. *Photii Myriobiblon sive bibliotheca*. Ed. David Hoeschel. Geneva: Oliva Pauli Stephani.
———, 1994. *The Bibliotheca, a selection translated, with notes*. Ed. Nigel Wilson. London: Duckworth.
Pingree, David, 1968. *The Thousands of Abu Ma'shar*. London: Warburg Institute.
Placcius, Vincent, 1708. *Theatrum anonymorum et pseudonymorum*. Hamburg: Sumptibus Viduae Gothofredi Libernickelii, Typis Spieringianis.
Plato, 1961. *The Collected Dialogues*. Ed. Edith Hamilton and Huntington Cairns. Bollingen Series 71. Princeton: Princeton University Press.
———. *Critias*. Trans. A. E. Taylor. 1212–24 in Plato 1961.
———, 1995. *Phaedrus*. Trans. Alexander Nehamas and Paul Woodruff. Indianapolis: Hackett.
———, 2000. *Timaeus*. Trans. Donald J. Zehl. Indianapolis: Hackett.
Plessner, M., 1954. "Hermes Trismegistus and Arab Science." *Studia Islamica* (Paris) 2: 45–59.
Pliny the Elder, 1472. *Caii Plinii Secundi Naturalis historiae libri*. Venice: Nicolaus Jenson.
———, 1942. *Natural History*. Books 3–7. Trans. H. Rackham. Reprint. Cambridge, MA: Harvard University Press, 1999.
———, 1945. *Natural History*. Books 12–16. Trans. H. Rackham. Reprint. Cambridge, MA: Harvard University Press, 2000.

Pliny the Younger, 1969. *Letters*. Vol. 1: Books 1–7. Trans. Betty Radice. Loeb Classical Library 55. Cambridge, MA: Harvard University Press.
Podany, Amanda H., 2019. "Kings and Conflict." 57–71 in Lassen et al. 2019.
Postel, Guillaume, 1538a. *Grammatica Arabica*. Paris: P. Gromors.
———, 1538b. *Linguarum duodecim characteribus differentium alphabetum introductio*. Paris: D. Lescuyer.
Potocki, Jan, 1995. *The Manuscript Found in Saragossa*. Trans. Ian Maclean. New York: Viking.
Preminger, Alex, and T. V. F. Brogan, eds., 1993. *The New Princeton Encyclopedia of Poetry and Poetics*. Princeton: Princeton University Press.
Quintilian, 1920. *The* Institutio Oratoria *of Quintilian*. Trans. H. E. Butler. Reprint. Cambridge, MA: Harvard University Press, 1989.
Rabelais, François, 1741. *Oeuvres, avec des remarques historiques et critiques de Mr. Le Duchat*. 3 vols. Amsterdam: J. F. Bernard.
———, 2006. *Gargantua and Pantagruel*. Trans. M. A. Screech. London: Penguin.
Reardon, B. P., ed., 1989. *Collected Ancient Greek Novels*. Berkeley: University of California Press.
Redford, Bruce, 2008. *Dilettanti: The Antic and the Antique in Eighteenth-century England*. Los Angeles: J. Paul Getty Museum.
Reimmann, Jacob Friderich, 1709. *Versuch einer Einleitung in die* Historiam literariam antediluvianam, *d. i. In die Geschichte der Gelehrsamkeit und derer Gelehrten vor der Sündfluth*. Halle: Renger.
———, 1728. *Idea systematis antiquitatis literariae generalioris et specialioris desiderati adhuc in republica eruditorum literaria*. Hildesheim: sumtibus Ludolphi Schöderi.
Reynolds, L. D., and N. G. Wilson, 2013. *Scribes and Scholars: A Guide to the Transmission of Greek and Latin Literature*. 4th ed. Oxford: Oxford University Press.
Richard of Bury, 1960. *The Philobiblon*. Trans. E. C. Thomas. Ed. Michael Maclagan. Oxford: Blackwell.
Richardson, Ernest Cushing, 1914. *The Beginnings of Libraries*. Princeton: Princeton University Press.
Roberts, J. M., 1972. *The Mythology of the Secret Societies*. Reprint. Frogmore, UK: Paladin.
Robinson, Andrew, 2007. *The Story of Writing: Alphabets, Hieroglyphs, & Pictograms*. 2nd ed. London: Thames and Hudson.
———, 2012. *Cracking the Egyptian Code: The Revolutionary Life of Jean-François Champollion*. Oxford: Oxford University Press.
Robson, Eleanor, 2013. "Reading the Libraries of Assyria and Babylonia." 38–56 in König et al. 2013.

Rocca, Angelo, 1591. *Bibliotheca Apostolica Vaticana a Sixto V. Pont. Max. in splendidiorem, commodioremq. locvm translata, et a fratre Angelo Roccha a Camerino . . . commentario variarvm artivm, ac scientiarum materijs curiosis, ac difficillimis, scituque dignis refertissimo, illustrata. Ad S. D. N. Gregorivm* XIV. Rome: Typographia Apostolica Vaticana.
Rodenbeck, John, 2004. "Travelers from an Antique Land: Shelley's Inspiration for 'Ozymandias.'" *Alif: Journal of Comparative Poetics* 24: 121–48.
Roper, Lyndal, 2016. *Martin Luther: Renegade and Prophet*. London: The Bodley Head.
Rossi, Azariah ben Moses dei, 2001. *The Light of the Eyes*. Trans. Joanna Weinberg. New Haven: Yale University Press.
Rossi, Paolo, 1984. *The Dark Abyss of Time: The History of the Earth and the History of Nations from Hooke to Vico*. Trans. Lydia G. Cochrane. Chicago: University of Chicago Press.
———, 1999. *Le sterminate antichità e nuovi saggi vichiani*. Scandicci, Florence: La Nuova Italia.
———, 2000. *Logic and the Art of Memory: The Quest for a Universal Language*. Trans. Stephen Clucas. Chicago: University of Chicago Press.
Rowland, Ingrid D., 2004. *The Scarith of Scornello: A Tale of Renaissance Forgery*. Chicago: University of Chicago Press.
Rowley-Conwy, Peter, 2007. *From Genesis to Prehistory: The Archaeological Three Age System and Its Contested Reception in Denmark, Britain, and Ireland*. Oxford Studies in the History of Archaeology. Oxford: Oxford University Press.
Ruska, J., 1926. "Arabische Nachrichten über Hermes Trismegistos und die Verwahrung seiner Bücher in Pyramiden und Schatzkammern." 61–68 in *Tabula Smaragdina, ein Beitrag zur Geschichte der hermetischen Literatur*. Heidelberg: Carl Winter's Universitätsbuchhandlung.
Ryck, Theodor, 1684. *Dissertatio de primis Italiae coloniis et Aeneae adventu*. Leiden: Apud P. Vander Aa.
———, 1692. *Lucae Holstenii, notae et castigationes in Stephanum Byzantium de urbibus. Omnia ex recensione Theodori Rykii*. Leiden: Apud P. Vander Aa.
Sabbadini, Remigio, 1967. *Le scoperte dei codici latini e greci ne' secoli XIV e XV*. 2 vols. Ed. Eugenio Garin. Florence: Sansoni. First published 1905–1914.
Said, Edward W., 2003. *Orientalism*. 25th Anniversary ed. New York: Vintage Books.
Sallbach [Saalbach], Christian, 1705. *De libris veterum schediasma*. Gryphiswaldiae: Starckii.
Sandys, John Edwin, 1967. *A History of Classical Scholarship*. 3 vols. Vol. 2, *From the Revival of Learning to the End of the Eighteenth Century (in Italy, France, England, and the Netherlands)*. Reprint. New York: Hafner. First published 1903–1908.

Schmandt-Besserat, Denise, 1992. *How Writing Came About*. Reprint. Austin: University of Texas Press, 1996.
———, 2007. *When Writing Met Art: From Symbol to Story*. Austin: University of Texas Press.
Schmitt, Charles B., 1966. "Perennial Philosophy: From Agostino Steuco to Leibniz." *Journal of the History of Ideas* 27: 505–32.
Scholem, Gershom, 1974. *Kabbalah*. Reprint. New York: Meridian, 1978.
Schoep, Ilse, 2018. "Building the Labyrinth: Arthur Evans and the Construction of Minoan Civilization." *American Journal of Archaeology* 122.1: 5–32.
Schreckenberg, Heinz, 1972. *Die Flavius-Josephus-Tradition in Antike und Mittelalter*. Leiden: Brill.
Schwarz, Christian Gottlieb, 1705–1717. *Disputationes de ornamentis librorum apud veteres*. Altdorf and Leipzig: various publishers.
Secret, François, 1959. "La Kabbale chez Du Bartas et son commentateur Claude Duret." *Studi Francesi* 7: 1–11.
Self, Will, 2006. *The Book of Dave: A Revelation of the Recent Past and the Distant Future*. New York: Bloomsbury.
Shelley, Mary Wollstonecraft, 2021. *The Last Man*. The Project Guttenberg eBook, https://www.gutenberg.org/files/18247/18247-h/18247-h.htm.
Sider, David, 2005. *The Library of the Villa dei Papiri at Herculaneum*. Los Angeles: J. Paul Getty Museum.
Sixtus of Siena [Fra' Sisto Senese, OP], 1742. *Bibliotheca sancta*. 2 vols. Ed. P. T. Milante. Naples. First published 1566.
Smith, Joseph, 1830. *The Book of Mormon: An Account Written by the Hand of Mormon: Upon Plates Taken from the Plates of Nephi*. Palmyra: Printed by E. B. Grandin, for the author.
———, 1981. *The Book of Mormon. An Account Written by the Hand of Mormon upon Plates Taken from the Plates of Nephi*. Reprint. Salt Lake City, UT: Church of Jesus Christ of Latter-Day Saints, 1986.
Speyer, Wolfgang, 1970. *Bücherfunder in der Glaubenswerbung der Antike. Mit einem Ausblick auf Mittelalter und Neuzeit*. Göttingen: Vandenhoeck & Ruprecht.
———, 1971. *Die Literarische Fälschung im Heidnischen und Christlichen Altertum: Ein Versuch Ihrer Deutung*. Munich: C. H. Beck'sche Verlagsbuchhandlung.
Stephens, Walter, 1984. "The Etruscans and the Ancient Theology in the Works of Annius of Viterbo." 309–22 in *Umanesimo a Roma nel Quattrocento*, ed. Paolo Brezzi and Maristella de Panizza Lorch. Rome: Instituto di Studi Romani; New York: Barnard College.
———, 1989. *Giants in Those Days: Folklore, Ancient History, and Nationalism*. Lincoln: University of Nebraska Press.

———, 1993. "Tasso's Heliodorus and the World of Romance." 67–87 in *In Search of the Ancient Novel*, ed. James Tatum. Baltimore: Johns Hopkins University Press.
———, 2004. "When Pope Noah Ruled the Etruscans: Annius of Viterbo and His Forged *Antiquities*, 1498." 201–23 in *Studia Humanitatis: Essays in Honor of Salvatore Camporeale. MLN* 119. 1, *Italian Issue*, Special Supplement.
———, 2005. "*Livres de haulte gresse*: Bibliographic Myth from Rabelais to Du Bartas." 60–83 in *La Littérature engagée aux XVIe et XVIIe siècles: Études en l'honneur de Gérard Defaux (1937–2004)*, ed. Samuel Junod, Florian Preisig, Frédéric Tinguely. *MLN* 120.1, *Italian Issue*, Special Supplement.
———, 2011. "Complex Pseudonymity: Annius of Viterbo's Multiple Persona Disorder." *MLN* 126.4: 689–708.
———, 2013. "From Berossos to Berosus Chaldæus: The Forgeries of Annius of Viterbo and Their Fortune." 277–89 in *The World of Berossos*. Proceedings of the 4th International Colloquium on The Ancient Near East between Classical and Ancient Oriental Traditions, Hatfield College, Durham University, UK, July 7–9, 2010, ed. Johannes Haubold, Giovanni B. Lanfranchi, Robert Rollinger, and John Steele. Wiesbaden: Harrasowitz Verlag.
———, 2016. "Discovering the Past: The Renaissance Arch-Forger and His Legacy." 66–84 in *Fakes, Lies, and Forgeries: Rare Books and Manuscripts from the Arthur and Janet Freeman Bibliotheca Fictiva Collection*. Ed. Earle Havens. Baltimore: Sheridan Libraries, Johns Hopkins University.
———, 2018. "Exposing the Arch-Forger: Annius of Viterbo's First Master Critic." 170–90 in Stephens and Havens 2018.
Stephens, Walter, and Earle A. Havens, eds., 2018. *Literary Forgery in Early Modern Europe 1450–1800*. Baltimore: Johns Hopkins University Press.
Sterling, Gregory E., 2011. "The Jewish Appropriation of Hellenistic Historiography." 231–43 in Marincola 2011.
Sterne, Laurence, 1965. *The Life and Opinions of Tristram Shandy*. Ed. Ian Watt. Boston: Houghton Mifflin.
Steuco, Agostino, 1540. *De perenni Philosophia*. Photographic reprint. Introd. Charles B. Schmitt. New York: Johnson Reprint, 1972.
Stewart, Susan, 1991. *Crimes of Writing: Problems in the Containment of Representation*. New York: Oxford University Press.
Stoneman, Richard, 2008. *Alexander the Great: A Life in Legend*. New Haven: Yale University Press.
Struve, Burchard Gotthelf, 1703. *Dissertatio historico litteraria de doctis impostoribus*. Jena, Germany: Müller.
———, 1754. *Introductio in notitiam rei litterariae et usum bibliothecarum*. 6th ed. Ed. Johann Christian Fisher. Frankfurt and Leipzig: Apud Henr. Ludovicum Broenner. First published 1704.

Tasso, Torquato, 1982. *Creation of the World*. Trans. Joseph Tusiani. Binghamton, NY: Center for Medieval and Early Renaissance Studies.
———, 1998. *Dialoghi*. Ed. Giovanni Baffetti. 2 vols. Milan: Rizzoli.
———, 2006. *Il mondo creato. Sacro poema*. Ed. Paolo Luparia. Edizione Nazionale delle Opere di Torquato Tasso (National Edition of the Works of Torquato Tasso). Alessandria: Edizioni dell'Orso.
Tertullian, 1971. *La toilette des femmes De cultu foeminarum*. Ed. Marie Turcan. Paris: Les Éditions du Cerf.
Thompson, David, and Alan F. Nagel, 1972. *The Three Crowns of Florence: Humanist Assessments of Dante, Petrarch, and Boccaccio*. New York: Harper and Row.
Thompson, Edward Maunde, 1912. *An Introduction to Greek and Latin Palaeography*. Oxford: Clarendon.
Thucydides, 1919. *Peloponnesian War*. 4 vols. Ed. Charles Forster Smith. Reprint. Cambridge, MA: Harvard University Press, 1969.
Todorov, Tzvetan, 1973. *The Fantastic: A Structural Approach to a Literary Genre*. Cleveland: Press of Case Western Reserve University.
Trithemius, Johannes, 1974. *In Praise of Scribes. De laude scriptorum*. Ed. Klaus Arnold; trans. Roland Behrendt. Lawrence, KS: Coronado Press.
Tuleja, Tad [Thaddeus], 1989. *The Catalogue of Lost Books: An Annotated and Seriously Addled Collection of Books That Should Have Been Written, But Never Were*. New York: Ballantine.
Turner, James, 2014. *Philology: The Forgotten Origins of the Modern Humanities*. Princeton: Princeton University Press.
Ursin, J. H., 1661. *De Zoroastro bactriano, Hermete Trismegisto, Sanchoniathone Phoenicio, eorumque scriptis*. Nuremberg: Michael Endter.
V.-David, Madeleine, 1965. *Le Débat sur les écritures et l'hiéroglyphe aux XVIIe et XVIIIe siècles: et l'application de la notion de déchiffrement aux écritures mortes*. Paris: S.E.V.P.E.N.
Valla, Lorenzo, 2007. *On the Donation of Constantine*. Trans. G. W. Bowersock. Reprint. Cambridge, MA: Harvard University Press, 2008.
Vanstiphout, Herman, 2004. *Epics of Sumerian Kings: The Matter of Aratta*. Ed. Jerrold S. Cooper. Leiden: Brill.
Vergil, Polydore, 2002. *On Discovery*. Trans. Brian Copenhaver. Cambridge, MA: Harvard University Press.
Vespasiano da Bisticci, 1926. *Renaissance Princes, Popes, and Prelates. The Vespasiano Memoirs: Lives of Illustrious Men of the XVth Century*. Trans. William George and Emily Waters. Reprint. New York: Harper Torchbooks, 1963.
———, 1951. *Vite di uomini illustri del secolo XV*. Ed. Paolo D'Ancona and Erhard Aeschlimann. Milan: Hoepli.

Vico, Giambattista, 1990. *Opere*. 2 vols. Ed. Andrea Battistini. Milan: Mondadori.

———, 2004. *La scienza nuova 1730*. Ed. Paolo Cristofolini and Manuela Sanna. Naples: Alfredo Guida.

———, 2020. *The New Science*. Trans. Jason Taylor and Robert Miner. Introd. Giuseppe Mazzotta. New Haven: Yale University Press.

Vincent of Beauvais, 1624. *Speculum quadriplex naturale doctrinale morale historiale*. 4 vols. Douai: Balthazar Bellerus.

Vockerodt, Gottfried, 1704. *Historia societatum et rei litterariae mundi primi*. 125–82 in his *Exercitationes academicae, sive commentatio de eruditorum societatibus, et varia re litteraria*. Gotha: Schal.

Voltaire, 1960. *Zadig, ou la destinée, histoire orientale*. 1–65 in *Romans et contes*. Ed. Henri Benac. Paris: Garnier.

———, 1962. *Philosophical Dictionary*. Trans. Peter Gay, 2 vols. New York: Basic Books.

———, 1981. *Voltaire:* Candide, Zadig *and Selected Stories*. Trans. Donald M. Frame. New York: Signet.

Von Däniken, Erich, 1970. *Chariots of the Gods? Unsolved Mysteries of the Past*. Trans. Michael Heron. New York: G. B. Putnam's Sons.

Walker, D. P., 1972. *The Ancient Theology: Studies in Christian Platonism from the Fifteenth to the Eighteenth Century*. London: Duckworth.

Watts, Edward J., 2017. *Hypatia: The Life and Legend of an Ancient Philosopher*. Oxford: Oxford University Press.

Wellard, James, 1973. *The Search for the Etruscans*. New York: Saturday Review Press.

Wendt, Herbert, 1968. *Before the Deluge*. Trans. Richard and Clara Winston. Reprint. London: Paladin, 1970.

West, Martin, ed., 2003. *Greek Epic Fragments*. Cambridge, MA: Harvard University Press.

Wilkins, John, 1694. *Mercury, or the Secret and Swift Messenger, Shewing How a Man May with Privacy and Speed Communicate His Thoughts to a Friend, at Any Distance*. 2nd ed. London: Richard Baldwin. First published 1641.

Winter, Irene J., 1985. "After the Battle Is Over: *The Stele of the Vultures* and the Beginning of Narrative in the Art of the Middle East." 11–32 in *Pictorial Narrative in Antiquity and the Middle Ages*, ed. Herbert L. Kessler and Marianna Shreve Simpson. Studies in the History of Art, vol. 16. Center for Advanced Study in the Visual Arts Symposium Series IV. National Gallery of Art, Washington, DC. Hanover, NH: University Press of New England.

Witt, Ronald G., 2000. *In the Footsteps of the Ancients: The Origins of Humanism from Lovato to Bruni*. Reprint: Leiden: Brill Academic, 2003.

Woods, Christopher, ed. 2015. *Visible Language: Inventions of Writing in the Ancient Middle East and Beyond*. With Emily Teeter and Geoff Emberling. 2nd corrected printing. Chicago: Oriental Institute of the University of Chicago.

Yates, Frances A. 1964. *Giordano Bruno and the Hermetic Tradition*. Reprint. New York: Vintage Books, 1969.

———, 1966. *The Art of Memory*. Chicago: University of Chicago Press.

Yeo, Richard, 2001. *Encyclopedic Visions: Scientific Dictionaries and Enlightenment Culture*. Cambridge: Cambridge University Press.

Zatti, Sergio, 2006. "Turpino's Role: Poetry and Truth in the *Furioso*." 60–94 in *The Quest for Epic: From Ariosto to Tasso*. Ed. Dennis Looney, trans. Sally Hill. Toronto: University of Toronto Press.

Index.

Tabula Hieroglyphica (quā descriptionem Mundi Nauticam designari censemus) cf. ea quidem plagam, ubi nostra quidem fert sententia, figura quævis ex præcipuis designes. e museo Jo: Georgij Horn Bavarici prin: &c:

memorandum that the plate or figure is printed the wro[ng way]
so that the paper looked upon through the back [...]

Sele 2da limbi — Sele 3tia — Sele 3tia limbi

Sud west versus west 31 — Sud west Sud — versus 30 — Sud west 29 — Sud versus west 28

west versus Sud fig. 33 — West Sud west 32

Polus Eccliptica Antarcticus occidenti proximus

Navigatio ab occidente in orientem Solamnis

Sud west Africus fig. 8 — Sud Auster f. 7 — Sud est Euchanus f. 6

Polus Mundi Antarcticus

west Nord west f. 21 — west Nord west f. 20 — Nord west versus west f. 19 — Nord west versus Nord f. 18 — Nord Nord west f. 17 — Nord versus west fig. 16

Polus Ecclipticæ Arcticus occasui proximus